TECHNICAL WRITING

John M. Lannon
Southeastern Massachusetts University

Little, Brown and Company
Boston   Toronto

Library of Congress Cataloging in Publication Data

Lannon, John M.
   Technical writing.

   Includes index.
   1. English language — Rhetoric.   2. English
language — Technical English.   3. Technical writing.
I. Title.
PE1475.L36   1985   808′.0666   84-23350
ISBN 0–316–51448–9

Library of Congress Catalog Card No. 84–23350

ISBN 0-316-51448-9

9 8 7 6 5 4 3 2 1

The author wishes to thank the students whose work
appears in this volume. All student work appears with
the kind permission of the students.

HAL

Published simultaneously in Canada
by Little, Brown & Company (Canada) Limited

Printed in the United States of America

TECHNICAL WRITING

Preface

This text offers a comprehensive and flexible introduction to technical communication. Designed specifically for heterogeneous classes, it speaks to a diversity of student interests. Rhetorical principles are explained, illustrated, and applied to a broad variety of assignments, from brief memos and summaries to formal reports and proposals. Exercises parallel the writing demands students will face in college and on the job.

The book is organized into four parts. Part I, Communicating with a Specified Audience, treats the process by which any document is written. The first four chapters analyze the writer's essential decisions about purpose, audience, content, organization, and style. Chapter 5 synthesizes these decisions.

Part II, Strategies for Technical Reporting, covers the principles of rhetoric, design, and research that are vital in the production of technical documents.

Part III, Specific Applications, adapts the earlier concepts and strategies to the composing of the varied documents that typically comprise technical writing.

To unify the concepts in Part I with the strategies in Part II and the applications in Part III, each sample document throughout the text is preceded by an audience-and-use analysis.

Finally, the appendixes contain a handbook for easy reference and the full text of interviews with four writers on the job.

The rationale for the sequence and substance of chapters is based on these assumptions:

1. Although no single, predictable sequence characterizes the writing process, it is in no way a random act; instead, the writing process is a deliberate act of problem solving. Beyond studying models of a particular document, students need guidance in the recursive process of decision making in order to generate their own useful documents.

2. Writers who lack rhetorical awareness begin "writing" too early, ignoring the initial decisions on which writing is based. Only by defining their writing situation, and by asking the right questions, can they find the right answers. The mistake of a too-quick start is compounded by the failure to write deliberately. Students begin to write deliberately only when they see writing as a form of problem solving.

3. All students can learn to recognize and incorporate into their work the essential rhetorical features of good writing: worthwhile content, sensible organization, and readable style.

4. Despite their knowledge differences, juniors and seniors generally face the same difficulties as freshmen or sophomores in planning, drafting, and revising a document.

5. The proliferation of technical writing courses has caused most classes to be grouped heterogeneously. This assortment of students with varied backgrounds calls for explanations that are accessible, examples and models that are broadly engaging and intelligible, and goals that are rigorous but collectively achievable.

6. The countless approaches to the teaching of technical and professional writing require that a textbook be as flexible as possible.

7. Class time in a writing course should not be wasted by lectures reiterating information readily found in a texbook. Instead, workshops can *apply* textbook principles by focusing on the texts composed by students in the course. The workshop approach, then, calls for a text that is both comprehensive and accessible. (Suggestions for workshop design are found in the Instructor's Manual.)

In line with the above assumptions, this book offers a pattern of cumulative skills, moving from assignments of summaries and expanded definitions to more complex tasks, ending with the formal report or proposal — an assignment that draws on most skills developed earlier. Within this structure, however, each chapter is self-contained for flexibility in course planning.

Ample exercises in each chapter offer practical applications at various levels of challenge and complexity. Thus the instructor who wishes to spend more time, for instance, on letters and short reports will find plentiful resources. Timely examples and models are drawn from a balance of student and on-the-job writing in a variety of fields, and are intelligible to students in all majors.

Individual chapters move from concept to practice in the following progression:

1. Defining each assignment in detail.

2. Explaining its purpose to students, whose implied question invariably is "Why are we doing this?"

3. Discussing the criteria the completed assignment should embody — with an emphasis on rhetorical purpose.

4. Providing models accompanied by explanations and audience-and-use analyses.

5. Offering guidelines for planning, drafting, and revising the document for a specified audience with specified needs.

6. Providing revision checklists for self- or peer evaluation.

The emphasis throughout is first on the process, then on the products of writing. Students are guided through inventing, developing, and organizing their materials recursively. They learn to make *all* the decisions necessary to produce a useful document.

Just about every page of this new edition is in some way improved, with the style more efficient, discussions more concise, and illustrations more vivid. Rhetoric theory is translated into sensible practice. Some specific additions include:

- in Chapter 1, discussion of the features unique to technical writing
- in Chapter 2, more on discovering and evaluating content for usefulness to a specified audience for a specified purpose
- a new Chapter 3, treating the standard paragraph as the structural model for any discourse of any length
- in Chapter 4, new sections on ambiguity, analogies, acronyms, and tone
- a new Chapter 5, giving an overview of the writing process in an actual work situation
- in Chapters 6–12, many new examples and models, with emphasis on high-tech and environmental subjects
- in Chapter 13, a new section on computer graphics
- in Chapter 14, a new section on electronic information searches, including a sample automated search
- in Chapter 15, two scentific/technical documentation systems and MLA's new in-text citations; a primary research report, with new primary research topics
- in Chapter 16, sections on claim and adjustment letters, and plenty of short-case exercises
- in Chapter 17, a major revision to emphasize persuasive memos, with many short-case exercises
- in Chapters 18 and 19, expanded discussions and new proposal and report models, with emphasis on primary research
- a new appendix containing interviews with working professionals who must write daily
- a new instructor's manual (8½″ by 11″ format), containing detailed suggestions and 60 master sheets to be reproduced as dittos or transparencies, or shown on an opaque projector. Some of the masters are designed for the instructor's convenience (syllabi, chapter quizzes, a final exam, and sample responses to various assignments). But to emphasize more than

the products of writing, many of the masters illustrate the writing process as a *thinking* process.

Many of the improvements in this edition were based on excellent advice from reviewers: Patrick Cheney, Pennsylvania State University; William J. Kelly, Bristol Community College; Thomas L. Warren, Oklahoma State University; Edith Weinstein, University of Akron; Valerie K. Wescott, Western Michigan University; and Sharon K. Wilson, Fort Hayes State University. Robert M. Hogge, U.S. Air Force Academy, gave suggestions of unsurpassed quality and detail. My thanks to you all.

At Southeastern Massachusetts University, Raymond Dumont, Jr., served as contributing editor for sizable sections, and acted as counselor, advisor, and all-around support system. Charles McNeil, instruction librarian, contributed vital material, advice, and editing for the section on automated literature searches. Once again, Jean Morgan tied all the loose ends of my teaching/administrative life. Shirley Haley did a superb job of editing the final manuscript, suggesting crucial changes, and contributing art, ideas, and energy.

At Little, Brown and Company, Molly Faulkner, Allison Hoover, and Joe Opiela gave gracious and unqualified support of all kinds; Victoria Keirnan made a difficult production job look and feel easy.

Finally, for all that matters: Chega, Daniel, and Sarah.

Contents

8
Dividing and Organizing   133

19
Analyzing Data and Writing a Formal Report   464

COMMUNICATING WITH A SPECIFIED AUDIENCE

1

Introduction to Technical Writing

TECHNICAL WRITING DEFINED

In technical writing, you report specialized information for the practical use of readers who have requested it. Readers may need your information to perform a task, answer a question, solve a problem, or make a decision. Here is a typical writing situation:

> Sarah Burnes is in her first month as a chemical engineer for Polsun, a leading manufacturer of cameras and multipurpose film. Her first assignment is to evaluate the quality of the main plant's incoming and outgoing water. (Chemicals in incoming water can damage negatives during production, and the production process itself can pollute outgoing water.) Upper management wants an answer to this question: How often should we change water filters? Sarah studies twenty-five pages of chemical analysis in computer printouts, and writes her report.

Technical writing is always done in response to some definite need. Any technical document is designed to give readers the information they need clearly and persuasively.

THE CONTRAST BETWEEN TECHNICAL AND NONTECHNICAL WRITING

Because readers *use* a technical document, it must be based on facts, and it must have one, single meaning. Poetry and fiction, then, would not be forms of technical writing because they are largely based on intuition, feelings, and imagination.[1] Also, a poem or story can suggest any number of meanings. Consider, for instance, this poem by Tennyson:

THE EAGLE: A FRAGMENT

He clasps the crag with crooked hands;
Close to the sun in lonely lands
Ringed with the azure world he stands.

The wrinkled sea beneath him crawls;
He watches from his mountain walls
And like a thunderbolt he falls.

The poet is sharing *his own* state of mind, his *impression* of the eagle. And impressions are by nature imprecise. So, no two readers of the poem are likely to share exactly the same meaning.

A nonfiction essay also is not technical writing because an essay, by definition, presents the writer's personal opinions, as in this piece:

The eagle is the most *noble* bird. This large and *awesome* creature perches on the highest cliffs, scanning the earth below. Against the sun, he presents a *dignified* and *formidable* silhouette, in *full command* of his world from his solitary perch. On sighting his prey, he dives with *swift* and *deadly accuracy,* *sure* and *self-reliant.* The eagle's *majestic demeanor, independence, pride,* and *invincible spirit* all symbolize American values. Thus it is no surprise that this *magnificent* bird was chosen as the national emblem of the United States.

This paragraph is highly personal; the writer describes his *impression* of the bird. Notice the many judgmental terms (in italics). Because these impressions might not be shared by other people, this is not technical writing. (Another writer, for instance, might see the eagle as vicious, ugly, or just another bird to be hunted.) Like the poem, this writing tells a good deal about the author's view of the eagle, but little about the bird itself.

In contrast to these nontechnical versions, here is a technical description of a bald eagle:

The bald eagle, *Haliaeetus leucocephalus,* is named for its snow-white head. One of the sea eagles, it nests along fresh or salt waters in polar regions of

---

[1] This section was inspired by Patrick M. Kelley and Roger E. Masse, "A Definition of Technical Writing," *The Technical Writing Teacher,* Vol. 4 (Spring 1977): 94–97.

the northern hemisphere, throughout most of the United States and south into Mexico. In recent years the number of bald eagles has been much reduced, and they are now most numerous in Alaska. The adult is blackish brown, with a snow-white head and tail. The bald eagle has unfeathered feet and toes. It is 30 to 40 inches long, and has a wingspan of 6 to 8 feet. It feeds mainly on fish; however, it catches very few itself, either pirating its food from other birds or picking up dead fish on the shore. In 1782, Congress adopted a design displaying the bird for the Great Seal of the United States, and the bald eagle became the national bird.[2]

This version is technical writing because it contains only precise factual information. Vague descriptive terms (such as *"large* and *awesome"*) are replaced by more precise terms ("30 to 40 inches long . . . a wingspan of 6 to 8 feet") that give readers a clear picture. As an expression of fact, this version allows only one interpretation.

We have just seen that a subject such as the eagle can be the theme of poetry, essay writing, or technical writing. So, technical writing is defined not by its subject, but by the author's purpose and the reader's needs. A technical subject is specialized, often mechanical or scientific in nature (e.g., a description of screw-thread gauging techniques). But many technical subjects — like the eagle — can be discussed from nontechnical points of view. Your responsibility in technical writing is to observe, interpret, and report from a technical point of view: on the basis of precise facts and verifiable evidence.

Besides having a purely *informative* purpose (to make readers understand something), a technical document can have a *persuasive* purpose (to encourage readers to do something). The next example is part of a report designed to persuade company executives to expand in-house computer capacity.

I recommend that our DEC 2060 hardware be upgraded by a maximum addition to main memory, a disk input/output control unit, and an additional disk storage unit. This expansion will (1) increase the number of simultaneous terminal users from 60 to 80, (2) increase the responsiveness of the system, and (3) provide sorely needed disk storage for word processing and company data bases.

Whether merely informative or persuasive as well, all useful technical documents share these features:

1. Each is the product of a writer who fully understands the subject.
2. Each focuses purely on the subject, *not* on the writer.
3. Each conveys *one* meaning, allowing one interpretation.
4. Each does more than merely record information. Instead of telling readers

---

[2] Reprinted with permission from *Collier's Encyclopedia* (© 1971, Crowell-Collier Educational Corporation).

everything they know, the writers above have been selective, tailoring their message to the specific needs of their audience.

5. Each is written at a level of technicality that will be understood by the specified audience (i.e., general information for the eagle entry in the encyclopedia, versus specialized information for the computer report to executives).

6. Each is efficient. Every word advances the writer's meaning. Nothing is wasted.

This book is designed to help you produce useful technical documents.

USES OF TECHNICAL WRITING SKILLS

Your value to any organization depends on how well you can convey to others what you know. In any field, almost anyone in a responsible position writes daily. Police and fire fighters write investigation reports that often serve as evidence in court. Medical personnel keep daily records that are crucial to patient welfare, and, increasingly, a basis for litigation. Managers write memos, personnel evaluations, requisitions, and instructions. Contractors write proposals, bids, and specifications for prospective customers. Engineers and architects plan, on paper, the structure of a project before contracts are awarded and construction begins. In a world of rapidly changing technology, good communication is more crucial than ever.

Thanks to the computer, we are now in the information revolution. No longer merely the knowledge stored in our heads, *information* now also consists of the computerized material to which we have access. To remain competitive in any field, people have to know about the latest developments. So, information itself becames the *ultimate* product. The successful professionals and businesses will be those who keep up with the rapid changes — those who receive, process, generate, and transmit information most efficiently.

Although automated office systems (word processors, electronic mail, teleconferencing) speed information flow, no computerized device can convert poor writing to good. When you sit down to write that first draft of a letter, memo, or report, it makes no difference whether you use quill-and-ink or the most sophisticated word processor. Good writing comes from hard work and constant practice. No automated shortcuts exist.

WRITING SKILLS IN YOUR CAREER

As college graduates, your duties will include planning, organizing, and controlling the work and information that flow in and out of the organization.

None of you can afford to write poorly. Consider, as evidence, these comments from authorities:

> One important reason working business people and professionals do not achieve . . . their potential . . . is their inability to communicate. . . . In every survey that asks business people what subjects they wish they had studied more carefully, their first or second answer is always communication.[3]

> One of the chief weaknesses of many college graduates is the inability to express themselves well. Even though technically qualified, they will not advance far with such a handicap.[4]

> The biggest untapped source of net profits for American business lies in the sprawling, edgeless area of written communication, where waste cries out for management action. Daily, this waste arises from the incredible amount of dull, difficult, obscure, and wordy writing that infests plants and offices.[5]

The message in these quotes is clear: expertise and motivation alone are not enough. Whatever your career plans, you can expect to be a "part-time" technical writer.

Employers first judge your writing by your application letter and résumé. If you join a large organization, your retention and promotion may be decided by executives you've never met. So, your letters, memos, and reports will be seen as a measure of the overall quality of your work. As you advance, your ability to communicate will become even more important, while your technical background may become less important. The higher your goals, the better communications skills you will need.

CHAPTER SUMMARY

Technical writing is the reporting of specialized information for the practical use of readers who have requested it. Nontechnical writing expresses the writer's feelings and imagination, and can have various meanings. In contrast, any technical document:

1. is the product of a writer who fully understands the subject,
2. focuses on the subject, *not* the writer,
3. conveys *one* meaning,

---

[3] Joseph Williams, *Style: Ten Lessons in Clarity and Grace* (Glenview, Ill.: Scott, Foresman and Co., 1981), p. x.

[4] Linwood E. Orange, *English: The Pre-Professional Major*, 2nd ed. (New York: Modern Language Association, 1973), p. 4.

[5] Langeley Carleton Keyes, "Profits in Prose," *Harvard Business Review* (January–February 1961): 105.

4. is tailored to the specific needs of an audience,

5. is at a level of technicality that will be understood by the specified audience, and

6. is efficient.

Your value to any organization depends on how well you communicate to others what you know. And much of the communication in the working world is carried out in writing. While computers have made information the ultimate product, no computer can transform bad writing to good. In fact, executives consistently rank writing and communication skills as the most vital employee qualities. So, in any career, you can expect to be a "part-time" technical writer.

EXERCISES

1. Locate a brief example of a technical document (or section) in your library or elsewhere. Make a photocopy, and bring it to class. Prepare to explain why your selection can be called technical writing (in terms of the features listed on pages 5–6).

2. Research the writing skills you will need in your career. (Begin by looking at the *Dictionary of Occupational Titles* in your library's reference section.) You might interview a successful person in your profession. Why and for whom will you write on the job? Explain in a memo to your instructor.

3. In a memo to your instructor, describe the specific skills you hope to acquire from your technical writing course. How, exactly, will you apply these skills in your career?

4. Find an example of a poorly written memo or letter. In a memo to your instructor, explain why and how the message fails. Attach a copy of the document to your memo. (Check around campus or look through unsolicited sales letters you or your family receive. Or ask a business acquaintance for a sample.) Now do the same for a well-written document, explaining instead how it succeeds.

5. Assume a friend of yours is skeptical about the need for writing skills in *your* major. This person argues that secretaries or word processors are available to "clean up" messy writing, and that computers soon will eliminate the need for most writing. Write your friend a letter, detailing why these assumptions are mistaken. Be as specific as possible, using vivid examples to support your point.

2

Writing for Readers

READERS DEFINED

All technical writing is for readers who will use your information for some purpose. You might write *to define* something — as to an insurance customer who wants to know what *variable annuity* means. You might write *to describe* something — as to an architectural client who wants to know what a new addition to her home will look like. You might write *to explain* something — as to a stereo technician who wants to know how to eliminate bass flutter in your company's new line of speakers. You might write *to persuade* someone — as to your vice-president in charge of marketing who wants to know if it's a good idea to launch an expensive advertising campaign for a new oil additive. Whatever your specific purpose, as a technical writer, you do not write for yourself, but to inform and persuade others.

In any factual writing, you must connect with the intended readers — sometimes one reader, sometimes several, sometimes many. Your goal is always to communicate useful information clearly and precisely.

THE CONCEPT OF "AUDIENCE"

Different readers have different backgrounds. So, something specialized that makes sense to you might confuse others, unless it is explained at their level of understanding. Information is useful only if it makes sense to its audience.

Examples of the differences in audience knowledge and needs are everywhere. Watching a TV football game with explayers or fans, for instance, is different from watching it with a novice. Terms such as *power sweep, blitz,* and *shotgun formation* are clear to informed viewers, but have to be defined or illustrated for the uninformed. Explaining how to repair a leaking faucet to someone with no mechanical background is different from explaining it to a do-it-yourselfer. Undoubtedly you can think of times when you've had to adjust a message for a less specialized audience.

Good writing connects with its readers by recognizing their differences in background and their specific needs. In some cases, a single message may appear in several versions for several audiences. A technical article about a new treatment for heart disease, for instance, might first appear in the *Journal of the American Medical Association,* which will be read by doctors and nurses. The same article may later be rewritten less technically for a cardiology textbook, which will be read by medical and nursing students. A nontechnical version might appear in *Good Housekeeping.* Though the three versions treat the same subject, each is adapted to the needs of its particular audience.

Technical writing is intended to be *used.* Whether giving instructions, describing a product, reporting research, or defining a specialized term, you become the teacher and the reader becomes the student. Because your readers know less than you do in this instance, you can expect certain questions:

- "What is it?"
- "What does it look like?"
- "How do I do it?"
- "How did you do it?"
- "Why did it happen?"
- "When will it happen?"
- "Why should we do it?"
- "How much will it cost?"
- "What materials and equipment does it require?"
- "What are the dangers?"

These are just some of the questions various reports are designed to answer. (Later chapters treat these questions in detail.) All readers need precise answers they can understand and use. Whatever specific questions your readers may have, they all add up to the one, big question any reader has of any message: What, exactly, do you mean? You always write to show a certain audience what you mean; and during the process of making your meaning clear, you often discover new meanings for yourself.

Think of your readers' needs as you think of your own. If chapter 1 of your introductory math textbook, for example, covered differential equations (advanced calculus), the author would have ignored your needs. For your background and purposes, the chapter would be useless. How clearly any message answers its readers' questions depends on how precisely it is adapted to their level of understanding.

ADAPTING MESSAGES TO AUDIENCES

When you write for a close acquaintance (friend, fellow computer hacker, classmate in experimental psychology, fellow dental technician, engineering colleague, chemistry professor who reads your lab reports, or your immediate boss), you know a good deal about your particular reader. Therefore, you automatically adapt your report to that particular reader's knowledge, interest, and needs. But sometimes you have to write for less clearly defined audiences, particularly when the audience is large (when you are writing for a magazine or professional journal, or asked to explain your company's stock-options program to members of many different departments). When you have only general knowledge about your audience's background, you must decide whether your message should be *highly technical, semitechnical,* or *nontechnical.*

The Highly Technical Message

Readers whose specialized training is more than or similar to your own expect technical data, without long explanations or interpretations. To illustrate, the following account of treatment given a heart attack victim is a highly technical message. The writer, a doctor on call in the emergency room, is reporting to the patient's doctor. This reader needs an exact record of his patient's symptoms and treatment.

> Mr. X was brought to the emergency room by ambulance at 1:00 A.M., September 27, 1984. The patient complained of severe chest pains, shortness of breath, and dizziness. Auscultation and electrocardiogram revealed a massive cardiac infarction and pulmonary edema marked by pronounced cyanosis. Vital signs were as follows: blood pressure, 80/40; pulse, 140/min.; respiration, 35/min. Lab tests recorded a wbc count of 20,000, an elevated serum transaminase, and a urea nitrogen level of 60 mg%. Urinalysis showed

4+ protein and 4+ granular cast/field, suggesting acute renal failure secondary to the hypotension.

The patient was given 10 mg of morphine stat, subcutaneously, followed by nasal oxygen and a 5% D & W IV. At 1:25 A.M. the cardiac monitor recorded an irregular sinus rhythm, suggesting left ventricular fibrillation. The patient was defibrillated stat and given a 50 mg bolus of Xylocaine IV. A Xylocaine drip was started, and sodium bicarbonate was administered until a normal heartbeat was established. By 3:00 A.M., the oscilloscope was recording a normal sinus rhythm.

As the heartbeat stabilized and cyanosis diminished, the patient was given 5 cc of Heparin IV, to be repeated every six hours. By 5:00 A.M. the BUN had fallen to 20 mg% and the vital signs had stabilized as follows: blood pressure, 110/60; pulse, 105/min.; respiration, 22/min. The patient was now conscious and responsive.

Written at the highest level of technicality, this report is clear only to the medical expert. The writer correctly assumes that her reader has the specialized knowledge to understand the message. Therefore, she does not define technical terms (pulmonary edema, sinus rhythm, etc.). Nor does she interpret lab findings (4+ protein, elevated serum transaminase, etc.). She uses abbreviations that she knows are familiar to her reader (wbc, BUN, 5% D & W IV). Because her reader knows the reasons for specific treatments and medications (defibrillation, Xylocaine drip, etc.), she includes no theoretical background. Her report is designed to provide concise answers to the main questions she can anticipate from her particular reader: "What happened?" "What course of treatment was given?" "What were the results?"

The Semitechnical Message

Readers whose specialized training is somewhat less than your own expect to have the background and interpretations of technical data spelled out. Semitechnical messages may address a broad array of readers. First-year medical students, for example, have a basic technical understanding, but second-, third-, and fourth-year students have increasingly higher levels of understanding. Yet in many cases, students in all four groups can be considered semitechnical readers. Therefore, when you write for a semitechnical audience, identify the *lowest* level of understanding among the group, and write to that level. You are wiser to give too much explanation than too little.

Here is a partial version of the earlier medical report. Written at a semitechnical level, it might appear in a textbook for first-year medical or nursing students, in a report for a medical social worker, in a patient history for the medical technology department, or in a monthly report for the hospital administration.

Examination by stethoscope and electrocardiogram revealed a massive failure of the heart muscle along with fluid build-up in the lungs, which produced a cyanotic discoloration of the lips and fingertips from lack of oxygen.

The patient's blood pressure at 80 mm Hg (systolic)/40 mm Hg (diastolic) was dangerously below its normal measure of 130/70. A pulse rate of 140/minute was almost twice the normal rate of 60–80. Respiration at 35/minute was over twice the normal rate of 12–16.

Laboratory blood tests yielded a white blood cell count of 20,000/cu mm (normal values: 5,000–10,000), indicating a severe inflammatory response by the heart muscle. The elevated serum transaminase enzymes (only produced in quantity when the heart muscle fails) confirmed the earlier diagnosis. A blood urea nitrogen level of 60 mg% (normal values: 12–16 mg%) indicated that the kidneys had ceased to filter out metabolic waste products. The 4+ protein and casts reported from the urinalysis (normal values: 0) revealed that the kidney tubules were degenerating as a result of the lowered blood pressure.

The patient was immediately given morphine to ease the chest pain, followed by oxygen to relieve strain on the cardiopulmonary system, and an intravenous solution of dextrose and water to prevent shock.

This version defines, interprets, and explains the significance of the raw data. Exact dosages here are not significant to these readers because they are not treating the patient; therefore, dosages are not mentioned. Normal values of lab tests and vital signs, however, make interpretation easier. (Highly technical readers would know these values, as well as the significance of specific dosages.) Knowing what medications the patient received would be especially important to the lab technician, because some medications affect blood test results. For a nontechnical audience, however, the above message needs even further translation.

The Nontechnical Message

Readers with little or no specialized training in your field expect technical data to be translated into their simplest form. Nontechnical readers are impatient with abstract theories but want enough background to help them make the right decision or take the right action. They are bored by long explanations but frustrated by bare facts not explained or interpreted. They want to understand a message without having to become students of the discipline. They expect a report that is clear on first reading, not one that requires review or study.

Following is a nontechnical version of the earlier medical report. The attending physician might write this version for the patient's spouse who is on a business trip overseas, or as part of a script for a documentary film about emergency room treatment.

Both heart sounds and electrical impulses were abnormal, indicating a massive heart attack caused by failure of a large part of the heart muscle. His lungs were swollen with fluid and his lips and fingertips showed a bluish discoloration from lack of oxygen.

The patient's blood pressure was dangerously low, creating the danger of shock. His pulse and respiration were almost twice the normal rate, indicating that the heart and lungs were being overworked in keeping oxygenated blood circulating freely.

Blood tests confirmed the heart attack diagnosis and indicated that waste products usually filtered out by the kidneys were building up in the bloodstream. Urine tests showed that the kidneys were failing as a result of the lowered blood pressure.

The patient was given medication to ease his chest pain, oxygen to ease the strain on his heart and lungs, and intravenous solution to prevent his blood vessels from collapsing and causing irreversible shock.

This nontechnical version translates all specialized information into simple terms. It mentions no specific medications, lab tests, or normal values because these would mean nothing to the nontechnical reader. The writer merely summarizes the events, and explains the causes of the crisis and the reasons for the particular treatment.

In some other situation, however (say, in a jury trial for malpractice), the nontechnical audience might need information about specific medication and treatment. Such a situation would call for a much longer report with elaborate explanations to simplify technical concepts — a kind of mini-course in emergency coronary treatment.

Each version of the medical report is useful *only* to readers at a particular level of understanding. Doctors and nurses have no need of the explanations in the two latter versions, but they do need the specialized data in the first. Beginning medical students and paramedics might be confused by the first version and bored by the third. Nontechnical readers would find both the first and second versions meaningless.

In your own job you will need to connect with audiences at various levels. As an architect, for instance, you might specify "800 board feet of 1×6 no. 2 white pine, rough sawn and kiln dried" for the building contractor (a wood expert). For the bank financing the project, you might describe the wood as "rough pine interior finish." For the client, you might write "new barn-board interior walls." Each version is clear and appropriately detailed for its intended audience.

Primary Versus Secondary Readers

When you write the same basic message for readers at different levels of technicality, classify those readers as *primary* or *secondary*. The primary readers usually are those who have requested the report and who probably will use it as a basis for decisions or actions. The secondary readers are those who will carry out the project, who will advise the primary readers about their decision, or who will somehow be affected by this decision. They will read your report for information that will help them to get the job done, to provide educated advice, or to remain abreast of new developments.

As often as not, the technical background of these two audiences will differ. Primary readers may require highly technical messages, while secondary readers may need semitechnical or nontechnical messages — or vice versa. When you must write for audiences at different levels, follow this rule:

1. If your report is short (a letter, memo, or anything less than two pages), rewrite it at various levels for various readers.

2. If your report is longer than two pages, maintain a level of technicality that connects with your primary readers. Then supplement the report with appendixes that address the secondary readers (technical appendixes when secondary readers are technical, or vice versa). Letters of transmittal, informative abstracts, and glossaries are other supplements that help less specialized audiences understand a highly technical report.

The following situation illustrates how certain reports must be adapted to both primary and secondary readers.

You are a metallurgical engineer in a Detroit consulting firm. Your supervisor has asked that you test the fractured rear axle of a 1981 Delphi pickup truck recently involved in a fatal accident. Your job is to determine whether the axle fracture was the cause or a result of the accident.

After testing the hardness and chemical composition of the metal and examining a series of microscopic photographs of the fractured surfaces (fractographs), you conclude that the fracture resulted from stress that developed *during* the accident.

Because your report may serve as evidence in a court case, you must explain your findings in meticulous detail. But your primary readers will be nonspecialists (attorneys, insurance representatives, possibly a judge and a jury), so you will have to translate your report, explaining the principles behind the various tests, defining specialized terms such as "chevron marks," "shrinkage cavities," and "dimpled core," and showing the significance of these features as evidence.

Secondary readers will include your supervisor and outside consulting engineers who will be evaluating your test procedures and assessing the validity of your findings. For these readers, you will have to include appendixes that spell out the technical details of your analysis: *how* hardness testing of the axle's case and core indicated that the axle had been properly carburized; *how* chemical analysis ruled out the possibility that the manufacturer had used inferior alloys; *how* light-microscope fractographs revealed that the origin of the fracture, its propagation direction, and the point of final rupture indicated a ductile fast fracture, not one caused by torsional fatigue.

In the above situation, the primary readers need to know *what your findings mean,* whereas the secondary readers need to know *how you arrived at your conclusions.* Unless the needs of each group are served independently, your information will be worthless.

FOCUSING ON READERS' NEEDS

In addition to choosing the appropriate level of technicality for your message, you need to make other decisions as well. Your choices of *what* you write (content) and *how* you write it (format, arrangement, style) are determined by your writing situation. Identify your purpose, and learn all you can about your audience *before* you write.

When you write for a particular reader or a small group of readers, you can focus sharply on your audience by asking yourself specific questions:

1. Who wants the report? Who else will read it?
2. Why do they want the report? How will they use it? What purpose do I want to achieve?
3. What is the technical background of the primary audience? The secondary audience?
4. How much does the audience already know about the subject? What material will have informative value?
5. What exactly does the audience need to know, and in what format? How much is enough?
6. When is the report due?

Follow the suggestions on the next few pages for answering these questions.

Reader Identification

Identify the primary readers by name, job title, and specialty (Martha Jones, Director of Quality Control, B.S. and M.S. in mechanical engineering). Are they superiors, colleagues, or subordinates? Are they inside or outside your organization? Are they apt to agree or disagree with your conclusions and recommendations? Will your report be taken as good or bad news?

Although your report must satisfy the needs of primary readers, it should not ignore the needs of any secondary readers. Identify those additional readers who are likely to be interested in or affected by your report, or who will affect the primary reader's perception or use of your report.

Purpose of the Request

Find out why your readers want the report and how they will use it. Ask them directly. Do they merely want a record of your activities or progress? Are you expected to supply only raw data or conclusions and recommendations as well? Will your readers take immediate action based on your report? Do they need step-by-step instructions? Are they simply collecting information for later use? Will the report be read and discarded, filed, or published? The more you learn about your audience's exact expectations and needs, the more useful you can make your report. In your audience's view, *what* is most important? On the basis of what you know of your audience, what purpose do you want your document to achieve?

The Reader's Technical Background

Colleagues on your project speak your technical language and can understand raw data. Supervisors responsible for several technical areas may want interpretations and recommendations. Managers may have only vague technical knowledge and thus will expect definitions and background explanations. Clients with no technical background will expect simplified versions that spell out what the data mean to *them* (to their health, pocketbook, business prospects, etc.). Keep in mind, however, that none of these generalizations may apply to your specific situation. Find out for yourself. By assessing your readers' technical background, you can avoid insulting their intelligence or writing over their heads. Make the level of technicality appropriate for your primary audience. For secondary audiences, rewrite the short report or include supplements with the long report. Remember, it's easier to make the mistake of being too technical than of being too simple. In any mixed audience, aim for the lowest level of technicality.

The Reader's Knowledge of the Subject

Do not waste time rehashing information readers already have. If preliminary reports on your subject have been written earlier, a brief summary of their major points will do.

Readers expect something *new* and *significant* from your message. Writing has informative value[1] when it:

1. conveys knowledge that is new *and* worthwhile to the intended audience, or

2. offers fresh insight about something familiar, causing readers to see things in a new way.

---

[1] Adapted from James L. Kinneavy's assertion that discourse ought to be unpredictable in *A Theory of Discourse* (Englewood Cliffs, N.J.: Prentice-Hall, 1971).

The informative value of any message is measured by the writer's purpose and the audience's needs. Say you have no computer background (only general knowledge), but you're trying to decide whether to take a computer course. And my purpose is to provide facts that will help you decide. In this situation, which of the following bits of information would you find useful?

1. Interest in computers has grown immensely in the last decade.
2. By 1990, at least 70 percent of businesses will be computer-dependent.
3. The first digital computer was built by Howard Aiken.
4. Information can be transmitted rapidly by computer.

The information in 2 is likely new to many novices, and is relevant to your needs and my purpose (stated above). Because 1 merely repeats common knowledge, it has no informative value here. Although 3 offers an unfamiliar fact, it is not relevant to my purpose and your needs. (But 3 could be relevant in another situation: if you were already in a computer course.) And everyone would agree with 4, so it offers you nothing worthwhile.

A message also has informative value if it offers fresh insight about a familiar fact. As this book's audience, for instance, you expect to learn about technical writing, and my purpose is to help you do so. In this situation, which of the following statements would you find useful?

1. Technical writing is hard work.
2. Technical writing is a process of deliberate decisions in response to a specific situation. In this process, you discover important meanings in your topic, and give your readers the information they need to understand your meanings.[2]

Statement 1 is no news to anyone who has ever picked up a pencil, so it has no informative value for you. But 2 offers a new way of seeing something familiar. Even if you've done much writing, and struggled through decisions about punctuation, organization, and so on, you probably have not viewed writing as entailing the many kinds of decisions discussed in this book (and illustrated in Chapter 5). Because 2 provides new insight into a familiar activity, you can say it has informative value. A writer *selects* only material that has informative value for a given audience.

Appropriate Details and Format

In the earlier medical reports we saw that dosages, drug names, and medical terms are significant to some readers but not others. The amount and kind of

[2] My thanks to Major Robert M. Hogge, U.S. Air Force Academy, for suggesting this definition.

detail in your report (How much is enough?) will depend on what you have learned about your readers and their needs. Were you asked to "keep it short" or to "be comprehensive"? Can you summarize certain material or does it all need to be spelled out? What length will your readers be willing to tolerate? Are your primary readers most interested in conclusions and recommendations or do they want a full description of your investigation as well? Have they requested a letter, a memo, a short report, or a long, formal report with supplements (title page, table of contents, appendixes, etc.)? Will visuals (charts, graphs, drawings, photographs, etc.) make certain information more accessible? What level of technicality will connect with your primary readers?

High Technicality   The diesel engine generates 10 BTUs per gallon of fuel, as opposed to the conventional gas engine's 8 BTUs.

Low Technicality   The diesel engine yields 25 percent better gas mileage than its gas-burning counterpart.

Every professional's challenge in the Age of Information is to keep abreast of the rapidly changing technology, to use it most effectively. What one has to know, therefore, can change quickly and often. For instance, quite possibly, a senior engineer (and supervisor) will know less about VLSI (Very-Large-Scale Integrated Circuits — the heart of microprocessors) than the recent graduate.

Sometimes, in writing for a large audience, you will be unable to identify all its members. In such cases, write for the least specialized. You are safer to risk boring some experts than to write over other people's heads.

Due Date

Is there a deadline for your report? Find out. Give yourself plenty of time to collect data and to write and revise the report. If possible, ask your primary readers to review an early draft and suggest improvements. A report dashed off at the eleventh hour impresses no one.

BRAINSTORMING FOR USEFUL MESSAGES

All readers expect a useful message, one that has three essential features:

1. *Content* that makes it worthwhile;
2. *Organization* that makes it easy to follow;
3. *Style* that makes it readable and clear.

Later chapters cover strategies for shaping various messages into organized memos, letters, and reports. Also, Chapter 3 shows how to organize paragraphs, and Chapter 4 how to sharpen your style. But for now, let's concentrate on *content* that is worthwhile.

When you begin working with an idea, content is raw material: ideas, insights, statistics, facts, or examples, anything that advances your meaning, that helps you answer this question: How can I find something worthwhile to say, something that will convey my meaning?

Outlines for discovering material about specific writing tasks are covered in later chapters. But the technique of *brainstorming* is your surest bet for finding worthwhile content. The aim is to get as much raw material as possible down *on paper*. Business and technical people, in fact, often use group brainstorming sessions to get ideas for new projects, marketing campaigns, and the like.

Some people brainstorm as a very first writing step, even before deciding about their purpose and their audience's needs, just as a way of getting started. Others save brainstorming until they have written a rough draft. But regardless of the sequence, good writers almost always brainstorm at some point in the writing process. They do so to insure they discover *all* the material readers might find useful.

The brainstorming procedure is simple: just concentrate on your writing situation, and jot down *every single thought*. Here are the steps:

1. Empty your head of anxieties (about bills, grades, or whatever). Sit with eyes closed for two minutes, thinking about *absolutely nothing*.

2. Now, concentrate on your situation, repeating this question: What can I say about my topic? (If you've already analyzed your purpose and audience, focusing on these will help.)

3. As ideas and phrases begin to flow, write every one down. Don't stop to judge relevance or worth, and don't worry about complete sentences or correct spelling. Simply get everything down on paper. The more ideas, the better chance you'll discover worthwhile content. Trust your imagination; even the wildest idea might lead to a valuable insight.

4. Keep pushing and sweating until you run out of ideas.

5. At the end of the session, you should have a chaotic mixture of nonsense, irrelevancies, and worthwhile content.

6. Take a break.

7. Now, confront your list. Strike out the useless material, and sort what's left into related categories. If other ideas crop up, include them as well. Your finished list should provide an ample supply of raw material.

As an illustration of how brainstorming works, assume you are having trouble balancing your budget in college. Bills are piling up, and your parents have written, asking how you plan to deal with the problem. So you decide to brainstorm for an answer to this question: How can I make ends meet? Here is how your brainstorming list might read:

get a roommate	transfer to a tuition-free school
move to a cheaper apartment	help from home
apply for a loan	write rich aunt Bertha
avoid expensive restaurants	make a strict budget
do less partying	sell skis and surfboard
find rich friend	turn in credit cards
buy used books	lock up checkbook
apply for work-study	do less dating
cook at home	work two jobs during summer
sell stereo	live on peanut butter
wash own shirts	trade car for old van, and live in it
apply for a scholarship	have phone removed
quit school and work full-time for a year	no weekend trips

As the next step, cross out ideas that seem impractical or unrealistic, and add any new ideas that emerge. Now, organize related items within broader categories (classification is discussed in Chapter 8). The six classes that seem to emerge from this list are:

Selling Things
Asking for Help
Changing Lifestyle
Making a Physical Move
Getting a Job
Managing Money

As you review your revised list, you ought to find the content you need for a persuasive letter to your parents.

Brainstorming helps you achieve the concentration needed for coming up with good ideas that otherwise you might never realize you had. This technique usually produces more information than you can use. So, after deleting some ideas you still have plenty of material to choose from. And chances are you will discover more while writing your letter, memo, or report. From this broad inventory, you will want to select only the material that is directly useful: namely, worthwhile content.

In Chapter 5, you will see how a working professional uses brainstorming during the process of writing an important memo to his superiors.

SAMPLE WRITING SITUATIONS

In many jobs you will have to ask daily questions about your audience's needs. Here are two scenarios where specialists who also must be "part-time" technical writers assess the needs of their audiences in different situations.

A DAY WITH A POLICE OFFICER

You are a police officer in the burglary division of an urban precinct.

– A local community college has asked you to lecture on the various fields of investigative police work during its career week. Your audience will be interested laypersons, possibly those considering a career in police work. You know they've been captive audiences at "lectures" for years, so you decide to liven things up with visuals. The police artist agrees to draw up a brightly colored flip chart with illustrations to accompany your summaries of each investigative area.

– Tomorrow in court you will testify for the prosecution of a felon you caught red-handed last month. His lawyer is a sly character who is known for making police testimony look foolish, so you are busy reviewing your notes and getting your report in perfect form. Your audience will consist of legal experts (judge, lawyers) as well as laypersons (jury). The key to a convincing testimony will be *clarity, accuracy,* and *precision,* with impartial and factual descriptions of *what* happened, as well as *when* and *where.*

– You are drafting an article about new fingerprinting techniques for a law enforcement journal. Your readers will be expert (veteran officers) and informed (junior officers). Because they understand "shoptalk" and know about the theory and practice of fingerprinting, they won't need extensive background information or definitions of specialized terms. Instead, they will read your article to learn about a new *procedure.* You will have to convince them your way is better. You decide to discuss the merits of your technique before explaining the process and its results. If your article is well received, you will later write a manual for using this technique.

– Your chief asks you to write a manual of investigative procedures for junior officers in your division. The manual will be bound in pocket size and carried by all junior officers on duty. Your audience is informed, but not expert. They have studied the theory at the police academy, but now they need the "how to." They will want rapid access to the various instructions for handling various problems, with each step spelled out. Most likely, they will use your instructions as a guide to *immediate* actions and responses. Therefore, you'll have to label clearly warnings and cautions *before* each step. Clarity throughout will be imperative.

– Next month you will speak before the chamber of commerce on "Protecting Your Business Against Burglary." The audience will be highly interested laypersons. Because burglary protection can be costly (alarm systems, guard dogs, etc.), the audience is apt to be most interested in the less expensive precautions they can take. They will want to remember your advice, so you decide to supplement your talk with visuals: specifically, a checklist for business owners that you will discuss item-by-item by using an overhead projector. The division secretary agrees to type the checklist and make copies for distribution at the end of your talk.

A DAY WITH A CIVIL ENGINEER

As a civil engineer in a materials-testing lab, you are in charge of friction studies on various road surfaces.

– Your vice-president calls to say your firm is competing for a state contract for a safety study of state bridge surfaces. Full reports of your testing procedures, progress, and results are needed for inclusion in your firm's contract bid. As engineer in charge, you are responsible for the quality of your firm's proposal. Your audience consists of experts (state engineers), informed readers (officers in the highway department), and laypersons (members of the state legislature). The legislature ultimately will decide which firm gets the contract, but they will act on the advice of engineers and the highway department. So you decide to keep your proposal at a level of technicality that will connect with the more specialized audience. The legislative committee will read only conclusions and cost data.

– During another call, the vice-president asks you to write safety instructions for paving crews on the new bay bridge. Your instructions will be read by all crew members, from project engineer to laborer, but when it comes to safety on this kind of job, even the specialized personnel may be considered laypersons. Therefore, you decide to keep the instructions simple and clear. To increase reader interest, you ask the graphics department to design a brochure for your instructions, including sketches and photographs of the hazardous areas and situations.

– At next month's engineering convention in Dallas you are scheduled to deliver a paper describing a new road-footing technique that reduces frost-heave damage in northern climates. (This paper will soon be published in an engineering journal.) Your audience will be civil engineering colleagues who are highly interested in your new procedure. This expert audience wants to know what it is and how it works — without lengthy background. You decide to compress your data by using conventional visuals (cross-sectional drawings, charts, graphs, maps, and formulas), as well as computer graphics.

– You draft a letter to a colleague you've never met, to ask about her new technique for increasing the durability and elasticity of rubberized asphalt surfaces. Your expert reader is a busy professional with little time to read someone's life story, but she will need *some* background about your own work in this area, along with clear, specific, and precise questions that she can answer quickly. As a gesture of goodwill, you close your letter with an offer to share *your* findings with her.

– You are putting the finishing touches on an article for a national magazine. The article describes the hazards encountered in building the trans-Alaskan highway. Your audience consists almost entirely of laypersons who are reading for general knowledge (and perhaps even for entertainment). To hold the reader's interest, you decide to focus on twelve dramatic incidents (avalanches, earthquakes, etc.). To further increase dramatic appeal, you decide to relate the factual details of each event in first-person narrative form, as in a journal or diary (e.g., "Friday, March 12: This morning's wind-chill factor lowered the temperature at our excavation site to −55°F").

In each situation above, the writers addressed their audience's need for clear, appropriate, and useful information. What they could learn about each audience beforehand guided their choice of what was appropriate and necessary.

Much of your writing in this course — and much of it on the job — will be aimed at general readers, but even when your readers are informed or expert, they will know less than you do about the subject. Whereas in college you write for an audience (the professor) who knows more than you, and who is *testing* your knowledge, on the job you write for an audience who knows less, and who is *using* your knowledge.

CHAPTER SUMMARY

Whenever you write, make your audience your first concern. Assume readers will use your information for practical purposes. Readers whose training is similar to yours expect a highly technical message. Those less specialized expect a semitechnical message. Readers with no technical training expect a nontechnical message.

When you write a single report for readers at different levels of specialization, classify them as *primary* or *secondary*. Rewrite short reports at various levels for various readers. Aim a long report at your primary audience, adding appendixes and other supplements for secondary readers.

Learn all you can about your audience, *before you write,* by answering these questions:

1. Who wants the report? Who else will read it?
2. Why do they want the report? How will they use it? What is my purpose?
3. How specialized is the primary audience? The secondary audience?
4. How much does the audience already know about the subject?
5. What exactly does the audience need to know, and in what format?
6. When is the report due?

Always brainstorm for useful messages.

EXERCISES

1. In a short report, identify and discuss the kinds of writing and speaking assignments you expect in your career. (If necessary, interview someone in your field for the information.) For whom will you be writing (colleagues, supervisors, clients, etc.)? Will you need to address both primary and sec-

ondary audiences? How will your readers use your information? For which audience are you likely to write most often?

2. Review the questions for audience analysis on page 16 and the scenarios on pages 22–24. Then ask a member of your field to describe five or six situations in which he or she has been a part-time technical writer. Finally, based on your notes from the interview, compose a scenario for a typical person in your field. Use the sample scenarios as models for your own.

3. Locate a short article from your field. (Or select part of a long article or a section from one of your textbooks for an advanced course.) Choose a piece written at the highest level of technicality you can understand and then translate the piece for a layperson. Exchange translations with a classmate from a different major. Read your neighbor's translation and write a paragraph evaluating the appropriateness of its level of technicality. Submit to your instructor a copy of the original, your translated version, and your evaluation of your neighbor's translation.

4. Assume a new employee is taking over your job because you have been promoted. Identify a specific problem in your old job that could cause difficulty for the new employee. Write the employee a set of instructions for avoiding or dealing with the problem. Before writing, perform an audience analysis by answering (on paper) the questions on page 16. Then brainstorm for details. Submit to your instructor your audience analysis, your brainstorming list, and your instructions.

5. Assume you live in the Northeast, and citizens in your state are voting on a solar-energy referendum that would channel millions of tax dollars toward solar technology. These two paragraphs are versions of a message designed to help you, as a voter, make an educated decision. Do both messages have informative value? Explain.

Solar power offers a realistic solution to the Northeast's energy problems. In recent years, the cost of fossil fuels (oil, coal, and natural gas) has risen sharply while supplies continue to decline. High prices and short supply will continue a worsening energy crisis. Because solar energy comes directly from the sun, it is an inexhaustible resource. By using this energy to heat and air-condition our buildings, as well as to provide electricity, we could decrease substantially our consumption of fossil fuels. In turn, we would be less dependent on the unstable Middle East for our oil supplies. Clearly, solar power is a good alternative to conventional energy sources.

Solar power offers a realistic solution to the Northeast's energy problems. To begin with, solar power is efficient. Solar collectors installed on fewer than 30 percent of roofs in the Northeast would provide over 70 percent of the area's heating and air-conditioning needs. Moreover, solar heat collectors are economical, operating for up to twenty years with little or no maintenance. These savings recoup the initial cost of installment within only ten years. Most important, solar power is safe. It can be transformed into electricity through photovoltaic cells (a type of storage battery) in a noiseless process that produces no air pollution — unlike coal, oil, and wood combustion. In sharp contrast to its nuclear counterpart, solar power produces no toxic wastes, and poses no catastrophic danger of meltdown. Thus, a massive conversion to solar power would insure abundant energy and a safe, clean environment for future generations.

3

Organizing Good Paragraphs

In Chapter 2, we saw that readers expect a message with worthwhile content. But the content also must be *accessible:* capable of being grasped easily by readers. Any reader approaches a message with definite expectations about its organization as well as its content.

THE SHAPE OF A MESSAGE

To follow a writer's line of thinking, readers need a message that is sensibly organized. And since thinking rarely occurs in a neat, predictable sequence, writers can't merely record their thoughts in their original order. Instead, writers must *shape* the material into an organized unit of meaning. So, in setting out to organize a message, writers face deliberate decisions:

- What do I want to emphasize?
- What do I say first?
- What comes after that?
- How do I stay on track?
- How do I end?

As with decisions about *content,* a writer's decisions about *organization* are guided by the audience's expectations. Most useful messages — whether in the form of a book, chapter, news article, letter, or memo — usually follow a common organizing pattern:

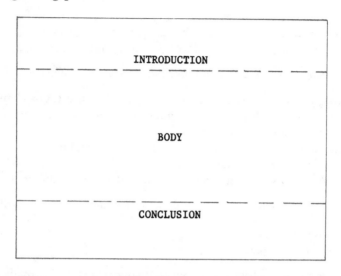

The *introduction* previews the discussion, and reveals the subject and purpose of the message. Here the writer makes a commitment to the readers about what will be said. Knowing immediately what to expect, readers can follow the message more easily.

The *body* delivers on the promise implied in the introduction, with a full explanation of the main point. This section has *unity* in that everything expands on the main point. And it has *coherence* in that the thought sequence is one continuous link from idea to idea.

The *conclusion* brings the message to a perceptible end, instead of just stopping. Often, the main point is reemphasized.

In just about any message, the shape of an organized unit of meaning is basically the same. This shape is best illustrated in the form of a standard support paragraph, a type of paragraph that develops a single main point in enough detail to convey the writer's exact meaning.

THE STANDARD SUPPORT PARAGRAPH

A standard support paragraph is a group of sentences focusing on one main organizing point. The main point, or viewpoint, is expressed as the *topic sentence.* Here are some sample topic sentences:

– Computer literacy soon will be a requirement for virtually all "educated" people.

– A video display terminal can endanger the operator's health.

– Chemical pesticides and herbicides are both ineffective and hazardous.

Each of these topic sentences only introduces a certain way of seeing a subject. Without supporting explanations, we couldn't possibly grasp the writer's exact meaning. Consider, for instance, the third statement:

– Chemical pesticides and herbicides are both ineffective and hazardous.

Imagine that you are a researcher for the Epson Electric Light Company, and you have been given this task: determine whether the company (1) should begin spraying pesticides and herbicides under its power lines, as many other utilities are doing, or (2) should continue with its manual ways of minimizing foliage and insect damage to lines and poles. If you simply responded with the preceding statement, your boss would have a number of questions:

– Why, exactly, are chemical pesticides and herbicides ineffective and hazardous?

– What are the problems?

– Can you explain?

By answering these questions while writing your report, you provide the necessary supporting details:

Introduction

Body (2–6)

Conclusion
(7–8)

1 Chemical pesticides and herbicides are both ineffective and hazardous. 2 Because none of these chemicals has permanent effects, pest populations invariably recover, and require respraying. 3 Repeated applications cause pests to develop immunities to the chemicals. 4 Furthermore, most pesticides and herbicides attack species other than the intended pest, killing off its natural predators, thus actually increasing the pest population. 5 Above all, chemical residues survive in the environment (and living tissue) for years, often carried hundreds of miles by wind and water. 6 This toxic legacy includes such biological effects as birth deformities, reproductive failures, brain damage, and cancer. 7 The ultimate victims of these chemicals would be our customers. 8 Therefore, I recommend we continue our present control methods.

Most paragraphs in technical writing follow this introduction-body-conclusion structure. The key is to begin with a clear topic (or orienting) sentence that states a generalization. Details in the body support the generalization.

Writing the Topic Sentence

Readers look to the first one or two sentences in a paragraph to orient themselves, to align their perceptions with the writer's. When readers know *what* to expect, they can follow more easily. By introducing your way of seeing something, the topic sentence gives readers a framework for understanding your message. Without this orienting framework, readers cannot possibly grasp your exact meaning. Consider, for instance, the following paragraph, whose topic sentence has been left out (read it *once,* only!):

> Besides containing several toxic metals, it percolates through the soil, leaching out naturally present metals. Pollutants such as mercury invade surface water, accumulating in fish tissues. Any organism eating the fish — or drinking the water — in turn, faces the risk of heavy metal poisoning. Moreover, acidified water can release heavy concentrations of lead, copper, and aluminum from metal plumbing, making ordinary tap water hazardous.

After one reading, are you able to provide a main point for the paragraph? Could you restate the message accurately in your own words? Probably not, even after a second reading. Without the orientation of a topic sentence, you have no framework for understanding the information in terms of its larger meaning. And because you don't know what to look for, you can't figure out where to place the emphasis: on polluted fish, on metal poisoning, on tap water? Without the organizing thread provided by a topic sentence, you have no way to tell.

Now, after inserting the following sentence at the beginning, reread the paragraph:

> Acid rain indirectly threatens human health.

In the light of this organizing point, the exact meaning of the message becomes obvious. The topic sentence gives us a framework by:

1. Naming the subject of the message (acid rain);
2. Stating the topic — the writer's specific viewpoint on the subject (that acid rain threatens human health);
3. Forecasting how the message will be developed (through an explanation of the process) in response to the reader's central question: *How exactly does acid rain threaten human health?*

In most support paragraphs, readers look to the topic sentence as the key to understanding the whole. For this reason, your topic sentence ought to appear *first* in your paragraph, unless you have good reason to decide otherwise.

Before you can write a good topic sentence, you must identify your purpose, based on what you know of your readers' needs. Then you can tailor your topic sentence to meet those needs. Assume, for instance, you're writing a report about whales, intended for readers you'd like to recruit as members of the Save-the-Whales Foundation. First, you must decide exactly what point you want to make about whales. And when that point becomes part of your topic sentence, it must provide enough direction for you to develop a worthwhile paragraph. Avoid topic sentences that lead nowhere:

> Whales are a species of mammal.
> Whales live only in salt water.

Also, the point made in your topic sentence must be focused enough to be covered in one paragraph. Avoid broad and abstract topic sentences like this one:

> Whales are interesting animals.

What is meant by "interesting"? Their breeding habits, their migration patterns, the way they exhibit intelligence, or something else?

Let's say you decide to discuss whale intelligence. What exact point do you want to make about whale intelligence?

> Whales seem to exhibit some intelligence.
> Whales are fairly intelligent.
> Whales are highly intelligent.

You decide that the final sentence above expresses your point most accurately. Now you think of ways to make this topic sentence more informative. After all, your readers will be asking: "Highly intelligent relative to what?" So you decide to relate whales' intelligence to that of other mammals:

> Whales are among the world's most intelligent mammals.

You now have a clear direction for developing support in the body section.

Let's look at some other directions your topic sentence might have taken:

> A good indication of whales' high intelligence is the way they play in game-like patterns.
>
> *or*
>
> Like children, a group of whales can spend hours playing tag.

Depending on your purpose and your reader's needs, you can make any main point more and more specific by focusing on smaller and smaller parts of it. The paragraph should then deliver what the topic sentence promises.

Writing the Body

The body of your paragraph contains the supporting details that explain and expand on your central idea. This support material answers the questions about your topic sentence you can expect from readers:

Says who?

What proof do you have to support your claim?

Can you give examples?

To answer these questions, you brainstorm your topic, listing everything you know about whale intelligence.

1. Whales help wounded fellow whales.
2. They communicate through sonar clicks and pings.
3. Scientists have studied whales.
4. They exhibit complex group behavior.
5. They teach and discipline their young.
6. They use marine vegetation, in addition to animal life, as a food source.
7. Scientists compare whales' intelligence with that of higher primates.
8. They follow fixed patterns of annual migration.
9. They engage in elaborate sexual foreplay.
10. They play in gamelike patterns.
11. They have a sophisticated communications system.

After selecting those facts that support the main idea,[1] you arrange them in related categories. Looking at your list, you notice that items 3, 4, 7, and 11 are major points, while the remaining items provide specific support for these points. You now have three categories of general evidence, which are further supported by specific details. When you combine and arrange this material, your paragraph looks like this:

> Whales are among the world's most intelligent mammals. Scientists studying whales rate their intelligence on a level with higher primates because of their complex group behavior. These impressive mammals have been seen teaching and disciplining their young, helping wounded fellow whales, engaging in elaborate sexual foreplay, and playing in definite gamelike patterns. They are able to coordinate such complex activities because of a highly effective communications system of sonar clicks and pings.

With your topic sentence and supporting details on paper, you are ready to write your conclusion.

Writing the Conclusion

Your concluding statement signals readers that the discussion of the central idea stated in your topic sentence is ending. It usually ties the paragraph together by summarizing, interpreting, or judging the facts. If the paragraph is part of a longer report, your conclusion can also prepare readers for a subse-

---

[1] Items 6 and 8 can be deleted from the list because they are not functions of a whale's intelligence. The former is a function of physiology, while the latter is instinct.

quent paragraph. Say that your paragraph on whale intelligence is to be followed by one describing abuses in the whaling industry. Here is an effective conclusion for the first paragraph:

> All in all, scientific evidence shows that whales have a high level of social organization. Unfortunately, the whales' intelligence is ignored by an industry that threatens them with extinction.

The final statement in this conclusion serves as a transition to the next major idea.

An introduction-body-conclusion structure should serve most of your paragraph needs in report writing.[2] Begin each support paragraph with a solid topic sentence and you will stay on target.

Structural Variation

Sometimes a central idea that needs detailed definition might call for a topic statement of two or more sentences (as shown in the needle-description paragraph, on page 35). Or your central idea might have several distinct parts, which would result in an excessively long paragraph. In this case you might break up the paragraph, letting your topic statement stand as a brief introductory paragraph and serve the separate subparts, which are set off as independent paragraphs for the readers' convenience.

> **The most common types of strip-mining procedures are open-pit mining, contour mining, and auger mining. The specific type employed will depend on the type of terrain covering the coal.**
>
> Open-pit mining is employed in the relatively flat lands in Western Kentucky, Oklahoma, and Kansas. Here, draglines and scoops operate directly on the coal seams. This process produces long, parallel rows of packed spoil banks, ten to thirty feet high, with steep slopes. Between the spoil banks are large pits that soon fill with water to produce pollution and flood hazards.
>
> Contour mining is most widely practiced in the mountainous terrain of the Cumberland Plateau and Eastern Kentucky. Here, bulldozers and explosives cut and blast the earth and rock covering a coal seam. Wide bands are removed from the mountain's circumference to reach the embedded coal beneath. The cutting and blasting result in a shelf along with a man-made cliff some sixty feet high, at a right angle to the shelf. The blasted and churned earth is pushed over the shelf to form a massive and unstable spoil bank that creates a danger of mud slides.
>
> Auger mining is employed when the mountain has been cut so thin that it can no longer be stripped. It is also used in other difficult-access terrain.

---

[2] Besides support paragraphs, of course, you will write introductory paragraphs, transitional paragraphs, and concluding paragraphs (as illustrated in the various sample reports that follow).

Here, large augers bore parallel rows of holes into the hidden coal seams to extract the embedded coal. Among the three strip-mining processes, auger mining causes the least damage to the surrounding landscape.

Each paragraph begins with a clear statement of the part of the subtopic it will discuss.

PARAGRAPH UNITY

A paragraph is unified when all its parts work toward the same end — when every word, phrase, and sentence explains, illustrates, and clarifies the central idea expressed in the topic sentence. Paragraph unity is destroyed when you drift away from your stated purpose by adding irrelevant details. Here is how the paragraph on whale intelligence could become disunified:

A Disunified Paragraph

Whales are among the most intelligent mammals ever to inhabit the earth. Scientists studying whales rate their intelligence on a level with higher primates because of their complex group behavior. For example, these impressive mammals have been seen teaching and disciplining their young, helping wounded fellow whales, engaging in elaborate sexual foreplay, and playing in definite gamelike patterns. **Whales continually need to search for food in order to survive. As fish populations decrease because of overfishing, the whale's quest for food becomes more difficult.**

This paragraph is not unified because it begins with whale intelligence and drifts into food problems. Stay on the track set by your topic sentence.

PARAGRAPH COHERENCE

A paragraph is coherent when it hangs together and flows smoothly in a clear direction — when all sentences are logically connected like links in a chain, leading toward a definite conclusion.

One way to damage paragraph coherence is to use too many short, choppy sentences. Two other ways to damage coherence are (1) to place sentences in the wrong order and (2) to use insufficient transitions and other connectors (pp. 569–572) to link related ideas. Here is how the paragraph on whaling intelligence could become incoherent:

An Incoherent Paragraph

Whales are among the most intelligent creatures ever to inhabit the earth. Scientists rate their intelligence on a level with higher primates. Whales exhibit complex group behavior. **The whaling industry ignores the whales'**

intelligence and threatens them with extinction. Whales have been seen teaching and disciplining their young, helping wounded fellow whales, engaging in elaborate sexual foreplay, and playing in definite gamelike patterns. **Scientific evidence shows that whales have a high order of social organization.** They are able to coordinate such complex group activities because of their apparently effective communications system of sonar clicks and pings.

This paragraph is not coherent because it fails to progress logically. The fourth sentence (in boldface) does not follow logically from the first three; it should be the last sentence. Also, the second-to-last sentence should follow the sentence that is now last. Finally, more transitions and connectors should be added to help the sentences flow smoothly from idea to idea. Here is the same paragraph with sentences written in correct sequence and with enough transitional expressions and other connectors (in boldface):

A Coherent Paragraph

Whales are among the most **intelligent** creatures ever to inhabit the earth. Scientists studying **whales** rate **their intelligence** on a level with higher primates **because** of **their** complex group behavior. **For example, these huge and impressive mammals** have been seen teaching and disciplining **their** young, helping wounded fellow whales, engaging in elaborate sexual foreplay, and playing in definite gamelike patterns. **They** are able to coordinate complex group activities **because** of **their** apparently effective communications system of sonar clicks and pings. **All in all,** scientific evidence shows that **whales** have a high order of social organization. **Unfortunately, the whales' intelligence** is ignored by an industry that threatens **them** with extinction.

The paragraph is now tighter and moves in a logical sequence. The sequence of development here can be called a *reasons sequence,* because reasons are given to support an observation and to lead to a recommendation. The logical sequence in the paragraph can be expressed like this:

1. a topic sentence about whale intelligence (*general*)
2. scientific documentation to support the thesis (*specific*)
3. specific examples and explanations of the scientific claims (*more specific*)
4. a statement that sums up and interprets the earlier evidence
5. a concluding statement that provides a bridge to the later discussion: abuses by the whaling industry

PARAGRAPHS DEVELOPED LOGICALLY

Once you have identified your reader and purpose, and gathered your supporting details, you will have to arrange these details in a way that makes the most sense. Following a logical sequence within a paragraph simply means that you decide on which idea to discuss first, which second, and so on. The sequence you select for any paragraph will depend on your subject, purpose, and readers' needs. Some possibilities follow.

Spatial Sequence

A spatial order of development begins at one location and ends at another. This order is most useful in a paragraph that describes a physical or geographical item or a mechanism. Simply describe the parts in the order in which readers would actually view them: left or right, inside to outside, etc. This writer has chosen a spatial order that proceeds from the needle's base (hub) to its point:

> A hypodermic needle is a slender, hollow, steel instrument used to introduce medication into the body (usually through a vein or muscle). It is a single piece composed of three parts, all considered sterile: the hub, the cannula, and the point. The hub is the lower, larger part of the needle that attaches to the necklike opening on the syringe barrel. Next is the cannula (stem), the smooth and slender central portion. Last is the point, which consists of a beveled (slanted) opening, ending in a sharp tip. The diameter of a needle's cannula is indicated by gauge number; commonly, a 24–25 gauge needle is used for subcutaneous injections. Needle lengths are varied to suit individual needs. Common lengths used for subcutaneous injections are ⅜, ½, ⅝, and ¾ inches. Regardless of length and diameter, all needles have the same functional design.

Chronological Sequence

A paragraph describing a series of events or giving instructions is most effective when its details are arranged according to a strict time sequence: first step, second step, etc.

> When you have collected all needed ingredients, prepare your cake batter. Begin by sifting 3½ cups of flour. Then measure into a mixing bowl the flour and sift it again, along with 2½ cups of granulated sugar, 5 teaspoons of baking powder, and 1 teaspoon of salt. To this mixture add ¾ cup of shortening, 4 eggs, 1½ cups of milk, and 2 teaspoons of vanilla extract. Finally, blend the ingredients by hand until the batter is mixed well enough to divide.

The paragraph explaining the process by which acid rain endangers human health (page 29) is another example of chronological sequence.

Examples Sequence

Often a topic sentence can best be supported by specific examples, usually arranged for greater emphasis.

Although strip mining is safer and cheaper than conventional mining, it is highly damaging to the surrounding landscape. Among its effects are scarred mountains, ruined land, and polluted waterways. Strip operations are altering our country's land at the rate of 5,000 acres per week. An estimated 10,500 miles of streams have been poisoned by silt drainage in Appalachia alone. If strip mining continues at its present rate, 16,000 square miles of United States land eventually will be stripped barren.

In the above paragraph, the most dramatic example is saved for the end.

Effect-to-Cause Sequence

A paragraph that first identifies a problem and then discusses its causes is typically found in problem-solving reports.

Modern whaling techniques have brought the whale population to the threshold of extinction. In the nineteenth century, the invention of the steamboat increased hunters' speed and mobility. Shortly afterward, the grenade harpoon was invented so that whales could be killed quickly and easily from the ship's deck. In 1904, a whaling station opened on Georgia Island in South America. This station became the gateway to Antarctic whaling for the nations of the world. In 1924, factory ships were designed that enabled 'round-the-clock whale tracking and processing. These ships could reduce a ninety-foot whale to its by-products in roughly thirty minutes. After World War II, more powerful boats with remote sensing devices gave a final boost to the whaling industry. The number of kills had now increased far above the whales' capacity to reproduce.

Cause-to-Effect Sequence

In a cause-to-effect sequence, the topic sentence identifies the cause(s), and the remainder of the paragraph discusses its effects.

Some of the most serious accidents involving gas water heaters occur when a flammable liquid is used in the vicinity. The heavier-than-air vapors of a flammable liquid such as gasoline can flow along the floor — even the length of the basement — and be explosively ignited by the flame of the water heater's pilot light or burner. Since the victim's clothing frequently ignites, the resulting burn injuries are commonly serious and extremely painful. They may require long hospitalization, and can result in disfigurement or death. *Never, under any circumstances, use a flammable liquid near a gas heater or any other open flame.*[3]

[3] From Fact Sheet No. 65, Consumer Product Safety Commission (Washington, D.C.: Government Printing Office, 1979).

Definition Sequence

For adequate definition, a term may require a full paragraph (as discussed in Chapter 7).

> The prisoner furlough program permits inmates to leave a state or county institution, unescorted, for no less than twelve hours and no more than seven consecutive days. The state corrections commissioner or the administrator of a county jail may grant furloughs to inmates for the following purposes: attending the funeral of a relative; visiting a critically ill relative; obtaining medical, psychiatric, or counseling services when such services are not available within the institution; contacting prospective employers; finding a suitable residence for use on permanent release; or for any other purpose that will help the inmate's reintegration into the community. Inmates released in 1973 after receiving at least one furlough had a one-year recidivism* rate of 17 percent. In contrast, inmates released in the same year after receiving no furloughs had a one-year recidivism rate of 25 percent. These figures are significant because convicts from all crime categories have been furloughed.

Reasons Sequence

A paragraph that provides detailed reasons to support a specific viewpoint or recommendation is often used in job-related writing, as in the pesticide/herbicide paragraph on page 28. For emphasis, the reasons usually are arranged in decreasing or increasing order of importance.

Problem-Causes-Solution Sequence

The problem-causes-solution paragraph is commonly used in daily activity or progress reports.

> On all waterfront buildings, the unpainted wood exteriors had been severely damaged by the high winds and sandstorms of the previous winter. After repairing the damage, we took the following protective steps against further storms. First, all joints, edges, and sashes were treated with water-repellent preservative to protect against water damage. Next, three coats of nonporous primer were applied to all exterior surfaces to prevent paint blistering and peeling. Finally, two coats of wood-quality latex paint were applied over the nonporous primer. To avoid future separation between coats of paint, the first coat was applied within two weeks of the priming coats,

---

* For the purpose of the study, a recidivist was defined as an ex-convict who was returned to any federal, state, or county jail for thirty days or more.

and the second within two weeks of the first. As of two weeks after completion, no blistering, peeling, or separation has occurred.

Comparison/Contrast Sequence

A paragraph discussing the similarities or differences (or both) between two or more items often is used in job-related writing.

> The ski industry's quest for a binding that ensures both good performance and safety has led to the development of two basic types. Although both bindings improve performance and increase the safety margin, they have different release and retention mechanisms. The first type consists of two units (one at the toe, another at the heel) that are spring-loaded. These units apply their retention forces directly to the boot sole. Thus the friction of the boot against the ski allows for the kind of ankle movement needed at high speeds over rough terrain, without causing the boot to release. In contrast, the second type has one spring-loaded unit at either the toe or the heel. From this unit extends a boot plate that travels the length of the boot to a fixed receptacle on its opposite end. With this plate binding, the boot plays no part in release or retention. Instead, retention force is applied directly to the boot plate, providing more stability for the recreational skier, but allowing for less ankle and boot movement before releasing. On the whole, the double-unit binding performs better in racing, but the plate binding is safer.

For the comparison and contrast of more specific data on each of these bindings, two lists would be most effective.

> The Salomon 555 offers the following features:
>
> 1. upward release at the heel and lateral release at the toe (thus eliminating 80 percent of all leg injuries)
> 2. lateral antishock capacity of 15 millimeters, with the highest available return-to-center force
> 3. two methods of reentry to the binding: for hard and deep powder conditions
> 4. five adjustments
> 5. maximum hold-down power for racers and experts
> 6. the necessity of boot alteration
> 7. release torque applied to the boot sole, which, under stress of normal use, alters its release characteristics, requiring readjustment and eventual replacement of boots
> 8. high durability and a combined weight of 74 ounces for $80.00
> 9. the most endorsements among alpine racers today
>
> The Americana offers these features:
>
> 1. upward release at the toe as well as upward and lateral release at the heel

2. lateral antishock capacity of 30 millimeters, with a moderate return-to-center force

3. two methods of reentry to the binding

4. two adjustments, one for boot length and another comprehensive adjustment for all angles of release and elasticity

5. hold-down power that is slightly compromised at the heel to provide lateral release potential

6. a boot plate that eliminates the need for boot alterations, wear on boots, and danger from friction

7. the durability of two moving parts and a combined weight of 56 ounces for $54.50

Instead of this block structure (in which one binding is discussed and then the other), the writer might have chosen a point-by-point structure (in which points in common for each item are listed together: e.g., "Reentry Methods"). The point-by-point comparison is particularly favored in feasibility and recommendation reports because it offers readers a meaningful comparison on the basis of common points rather than by cataloguing items separately.

PARAGRAPH LENGTH

Paragraph length depends on the writer's purpose and audience's needs. To guide your decision about length, answer this question:

> How much and what kind of support do I need, to convey my exact meaning to *these* readers?

When deciding about paragraph length, consider these guidelines:

1. Too many short paragraphs make a message seem choppy and poorly organized. A series of short paragraphs, however, *is* effective in a set of step-by-step instructions.

2. Too many long paragraphs can be hard to follow. Sometimes, important ideas can get buried in the middle of a long paragraph. Unless your paragraph is patterned as a list (like this one), it should generally be no longer than fifteen lines.

3. A short paragraph can attract readers' attention to an important idea, setting it off from the rest of the discussion. Sometimes, even a single-sentence paragraph can do the same thing:

> In summary, we can prevent further damage from mud slides only by building a retaining wall behind the #3 construction site immediately.

4. A combination of shorter and longer paragraphs is best for emphasis, if your subject allows this arrangement.

CHAPTER SUMMARY

Most useful messages have a definite shape: introduction, body, and conclusion. This shape is illustrated by a standard support paragraph. The writer's main point is usually expressed first in a topic (or orienting) sentence, then explained in the paragraph body, and reemphasized in the conclusion.

Good paragraphs follow principles of unity, coherence, development, and length. A paragraph is unified when all its support relates to the main point. It is coherent when the main and supporting points are all clearly linked. Depending on your subject, purpose, and readers' needs, you might develop your paragraph in any of several logical sequences, and in various lengths.

EXERCISES

1. The following topic sentences either provide no direction, are unfocused, or are not sufficiently informative. Revise them.

Examples
A. Our firm employs twelve people. (leads nowhere)
   Revised: Working in a small architectural firm has helped me appreciate the importance of being versatile.
B. Writing is a complex skill. (unfocused)
   Revised: Writing is a process that involves a series of deliberate decisions about audience needs, purpose, content, arrangement, and style.
C. A Mercedes-Benz is a great all-around car. (not sufficiently informative)
   Revised: A Mercedes-Benz offers safety, performance, durability, and luxury.

a. Women have changed radically in the last decade.
b. My colleague's name is Bill.
c. My job is (awful, great).
d. Professor Jones is unfair to students.
e. Nursing is a popular profession.
f. My company hires more men than women.

2. Below are assorted writing situations. Select one (or more, if assigned), and respond with a unified, coherent, and logically developed paragraph. Be sure to identify (on paper) your audience's needs, and its uses of your information. Begin with a definite topic sentence. Make a list of anticipated reader questions about your topic sentence. Brainstorm for worthwhile content. (You might read pages 82–84 in Chapter 5, to see how a writer brainstorms in an actual working situation.)

Submit your final paragraph to your instructor, with your audience-and-use analysis, audience questions, brainstorming list, and early drafts attached.

a. When you applied to this college, you requested a specific major (nursing, computer science, or some such). The college later notified you that you had been accepted as a General Studies student because qualified applicants for your desired major far exceeded openings. But the letter went on to say this:

A limited number of General Studies students will be allowed to transfer into the _____ program at the beginning of second semester. Any student who wishes to be considered for transfer should submit a brief statement (no more than one 200-word paragraph) explaining his/her reasons for choosing this major. Statements will be evaluated by a committee of seniors and faculty in the _____ program.

Write the paragraph.

b. From the start, you've had major complaints about inadequate facilities in your dorm or apartment (leaking faucets, no parking space, or the like). You decide to take action by writing a one-paragraph note to the president of your dorm council (or the supervisor of your apartment building), spelling out the problems and demanding a solution.

c. The student senate has published a request for nominations for the Teacher-of-the-Year award. This request stipulates that all nominations be accompanied by a paragraph of no more than 200 words, showing why your candidate should win. You decide to nominate Professor X. Write the paragraph.

d. You've decided to apply for a scholarship offered by your college. Among other materials, the scholarship committee requests a paragraph explaining your reasons for attending this college. Write the paragraph.

e. Think about a place in your town or on campus that needs improvement. Describe the problem in one paragraph to a specified audience who will use your information as a basis for action.

f. Assume it's time for end-of-semester course evaluations by students. Write a one-paragraph evaluation of your favorite course, to be read by the department chairperson.

g. Describe the job outlook in your chosen field. Write one paragraph for a high-school senior interested in your major. Your paragraph will be included in the Career Handbook published by your college.

4

Revising for a Readable Style

Readability Defined

Writing Efficient Sentences

Choosing the Exact Words

Adjusting the Tone

Exercises

READABILITY DEFINED

A technical writer's first task is to inform or persuade through efficient sentences and exact language, not to impress or entertain through fancy verbal displays. So, technical writers express worthwhile information — even highly specialized information — in the most straightforward way for their audience.

You might write for a broad or specific audience, or for an expert or non-expert audience. (We've already seen how information inappropriate for a particular audience is useless.) But no matter how appropriate your information, the audience's needs will not be served unless your writing is easy to follow and understand — in a word, *readable*. And hardly any writer ever "says it right" the first time. So, revise *any* writing until its sentences are efficient and its words exact. To help your audience spend less time reading, spend more time revising.

Readers of technical documents are busy and impatient. They don't wish to put more into reading a document than they can get from it. They hate waste and expect efficiency. Every sentence and word in a document should carry its own weight, and advance the writer's meaning.

WRITING EFFICIENT SENTENCES

Observe the same rule with sentence style as with document content: make it long enough to be understood, yet short enough to be tolerated. For instance, if you're responding to someone's job application, don't write like this:

> We are in receipt of your recent correspondence indicating your interest in the position listed below. Your correspondence has been duly forwarded to the office with candidate selection responsibility for consideration. You may expect to hear from the aforementioned office relative to your application as the selection process progresses.

Notice how hard you've worked to extract information that could be expressed like this:

> We've received your application for the position listed below, and forwarded it to the office that will select candidates. At each stage of the selection process, we will inform you of the status of your application.

Never make readers work harder than they have to.

Generally, sentences in a technical document should be no longer than twenty-five words. But (as we've seen above) word or syllable count are not the sole measures of good sentences. An efficient sentence emphasizes relationships, offers solid information, and makes for easy reading. In short, an efficient sentence is *clear, concise,* and *fluent.*

Making Sentences Clear

A clear sentence communicates its precise meaning on first reading. It signals relationships among its parts, and emphasizes the key thought. Assuming that their level of technicality is appropriate for the audience, unclear sentences generally exhibit one of three types of bad style:

1. Sentences lacking any sensible meaning at all:

> – Aging in the bottle for six months is desirable before drinking.
> – This design may become a possibility because of no other choices rather than a decision.

Such sentences are hopeless, often the result of hasty writing, poor concentration, or no proofreading. Writers can only discard them, and start over.

2. Sentences lacking one definite meaning, suggesting instead at least two possible meanings:

> – Most city workers strike on Friday.
> – Visiting managers can be boring.

Sentences of this type are ambiguous. Revisions usually can rescue them.

3. Sentences that do express a single meaning, but that have to be read more than once to be understood:

> — Bill is well qualified for the job in our shipping department, having a forklift operator's license which was mandatory at the Acme Warehouse, where he was laid off, and at the Post Office in Woburn, where he was approved by the government to handle and package important material.

Sentences of this type can be unscrambled for better readability. The following guidelines will help you revise for clarity.

Avoid Telegraphic Writing

Missing function words (articles, prepositions, linking verbs) often leave readers grasping for your meaning.

> Ambiguous   New housing for elderly not yet dead.
>
> Revised   New housing for elderly *is* not yet dead.
>
> Ambiguous   Side test tube cracked at 30° F.
>
> Revised   *The* side *of the* test tube cracked at 30° F.
>
> *or*
>
> *The* side test tube cracked at 30° F.

Avoid Ambiguous Phrasing and Word Choice

In technical writing, a sentence should have only *one* meaning. Make sure word choice and word order are absolutely clear.

> Ambiguous   I cannot recommend this candidate too highly. (Are you recommending or not?)
>
> Revised   This candidate has my highest recommendation.
>
> *or*
>
> I cannot recommend this candidate highly.
>
> Ambiguous   The Conservation Board will confer on Nantucket Island Tuesday. (Will Nantucket be the site or the subject?)
>
> Revised   The Conservation Board will confer *about* Nantucket.
>
> *or*
>
> Nantucket will be the site of the Board's meeting.

Keep Pronoun References Unambiguous

Whatever noun (referent) a pronoun replaces must be identified. Otherwise, sentences like this one result:

Ambiguous Referent   Our patients are enjoying the warm days while they last. (Are the patients or the warm days on the way out?)

Depending on whether the referent for *they* is *days* or *patients,* the sentence can be clarified as follows:

Clear Referent   While these warm days last, our patients are enjoying them.

or

Our terminal patients are enjoying the warm days.

Ambiguous Referent   Janice dislikes working with Claire because she's the nervous type. (Who's the nervous type? Janice or Claire?)

Clear Referent   Because Janice is the nervous type, she dislikes working with Claire.

(See pages 576–577 for a detailed discussion of pronoun references.)

Avoid Punctuation Ambiguities

Too often, a missing hyphen, comma, or other punctuation mark can fog your meaning:

Missing Hyphen   Repack the trailer's inner wheel bearings. (Inner-wheel bearings or inner wheel-bearings?)

Our president is a high fidelity fanatic. (Someone who likes drugs, but refuses to fool around?)

Missing Commas   Is liquid hydrogen being mass produced today? If so how[,] and where is it stored?

(See how the meaning changes with a comma after "how"?)

Police fired on the crowd[,] killing four protesters. (Without the comma, it appears that the crowd, not the police, killed the four protesters.)

While missing hyphens and commas are prime causes of punctuation ambiguity, other omissions can cause similar problems. For example, a missing colon after "kill" gives us this headline: "Moose Kill 200." Or a missing apos-

trophe after Myers creates this gem: "Myers Remains Buried in Portland."
Punctuation *does* affect meaning! (See pages 553–569 for punctuation rules.)

Use "That" Wisely

A needless *that* adds clutter:

> Faulty   This [is a] problem [that] bothers me.

> Faulty   I found the diskette [that] I misplaced.

On page 53, we'll look at other examples of needless use of "that." But in
certain cases, a well-placed *that* gives readers a cue to your meaning by signaling a definite relationship. Consider, for example, the following sentence
opener:

> Executives in our company fear lower-level managers. . . .

Thus far along in the sentence, we might wonder *why* executives fear lower-
level managers. Now look at the whole sentence:

> Misleading   Executives in our company fear lower-level managers
> will resist office automation.

Ah! The writer wasn't talking about executives *fearing* managers, but about
executives *fearing resistance* from management. So we've had to correct the
initial impression received from the early part of the sentence. A simple *that*
would have saved us from reading the sentence twice.

> Revised   Executives in our company fear that lower-level man-
> agers will resist office automation.

In the sentence below, a missing *that* makes the sentence ambiguous, no matter how many times we read it.

> Ambiguous   The reactor operator told management several times
> he had feared a loss-of-coolant accident.

Did the operator tell management several times or once? An appropriate *that*
would clarify the meaning.

> Revised   The reactor operator told management *that* several
> times he had feared a loss-of-coolant accident.

> *or*

> The reactor operator told management several times
> *that* he had feared a loss-of-coolant accident.

Avoid Ambiguous Modifiers

Modifiers explain or define other words. But unless you keep a modifier close to the word it modifies, your sentence might read like this:

Misplaced Modifier	Many executives are skeptical about office automation **as well as managers.** (Are the executives skeptical about managers?)
Revised	Many executives **as well as managers** are skeptical about office automation.

Another problem with ambiguity occurs when the modifier has no word to modify.

Dangling Modifier	**Being well-known in the computer industry,** I would appreciate your advice.

The writer was trying to say that the *reader* is well-known, but with no word to modify, the modifier (in bold print) dangles. By inserting a subject, we can repair the confusing message.

Revised	Because **you** are well-known in the computer industry, I would appreciate your advice.

Dangling modifiers can cause hopelessly ambiguous messages:

Ambiguous	Before ordering the new laser printer for the publication manager, the accounting department must approve the purchase.

Who will put in the order: the accounting department, the publication manager, or someone else?

Revised	Before Ms. Washington can order the new laser printer for the publication manager, the accounting department must approve the purchase.

Unstack the Modifiers

Too many nouns stacked up as modifiers in front of another noun make for hard reading, and they can be hopelessly ambiguous as well. For instance, is a "training session participant evaluation form" designed for participants to evaluate the session or for the trainers to evaluate the participants? Stacked nouns also deaden your style. Bring your style and your reader to life by using action verbs.

Faulty	Her job involves **fault-analysis systems troubleshooting handbook** preparation.

Revised   In her job, she prepares handbooks for trouble-shoot-
ing fault-analysis systems.

Faulty   Our **vehicle air conditioner compression cut-off** device will reduce fuel consumption by 5 percent.

Revised   Our compression cut-off device for vehicle air condi-
tioners will reduce fuel consumption by 5 percent.

No such problem with readability occurs when adjectives are stacked.

Correct   He was a **perplexed, angry, confused,** but **dedicated**
employee.

Emphasize Key Words

Just as any paragraph has a key sentence, any sentence has a key word or
phrase that sums up its meaning. For emphasis, place the key word or phrase
at the beginning or end of the sentence.

Faulty   I expect a **refund** because of your error in my shipment.

Correct   Because of your error in my shipment, I expect a
refund.

Faulty   These days, "**quality**" means little to **workers** in some
industries.

Correct   "**Quality**" in some industries these days means little
to **workers.**

To provide readers with a forecast, place the action verb at the beginning of
any instruction.

Correct   **Insert** the diskette before activating the disk drive.

Use Active and Passive Voice Selectively

Most often, the active voice ("I did it") is better than the passive voice ("It
was done by me"). But for certain kinds of emphasis, the passive voice is preferred. Let the following guidelines govern your choice.

The active voice follows an actor-action-recipient pattern:

actor	*action*	*recipient*
Joe	lost	the check
subject	*verb*	*object*

The passive voice reverses the pattern, making the recipient the subject and
either omitting the actor or naming him/her in a prepositional phrase:

recipient	*action*	*actor*
The check	was lost	by Joe
subject	*verb*	*prepositional phrase*

In each version, the word or phrase in the *subject* position receives the emphasis.

Active	**A falling girder** injured the foreman.

Passive	**The foreman** was injured by a falling girder.

In most writing, give preference to the active voice. It is more direct and economical. Use the passive voice only to emphasize the *recipient* rather than the *actor*.

Passive	**Mr. X** was brought to the emergency room by ambulance.

Passive	**The hijacking story** was broadcast by all stations.

Use the passive voice to emphasize events or results when the *actor* is unknown, not apparent, or unimportant.

Passive	**The victim** was badly beaten.

Passive	**The leak in the nuclear-core housing** was finally discovered.

Passive	**Parts of the plane** were found as far as two miles from the crash site.

One danger of passive construction is that it camouflages the person responsible for the action:

Passive	A mistake was made in your shipment. (By whom?)
"Irresponsible"	The manager was kissed.
	It was decided not to offer you the job.
	These offices were designed poorly.

Don't hide behind the passive unless, of course, the person behind the action has reason for being protected.

Correct	The criminal was positively identified.

In reporting errors or bad news, use the active voice for greater sincerity.
The passive voice is a weaker construction. It creates an impersonal tone.

Weak and Impersonal	An offer will be made by us next week.

Stronger	We will make you an offer next week.

Ordinarily, use the active voice in giving instructions:

Faulty   The lid should be sealed with wax.

Correct   Seal the lid with wax.

Faulty   Care should be exercised.

Correct   Be careful.

Use the active voice when you want action. Otherwise your statement will have no impact.

Faulty   If the aforementioned claim is not settled immediately, the Better Business Bureau will be contacted,
and their advice on legal action will be taken.

This statement is unlikely to move readers to action. Who is making the claim? Who should be doing the settling? Who will be contacting the Bureau? Who will take action? There's nobody here!

Correct   If you do not settle my claim by May 15, I will contact the Better Business Bureau. Upon their advice,
I will take legal action.

Making Sentences Concise

Remember that most of us are far more interested in what we have to say than our audience may be. As a result, we can sometimes fall into two kinds of wordiness: one kind gives more information than readers need (see, for instance, pages 178–179); the other kind just uses too many words to express the information that *is* needed. Every word in the message should advance your meaning.

A concise message is brief but informative. It gets right to the point, without clutter. A brief but vague message, on the other hand, is useless.

Brief but Vague   These structural supports are too heavy.

Cluttered   These structural supports weigh 300 pounds, thereby
exceeding the load tolerance specified by the building inspector for this project by a total of 50 percent.

Brief but   These structural supports weigh 300 pounds, thereby
Informative   exceeding our specified load tolerance by 50 percent.

Avoid wordiness. Use fewer words when fewer will do.

Cluttered   At this point in time I would like to say that we are
ready to move ahead with the project.

Concise   We are now ready.

The passage below is both wordy and poorly detailed.

Low-information Sentences   The lawn tennis court is bounded at each end by a screen. This *high* steel or wooden fence is placed *so that it is* at an *appropriate* distance beyond the baseline *of the court. Its function is significant in that* it prevents the ball from bouncing out of the playing area. The screen *is located so as to* mark the distal boundaries of the playing surface. Each segment *of the screen* is supported by *sturdy* poles *which are* set at *measured lengths.* (82 words)

There is no density here, no solid information. The words in italics either cause clutter or are too vague. Here is a concise version — informative and to the point.

High-information Sentences   The lawn tennis court is bounded at each end by a screen. This ten-to-twelve-foot steel or wooden fence is set exactly 21 feet beyond each baseline. The screen prevents the ball from bouncing out of the playing area, and marks the rear boundaries of the playing surface. Each segment is supported by 4-inch diameter poles set at 3-foot intervals. (65 words)

First drafts rarely are concise. Always revise parts that are wordy, repetitious, or vague. Get rid of anything that adds no meaning.

Eliminate Redundancy

Avoid using a phrase when a word will do. Each redundant phrase below can be reduced to a single word — without any loss in meaning.

at a rapid rate	=	rapidly
has the ability to	=	can
in this day and age	=	today
in regard to	=	about
in close proximity	=	near
the majority of	=	most
due to the fact that	=	because
aware of the fact that	=	know
on a personal basis	=	personally
take the place of	=	substitute

Another form of redundancy occurs when the same thing is said twice, in different words. Each bracketed word or phrase below merely adds clutter because its meaning is contained in the other word.

a [dead] corpse	[totally] monopolize
the [final] conclusion	[very] vital

[utmost] perfection	[past] experience
[mental] awareness	[close] scrutiny
[the month of] August	mix [together]
[valuable] asset	audible [to the ear]
[the color] green	correct [amount of] change
[mutual] cooperation	stands out [the most]
[fellow] colleagues	[the person,] Jim
free [of charge]	a [circular] disk

Avoid Overstuffing

A sentence that crams in too many ideas forces readers to struggle over its meaning.

Overstuffed   A smoke-filled room causes not only teary eyes and runny noses but also can alter auditory and visual perception as well, which is irritating but not associated with any serious disease, except for people with heart and lung ailments, who are threatened with major problems from smoke.

Correct   A smoke-filled room not only causes teary eyes and runny noses but can alter auditory and visual perception as well. Although the smoke itself does not produce serious disease, it does pose a threat to people with heart and lung ailments.

Avoid Needless Repetition

Much repetition in the following passage can be eliminated when sentences are combined.

Repetitious   Breathing is restored by artificial respiration. Artificial respiration means that breathing is being maintained by artificial means. Techniques of artificial respiration are mouth-to-mouth and mouth-to-nose. Artificial respiration must always be performed when external cardiac massage is being carried out. (43 words)

Revised   Breathing is restored by artificial respiration, either mouth-to-mouth or mouth-to-nose. Always give artificial respiration when performing external cardiac massage. (23 words)

Use Needed Repetition

Don't be afraid to repeat, or at least rephrase, material (even whole paragraphs) if you feel that readers need reminders at points in your report. Effective repetition helps avoid cross-references like this: "See page 903."

Avoid "There" Sentence Openers

Save words and improve your emphasis by avoiding "There is" and "There are" at the beginning of sentences.

> Weak   There are several reasons why Joe left the company.
>
> Revised   Joe left the company for several reasons.
>
> Weak   There is a danger of collapse in number 2 mine shaft.
>
> Revised   Number 2 mine shaft is in danger of collapsing.

Avoid Certain "It" Sentence Openers

Eliminate any "It" that does not refer to something specific.

> Wordy   [It was] his negative attitude [that] caused him to be fired.
>
> Wordy   It gives me great pleasure to introduce our new colleague.
>
> Improved   I am pleased to introduce our new colleague.

Delete Needless "To Be" Constructions

Forms of the verb "to be" ("is," "was," "are," etc.) often add clutter without adding meaning.

> Wordy   She seems [to be] upset.
>
> Wordy   I find some employees [to be] incompetent.

Avoid Excessive Prepositions

Although they *can* help break up stacked modifiers, prepositions (especially "of") can combine with forms of the verb "to be" to make wordy sentences.

> Wordy   These are the recommendations of some of the members of the committee.
>
> Revised   Some committee members made these recommendations.
>
> Wordy   An engineer by the name of Snopes just called.
>
> Revised   An engineer named Snopes just called.

Use "That" and "Which" Sparingly

Excessive use of "is" usually drags along a needless "that" or "which."

> Wordy   The report [, which is] about consumer attitudes[,] is three hundred pages long.

Wordy   This [is a] problem [that] requires immediate atten-
         tion.

Fight Noun Addiction

Excessive nouns make sentences awkward and wordy. They are often found
with needless prepositions or weak verbs, such as "is," "has," or "make." Break
up noun clusters by substituting action verbs.

Wordy   His memo is a request that we conduct a study of
         the problem.

Revised His memo requests we study the problem.

Wordy   I am giving consideration to the possibility of a
         change in career.

Revised I am considering changing careers.

Make Negatives Positive

Save words and get to the point by eliminating negative constructions.

Weak and Wordy   I did not gain anything from this course.

Revised   I gained nothing from this course.

Wordy   I don't see any difference between these two models.

Revised   I see no difference between these two models.

Clear Out the Clutter Words

Clutter words stretch a message without adding meaning. Here are some of
the most common: "very," "definitely," "quite," "extremely," "rather," "somewhat," "really," "actually," "situation," "aspect," "factor." If you must use
them, be sure they serve a purpose in your message.

Cluttered   Actually, one aspect of a business situation that could
             definitely make me quite happy would be to have a
             somewhat adventurous partner who really shared my
             extreme love of taking risks.

Revised   I seek an adventurous business partner who enjoys
             risks.

Delete Needless Prefaces

Get to the point. Deliver the pitch without a long wind-up.

Wordy   [I am writing this letter because] I wish to apply for
         the position of copy editor.

[The conclusion we can draw is that] writing is hard work.

Delete Needless Qualifiers

Qualifiers are expressions such as "I feel," "it seems," "I believe," "in my opinion," and "I think." Use them only to emphasize that your assertion is one of opinion, not of fact — an assertion subject to change.

> Appropriate Qualifier   Despite Frank's poor academic performance last semester, he will, I think, do well in college.

If you are sure of your assertion, however, eliminate the qualifier. Otherwise, it waters down your meaning and sounds evasive.

> Needless Qualifiers   [It seems that] I've wrecked the company car.
> [It would appear that] I've lost your credit card.
> [In my opinion,] you have a valid argument.

Making Sentences Fluent

Fluent sentences are polished, graceful, easy to read. Varied length and word order make them free from choppiness and monotony. The following suggestions will help you write fluent sentences.

Combine Related Ideas

Use coordination and subordination (pages 542–544) to combine a series of short, choppy sentences.

> Choppy   Jogging can be healthful if you have the right equipment. Most important are well-fitting shoes. These are important because without them you take the chance of injuring your legs. Your knees are especially prone to injury.
>
> Fluent   Jogging can be healthful if you have the right equipment. Well-fitting shoes are most important because they prevent injuries to your legs, especially your knees.

Most sets of information can be combined in various ways — depending on where the emphasis belongs. Consider the following sets of information.

1. Roland James is the third candidate for our engineering position.
2. He has graduated from an excellent program.

3. He has no practical experience.
4. He comes highly recommended.

Assume you are writing your colleagues a memo that records your impression of this candidate. If you want to emphasize a negative impression, you might combine the above data this way:

> Although Roland James, the third candidate for our engineering position, has graduated from an excellent program and comes highly recommended, **he has no practical experience.**

The thought in the independent clause is the one that receives the emphasis.

If, on the other hand, you are undecided about this candidate, but are leaning in a negative direction, you might combine the data this way:

> Roland James, the third candidate for our engineering position, has graduated from an excellent program and comes highly recommended, but he has no practical experience.

In the above example, the two independent clauses are joined by a coordinating conjunction ("but"), which suggests that both sides of the issue are of equal importance. Placing the negative feature last, however, gives it slightly more emphasis. If instead you were leaning toward the positive, you could reverse the order, placing the positive feature after the "but."

To emphasize strong support for the candidate, you might use the following combination:

> Although Roland James, the third candidate for our engineering position, has no practical experience, **he has graduated from an excellent program and comes highly recommended.**

Our meaning is reflected not only by what we say but also by how we arrange our information.

Vary Sentence Construction and Length

Long or short sentences have a purpose: to express ideas logically or forcefully.[1] Ideas that are linked need to be in the same sentence, to show their relationship.

> Weak   The loss-of-coolant alarm sounded. The operator shut down the reactor.
>
> Revised   When the loss-of-coolant alarm sounded, the operator shut down the reactor.

We combine ideas above to show that one action resulted from another.

---

[1] My thanks to Professor Edith K. Weinstein, University of Akron, for suggesting this distinction.

Conversely, an idea that should stand alone for emphasis needs its own sentence.

> Correct   Cigarettes are one of America's biggest public-health threats.

Too much of anything loses its significance. An unbroken series of short or long sentences makes for confused reading, as does a series with identical openings.

> Dreary   There are some drawbacks about diesel engines. They are much noisier than standard engines. They are difficult to start in cold weather. They cause considerable vibration. They also give off an unpleasant odor. They cause sulphur dioxide pollution. For these reasons, some auto makers are limiting their diesel models to light trucks only.

Combine and rephrase for significance:

> Revised   Diesel engines have some drawbacks. Most obvious are their noisiness, cold-weather starting difficulties, vibration, unpleasant odor, and sulphur dioxide emissions. Therefore, some auto makers are limiting their diesel models to light trucks only.

Opening most sentences with "The," "This," "He," "She," "It" or "They" creates monotony. When you write in the first person, overusing "I" suggests you are self-centered. Do not, however, avoid personal pronouns if they make the writing more readable (for instance, by eliminating passive constructions). Instead, to avoid repetition, combine ideas and shift word order.

Use Short Sentences for Special Emphasis

With all this talk about combining ideas, one might conclude that short sentences have no place in writing. Wrong. Whereas long sentences combine ideas to clarify relationships, short sentences (even one-word sentences) isolate a single idea for special emphasis. They stick in readers' minds.

CHOOSING THE EXACT WORDS

Your choice of words ultimately determines the quality of your writing. Keep your expressions simple, jargon-free, original, convincing, precise, and concrete and specific.

Keeping the Language Simple

Overblown diction and needless jargon obscure your message and make readers
work too hard.

Deflate the Diction

Say it in plain English. Avoid inflated diction. Don't use three syllables when
one will do. Trade for less:

utilize	=	use
approximately	=	about
to be cognizant	=	to know
to endeavor	=	to try
endeavor	=	effort
securing employment	=	finding a job
demonstrate	=	show
phenomenon	=	event
determine	=	find
multiplicity of	=	many
necessary	=	needed
effectuate	=	do
component	=	part
terminate	=	end

Count the syllables. Trim when you can. Don't write like the author of a report
from the Federal Aviation Administration who suggested that manufacturers of
the DC-10 be directed "to re-evaluate the design of the entire pylon assembly
to minimize design factors which are resulting in sensitive and/or critical maintenance and inspection procedures" (25 words, 50 syllables). Here is a plain-
English translation: "Redesign the pylons so they are easier to maintain and
inspect" (11 words, 18 syllables). Here is another example:

> Inflated   Upgrade your present employment situation. (5
> words, 12 syllables)

> Revised   Get a better job. (4 words, 5 syllables)

These are examples of the worst kind of flab: too many words, and words that
are bigger than needed. Keep it lean.

> Inflated   Accoustical attenuation for the food consumption area
> is needed.

> Revised   The cafeteria needs soundproofing.

> Inflated   Replacement of the weak battery should be effectu-
> ated.

Revised	Replace the weak battery.
Inflated	Make an improvement in the clerical situation.
Revised	Hire more (or better?) secretaries. (Inflated language also can be ambiguous!)

Of course, now and then the fancier word can serve best, if it expresses your exact meaning. For instance, don't substitute *end* for *terminate* if you're referring to something with an established limit.

Correct	Our warehouse lease terminates this month.

Also, if a single fancy word can replace a handful of simple words, and can sharpen your meaning, use the fancy word.

Weak	Six rectangular grooves around the outside edges of the steel plate are needed for the pressure clamps to fit into.
Revised	Six rectangular grooves on the steel plate perimeter accommodate the pressure clamps.

Avoid Needless Jargon

Various professions have their own "shorthand." Among specialists, certain technical terms do save time, and they communicate precise meaning economically. For example, "stat" on page 12 is shorthand for "Drop whatever else you're doing and deal immediately with the emergency." For computer buffs, a "glitch" is a momentary surge in current that can erase the contents of internal memory; a "bug" is an error that causes a program to run incorrectly; a computer that is "down" is out of service. These are just a few examples of computer jargon that communicates clear meaning to the initiated.

There are two ways, however, to use technical language — appropriately and inappropriately — and the latter is what we call needless jargon, meaningless to insiders as well as outsiders.

Needless Jargon	Unless all parties to the contract interface within the same planning framework at an identical point in time, the project will be rendered inoperative.
Revised	Unless we coordinate our efforts, the project will fail.
Needless Jargon	Intercom utilization will be used to initiate substitute teacher operative involvement.
Revised	Teachers who have to substitute will be notified on the intercom.

Needless Jargon   For the obtaining of the X-33 word processor, our firm will have to accomplish the disbursement of funds to the amount of $6,000.

Revised   Our firm will have to pay $6,000 for the X-33 word processor.

The practice in commercial jargon of adding "wise" to nouns as shorthand for "with reference to" is unacceptable: "saleswise," "timewise," "businesswise," "costwise," "moneywise," etc. Jargon-ridden writing makes you seem as if you are trying to put something over on your reader or as if you doubt the validity of your own statements. Before using any jargon, think about your specific readers, and use jargon only if it helps you communicate better.

The following letter, an unfortunate mix of 65-cent words, jargon, and needless use of passive construction, is a version of one published in a newspaper.

In the absence of definitive studies regarding the optimum length of the school day, I can only state my personal opinion based upon observations made by me and upon teacher observations that have been conveyed to me. Considering the length of the present school day, it is my opinion that the day is excessive length-wise for most elementary pupils, certainly for almost all of the primary children.

To find the answer to the problem requires consideration of two ways in which the problem may be viewed. One way focuses upon the needs of the children, while the other focuses upon logistics, transportation, scheduling, and other limits imposed by the educational system. If it is necessary to prioritize these two ideas, it would seem most reasonable to give the first consideration to the primary reason for the very existence of the system, i.e., to meet the educational needs of the children the system is trying to serve.

Here is a plain-English translation:

Although no studies have defined the best length for a school day, my experience and teachers' comments lead me to believe that the school day is too long for most elementary students — especially the primary students.

We can view this problem from the children's point of view (health, psychological welfare, and so on) or from the system's point of view (scheduling, transportation, utilities costs, and so on). But our primary concern is the children, because the system exists to serve their needs.

Use Acronyms Selectively

Acronyms are another form of specialized shorthand or jargon. They are formed from the first letters of a phrase (e.g., "LOCA," from Loss-of-Coolant Accident) or from a combination of first letters and parts of words (e.g., "bit," from *b*inary dig*it*). Computer science especially has given rise to countless acronyms. Here are two of the more common ones:

Bit: A binary digit whose value is 0 or 1. Within the computer, all data are expressed as a combination of bits in various sequences.

CPU (Central Processing Unit): The part of the computer which controls the transfer of information and carries out arithmetic and logical instructions.

Other computer-related acronyms include RAM (random access memory), CRT (cathode ray tube), and DOS (disk-operating system).

Acronyms *can* communicate concisely, but only if your audience knows their specific meanings. Unless you're certain of your audience's knowledge, define the acronym on first use. The less specialized your audience, the fewer acronyms you should use. Use acronyms only to *express*, not *impress*.

Being Convincing

Observe the following guidelines to give your writing the qualities of original expression and convincing tone.

Avoid Triteness

Writers who rely on tired old phrases (clichés) seem either too lazy or careless to find convincing or exact ways to say what they mean. The expressions below are just a few of the thousands that have become worn out through overuse.

first and foremost	not by a long shot
in-depth study	last but not least
in the final analysis	in a manner of speaking
consensus of opinion	the bottom line
it is interesting to note	welcome aboard
it has come to my attention	over the hill
needless to say	water under the bridge
as a matter of fact	take it in stride
to all intents and purposes	holding the bag
close the deal	bite the bullet

If it sounds like something you've heard before, don't write it.

Avoid Overstatement

Overstatement is a sure way for writers to lose credibility. Don't exaggerate to make your point.

Overstated   If you try skiing, you will find it to be one of the most memorable experiences of your life.

Revised   If you try skiing, you will enjoy it.

Overstated   If you hire me, I will be the best worker you've ever had.

Revised   If you hire me, I will do my best.

Avoid Sweeping Generalizations

Sweeping generalizations harm credibility because they lack factual support.

Sweeping   Television is rotting everyone's brain.

Revised   Many authorities argue that television is one cause of declining literacy.

Sweeping   Democracy in America is collapsing.

Revised   American democracy is threatened by fanatical organizations — both from the left and right.

Avoid Euphemisms

Euphemisms are expressions that aim at politeness, that make unpleasant or delicate subjects seem less offensive. Thus, one "powders one's nose" instead of using the toilet; "passes on," "passes away," "finds eternal rest," or "goes to meet one's maker" instead of dying. To the extent that euphemisms save hurt feelings or embarrassment, they are by all means legitimate.

One danger with euphemisms, however, is that they can sugar-coat the hard truth when only the truth will serve: criminals become "offenders"; illiterates become "disadvantaged students"; failing students become "underachievers"; political corruption becomes "lobbying"; rape becomes "sexual assault"; wars become "conflicts"; the nuclear disaster at Three Mile Island becomes an "incident"; and the prospect of nuclear holocaust is translated by one United States senator into "in the event of a nuclear exchange."

But as benign as conflicts, nuclear incidents, and nuclear exchanges may sound, they kill people just as dead as do wars, nuclear disasters, and a nuclear holocaust. Plain talk is better.

Being Precise

Be sure what you say is what you mean. Even words listed as synonyms contain a different shade of meaning. Do you mean to say, "I'm slender; you're thin; he's lean; and she's scrawny"? The wrong choice could be disastrous. Is your nose reacting to a "smell," "essence," "odor," "scent," "fragrance," "stench," or "stink"? Do not use one word when you mean another. Don't, for instance, write to apply for a job with a statement like this:

> Another attractive feature of X corporation is its **adequate** training program.

The sense conveyed by "adequate" clashes with the whole notion of "attractive." The above writer later explained his choice of *adequate:* the program, he said, was not highly ranked, so he tried to choose a word that would be sincere. But his choice has two liabilities:

1. *Adequate,* in this context, is insulting. Who likes to be called adequate?
2. The word suggests that the writer has placed himself in a position of judgment, overstepping his bounds as an applicant.

Any of several alternatives (*solid, respectable, growing*) retains sincerity without being offensive — or don't use any modifier!

Precision is not only important to the tone of your writing but also to its informative value. For instance, how would you like to be the executive who had to act on the following recommendation?

> Imprecise   I recommend a solar heating system for our new building because the price of solar will soon far exceed that of fossil fuels economically. (This writer meant to say that solar would be cheaper than other fuels.)

Always seek the most exact word for your meaning. In defining, say, *resistor,* for a nontechnical audience, you might call it "an electrical device that regulates current flow through a circuit." *Regulates,* however, is not as precise as *restricts* in explaining a resistor's function.

Learn the differences between words that sound alike. Here are some of the most commonly confused:

affect/effect	healthy/healthful
continual/continuous	imply/infer
fearful/fearsome	sensual/sensuous
formally/formerly	worse/worst

Do not write "Skiing is healthy" when you mean that skiing is good for one's health (healthful). Healthful things help keep us healthy.

Avoid *due to* when you mean *because.* (A train can be due to arrive!) Even worse is *due to the fact that* (imprecision burdened by flab). Don't write about *getting used to* something when you mean *becoming accustomed.* Tools *get used to* do a job; cars *are used to* travel, but you *become accustomed to* new circumstances.

Be on the lookout for imprecise (and therefore illogical) comparisons.

> Imprecise   Your rate of interest is higher than the First National Bank. (Can a rate be higher than a bank?)

> Revised   Your rate of interest is higher than that of the First National Bank.

Being Concrete and Specific

Informative writing *shows* as well as *tells*. Consider an abstract and general assertion like this one:

> Crossing the street in front of our office, pedestrians place their lives in danger.

You need to support the assertion — to *show*, with concrete and specific examples. Do not say simply,

> For example, **an office worker** was **injured** there by a **vehicle recently.**

Because they are much too general, the boldface words only *tell*. Instead, say,

> Alan Hill was hit by a speeding truck last Tuesday and suffered a broken leg.

Notice how the informative value of the following terms increases as we move down the scale.

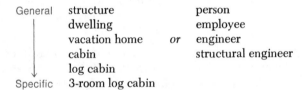

General	structure		person
	dwelling		employee
	vacation home	*or*	engineer
	cabin		structural engineer
	log cabin		
Specific	3-room log cabin		

Choose high-information words that show exactly what you mean. Don't write "thing" when you mean "lever," "switch," "micrometer," or "compass." Instead of evaluating an employee as "swell," "great," "terrific," "lousy," "terrible," or "awful," use informative modifiers such as "reliable," "skillful," "competent," "dishonest," "irritable," or "incompetent." The earlier modifiers reveal your attitude, but nothing about the person. Readers have to know what we mean.

Abstract	Professor Jones's office looks like a disaster area.
Concrete	Professor Jones's office has a floor strewn with books, a desk buried beneath a mountain of uncorrected papers, and ashtrays overflowing with cigarette butts.
General	Industrial emissions are causing lakes to die.
Specific	Sulphur dioxide emissions from coal-burning plants combine with atmospheric water to produce sulfuric acid. The resultant "acid rain" increases the acidity of lakes to the point where they no longer can support life.

When applicable, provide solid numbers and statistics that get your point across:

General	In 1972, thousands of people were killed or injured on America's highways. Many families had at least one relative who was a casualty. After the speed limit was lowered to 55 miles per hour in late 1972, the death toll began to drop.
Specific	In 1972, 56,000 people died on America's highways; 200,000 were injured; 15,000 children were orphaned. In that year, if you were a member of a family of five, chances are that someone related to you by blood or law was killed or injured in an auto accident. After the speed limit was lowered to 55 miles per hour in late 1972, the death toll dropped steadily to 41,000 deaths in 1975.

Concrete and specific information is not only more informative; it is more convincing as well.

Using Analogies to Sharpen the Image

Ordinary comparison shows similarities between two things *of the same class* (two computer keyboards, two technicians, two methods of cleaning dioxin-contaminated sites). Analogy, on the other hand, shows some essential similarity between two things of *different classes* (report writing and computer programming, computer memory and post-office boxes).

Analogies are good for emphasizing a point (for example, that some rain is now as acidic as vinegar). But they are especially useful in translating something abstract, complex, or unfamiliar to laypersons, as long as the easier subject is broadly familiar to readers. For instance, to translate how power dissipation in a resistor (as it restricts current flow) produces heat, you might use a kitchen toaster as an analogy. However, to understand that the toaster coil serves as a resistor (so it can toast bread), readers would have to be familiar with toasters in general. Analogy therefore calls for particularly careful analyses of audience.

Analogies are one reason computers have become less intimidating to the public. Think of the lifelike characteristics computers have been assigned. For instance, a word processor is sometimes called a "typewriter with a memory." If the hardware and software are easy to use, they are termed "user friendly." If we need to process lots of words or numbers, we need a computer with an adequate "memory" for "word- or number-crunching." In a power loss, the computer will "crash" (or lose everything in its memory).

Analogies can save words and convey vivid images. *Collier's Encyclopedia,* for instance, describes the tail of an eagle in flight as "spread like a fan." On page 185 of this book, the description of a trout-feeder mechanism uses this analogy to clarify the positional relationship of two working parts:

> Analogy   The metal rod is inserted and centered, *crosslike* between the inner and outer section of the clip. . . .

Without the analogy, *crosslike,* we would need something like this to visualize the relationship:

> Missing Analogy   The metal rod is inserted, *perpendicular to the long plane and parallel to the flat plane,* between the inner and outer sections of the clip. . . .

This second version is doubly inefficient: more words are needed to communicate, and more work is needed to understand the meaning.[2]

Besides naming things in more vivid ways, analogies help *explain* things. Here is an extended analogy that helps readers understand something unfamiliar (computer programming) through a comparison to something more familiar (writing).

EXTENDED ANALOGY

Like report writers, computer programmers rely on problem-solving skills and procedural thinking to analyze a problem and generate a solution. Both the writing process and the programming process entail a deliberate set of deliberate decisions.

Writers and programmers define their process and refine their product. First, they decide about their purpose and their audience's needs. Both must define and express complex thoughts within the constraints of a particular language. They must formulate, organize, and articulate the details of a logical map that allows for no ambiguity and requires absolute precision. Writers revise their reports and programmers debug their programs in a cyclical (recursive) process, continuously refining their meaning, rearranging parts, and sharpening their expression.

Both writers and programmers must generate messages that have *one* interpretation only. The writer's instructions, as well as the programmer's, must yield the exactly specified results. Both writers and programmers must be sure of their exact meaning, and must communicate that meaning clearly. Both reports and programs should be written in a style and format that allow readers to follow the line of reasoning clearly, and to identify the significant parts.

---

[2] Analogy is, of course, a form of metaphor. For an inspired discussion of metaphor's role, see John S. Harris, "Metaphor in Technical Writing," *The Technical Writing Teacher,* Vol. 2 (Winter, 1975): 9–13.

To make reports easy for readers to follow, writers provide cues through sensible arrangement (transitions, topic sentences, paragraphs), formatting devices (headings, white space, underlining, varying typeface) and report supplements (title page, table of contents, abstract, appendixes). To make programs easy for users to follow, programmers provide documentation: they explain why the program was developed; they describe what the program will do and explain how it will do it; they give instructions for using the program and interpreting its results.

Written reports and computer programs can have similar things go wrong. Writers might give inappropriate or unclear information, while programmers might include documentation computer users can't follow. Missing or incorrect punctuation might cause the reader or computer to misinterpret a sentence or a program instruction. A misspelled or imprecise word will be rejected by the reader or unrecognized by the computer. And, worst of all, a misleading message (a wrong instruction, something left out, or incorrect data) might cause the reader to act or the computer to run, but with the *wrong* results.

ADJUSTING THE TONE

Your tone is your personal mark — the personality that emerges from between the lines. The tone you create in any writing depends on (1) the distance you impose between yourself and the reader, and (2) the attitude you express toward the subject. Assume a friend is going to take over a job you've held. You've decided to write instructions for your friend. Here is your first sentence.

> Now that you've arrived in the glamorous world of office work, put on your track shoes; this is no ordinary manager-trainee job.

What kind of tone emerges in this sentence? Notice that the writer imposes little distance between herself and the reader (she uses the direct address, "you," and the humorous suggestion to "put on your track shoes"). And the ironic use of "glamorous" suggests just the opposite.

For a different reader (let's say, a company training manual written for a stranger) the writer would have chosen some other way to open. For example:

> As a manager trainee at GlobalTech, you will work for many managers. In short, you will spend little of your day at your desk.

The tone has changed; it is no longer intimate, and the writer expresses no distinct attitude toward the job. For yet another audience (say, in a company pamphlet for new trainees), the writer might have altered her tone again:

> Manager trainees at GlobalTech are responsible for duties that extend far beyond desk work.

Here the writer imposes even more distance between herself and her audience, especially through the shift from second- to third-person address.

Your tone changes in response to the situation and audience, even though the subject may be the same. For instance, letters to your professor, your grandmother, and your friend, each about a disputed grade, should have noticeably different tones:

1. Dear Professor Snapjaws:
   I am convinced that my failing grade in Professional Writing did not reflect a fair evaluation of my work over the semester. . . . [a formal tone]
2. Dear Grandma,
   Thanks for your letter. I'm doing well in school, except for the failing grade I unjustly received in Professional Writing. . . . [a semiformal tone]
3. Dear Fred,
   Boy, have I been shafted! That old turkey, Snapjaws, gave me an F in Professional Writing. . . . [informal tone]

In each version, the writer expresses disapproval. But as the distance between writer and audience decreases, his statement becomes increasingly informal. The writer adjusts tone to suit the audience. He also uses tone to control the distance from his audience. Clearly the intimate tone and attitude of outrage in the letter to Fred would be inappropriate for Professor Snapjaws, to whom the writer expresses firm disapproval. And the attitude of polite disapproval in the letter to Grandma would be inappropriate for Fred or the professor.

Establishing Appropriate Distance

When you meet someone, you respond in a tone that defines your relationship:

- Honored to make your acquaintance. [formal tone — greatest distance]
- This is indeed an honor. [formal]
- Nice to meet you. [semiformal — medium distance]
- Hello. [semiformal]
- How do you do? [semiformal]
- Hi. [informal — least distance]
- What's happening? [informal — slang]

Your greeting is based on how much distance you decide is appropriate, and, in turn, the tone determines how you come across. Each of these responses is appropriate in certain situations, inappropriate in others. "What's happening?" might be OK for another student, but not the college president.
    As a rule,

- Use a formal or semiformal tone with superiors or dignitaries (depending on what you think the reader expects).

– Use a semiformal or informal tone in technical writing (depending on how close you feel to your readers).

– Use an informal tone when you want your writing to be conversational, or when you want your writing to sound like people talking to other people.

Whichever tone you decide on, be consistent throughout your message.

> Inconsistent Tone   My office isn't fit for a pig [too informal];
> it is ungraciously unattractive [too formal].
>
> Revised   The shabbiness of my office makes it unfit for work.

In general, lean toward an informal tone without falling into slang. Keep your writing conversational by following these suggestions:

1. Keep the language simple and free of needless jargon (pages 58–61).
2. Use an occasional contraction.
3. Address readers directly, when possible.
4. Use "I" when appropriate.
5. Prefer active to passive voice (pages 48–50).

Use an Occasional Contraction

Use a contraction to loosen the tone. Balance an *I am* with an *I'm*, a *you are* with a *you're*, an *it is* with an *it's* (as we've done throughout this book). Be careful, though, not to overuse contractions.

> Excessive   It's a shame that Clementine's been crying since she's
> Contractions   learned her date'll be late because his tire's flat and
> his wallet's lost.

Generally, use contractions only with pronouns, not with nouns or proper names. Otherwise, the constructions are awkward or ambiguous.

> Awkward   Barbara'll be here soon.
> Contractions   Health's important.
> Love'll make you happy.
>
> Ambiguous   The dog's barking.
> Contractions   Bill's skiing.

These ambiguous contractions can easily be confused with possessive constructions.

Address Readers Directly

Whenever appropriate, use direct address. Otherwise, your writing sounds like this: "Whenever appropriate, a writer should use direct address." Notice how distance *and* words increase.

Distant   Students at the college will find the faculty always willing to help.

Closer   As a student at our college, you will find the faculty always willing to help.

Caution: Use "you" only to correspond *directly* with the audience, such as in a letter, memo, instructions, or some form of advice, encouragement, or persuasion. Do not use second-person address when your subject and purpose call for first or third person. Otherwise, you might write something wordy and awkward like this:

Weak   When you are in northern Ontario, you can see wilderness lakes everywhere around you.

Revised   Wilderness lakes abound in northern Ontario.

Use "I" When Appropriate

Don't disappear behind your writing. Use "I" when referring to yourself.

Distant   This writer would like a refund.

Revised   I would like a refund.

Distant   The fear was awful until the police arrived.

Revised   I was terrified until the police arrived.

Avoid opening too many sentences with "I." Combine ideas and shift word order instead.

Expressing a Clear and Appropriate Attitude

Earlier chapters emphasize the need for technical documents to be based on facts, *not* on uninformed opinion. But in situations where your attitude and informed opinion *are* expected, let readers know where you stand. Don't force them to translate. Say "I enjoyed the fiber optics seminar" instead of "My attitude toward the fiber optics seminar was one of high approval." Say "Let's liven up our dull relationship" instead of "We should inject some rejuvenation into our lifeless liaison." If, however, your job is to report impartially, do not give a biased view by inserting your attitude.

Remain Unbiased

Your responsibility as a writer is to report accurately and fairly without distorting the evidence or injecting "loaded" words that reflect personal biases. If you *are* asked to include your interpretations and conclusions, base them on facts. Even controversial subjects deserve unbiased treatment. For instance, imagine you have been sent to investigate the causes of an employee-management confrontation at your company's Omaha branch. Your initial report, written for the New York central office, is intended simply to describe the incident. Here is how an unbiased description might read:

> At 9:00 A.M. on Tuesday, January 21, eighty women employees entered the executive offices of our Omaha branch and remained for six hours, bringing business to a virtual halt. The group issued a formal statement of protest, claiming that their working conditions were repressive, their salary scale unfair, and their promotional opportunities limited. The women demanded affirmative action, insisting that the company's hiring and promotional policies and wage scales be revised. The demonstration ended when Garvin Tate, vice-president in charge of personnel, promised to appoint a committee to investigate the group's claims and to correct any inequities.

Notice the absence of implied judgments; the facts are presented objectively. A less impartial version of the event, from the women's side, might read as follows:

> Last Tuesday, sisters struck another blow against male supremacy when eighty women employees paralyzed the pinstriped world of solidly entrenched sexism for more than six hours. The timely and articulate protest was aimed against degrading working conditions, unfair salary scales, and lack of promotional opportunities. Stunned executives watched helplessly as the group occupied their offices. The women were determined to continue their occupation until their demands for equal rights were met. Embarrassed company officials soon perceived the magnitude of this protest action and agreed to study the group's demands and to revise the company's discriminatory policies. The success of this long overdue confrontation serves as an inspiration to oppressed women employees everywhere.

Notice how the use of judgmental words and qualifiers ("male supremacy," "degrading," "paralyzed," "articulate," "stunned," "discriminatory," etc.) injects the writer's personal attitude, even though it's not called for. In contrast to the above bias, the following version defends the status quo:

> Our Omaha branch was the scene of an amusing battle of the sexes last Tuesday, when a Women's Lib group, eighty strong, staged a six-hour sit-in at the company's executive offices. The protest was lodged against alleged inequities in hiring, wages, working conditions, and promotion for women in our company. The libbers threatened to remain in the building until their demands for "equal rights" were met. Bemused company officials responded to this carnival demonstration with patience and dignity, assuring the militants that their claims and demands — however inaccurate and immoderate — would receive just consideration.

Again, the use of qualifying adjectives and superlatives slants the tone of the report. Let your facts speak for themselves.

Avoid Sexist Language

Another element of bias occurs when people use sexist language, stereotyping people on the basis of sex. Your usage is sexist if you refer in general to doctors, managers, lawyers, company presidents, engineers, and other professionals as *he* or *him* while referring to nurses, secretaries, homemakers, and the like as *she* or *her*. Our goal as communicators is to *identify* with our audience, not to exclude half the population. Sexist language is demeaning and misleading.

Follow these guidelines to eliminate sexism from your communications:

– Use neutral terms. For example,

chair, or chairperson	rather than chairman
businessperson	rather than businessman
supervisor	rather than foreman
police officer	rather than policeman
letter carrier	rather than postman
homemaker	rather than housewife

– Use plural forms. For example, instead of using "The manager ... he," use "The managers ... they."

– When possible (as in direct address), use "you." For example, "You can begin to eliminate sexual bias by becoming aware of the problem."

– When possible, use a person's name, then refer to that person with the appropriate pronoun. For example, "Ms. Williams..., she...." Mr. Williams..., he...."

– Drop endings such as "-ess," and "-ette" used to denote females (e.g., poetess, authoress, bachelorette, majorette).

– Avoid the overuse of pairings ("him or her," "she and he," "his or her" "he/she" and the combined form, "s/he"). Too many such pairings are awkward, call attention to themselves, and do little to eliminate sexist language.

– See Chapter 16 for guidelines on how to avoid sexist salutations (e.g., Dear Sirs, Gentlemen, etc.).

EXERCISES

1. The following sentences are unclear because of missing function words, ambiguous phrasing or punctuation, or overstuffing. Revise them so their meaning is clear. (For sentences that suggest two meanings, write two versions.)

a. State law requires that restaurants serve food with a sanitation certificate.
b. Bring dictionary to exam as reference.
c. Making the shelves look neater should be another one of our priorities for increasing sales, because if the merchandise is not always neatly arranged, customers will not have a good impression, whereas if it is neat, they probably will return.
d. His constant aggravation will someday cause a nervous breakdown.
e. Along with losing weight, swimming tones muscles and improves lung capacity.
f. While camping, we live in different surroundings from Friday to Sunday.
g. A man eating shark was spotted in Boca Grande Harbor.
h. Wearing high boots, the snake failed to injure the supervisor.

2. Improve clarity and emphasis in the following sentences by unstacking the modifiers or by rearranging word order.

a. Our students are tested by an incoming freshman mandatory writing proficiency exam.
b. Education enables us to recognize excellence and to achieve it.
c. The new diesel engine trailer truck driver training school is now enrolling students.
d. In a business relationship, trust makes it work.
e. A densely packed several-inch-thick layer of pine needles covered the ground.
f. Henry is enrolled in a home entertainment electronic system home study course.

3. Some of the following sentences need to be rewritten in the active or passive voice for better emphasis, less awkwardness, more directness, or greater economy. Make the necessary changes and be prepared to give reasons for each. Mark an E by the sentences that are already effective.

a. It is believed by us that this contract is faulty.
b. The tall model wore the $50,000 mink coat.
c. Joe has been fired.
d. Hard hats should be worn at all times on this job.
e. A tornado destroyed our brand new tractor.
f. It was decided not to accept your invitation.
g. A check for full payment will be sent next week.
h. This package should be kept cool.
i. A rockslide buried the mine entrance.
j. Searchers found the victim almost dead.
k. It is my hope that you succeed.

4. Get rid of clutter in the following sentences by eliminating redundancies and needless repetition.

a. He is a man who works hard.
b. This report is the most informative report I've read in months.
c. I am aware of the fact that Sam is a trustworthy person.
d. On previous occasions, we have worked together.
e. I have admiration for Professor Jones.
f. I'll meet you in the downtown area of the city.
g. She has the talent to restore old furniture back to life.
h. In the event that you need help, call me.
i. Sally is an associate of mine.
j. Stretching is very vital before exercising.

5. Make the following sentences more concise by eliminating "There is" and "There are" sentence openers and the needless use of "it" "to be," "is," "of," "that," and "which."

a. I consider George to be a talented technician.
b. The sales volume for last month, which was dreadfully low, disappointed everyone.
c. This step must be practiced in order for it to become effective.
d. There are certain people whom I enjoy traveling with.
e. Another reason the job is attractive is because the salary is excellent.
f. Smoking of cigarettes is considered by many people to be the worst habit of all habits of human beings.
g. Our summer house, which is located on Cape Cod, is for sale.
h. Many of the jobs that I have held have been interesting.
i. There are many employees who always do excellent work.
j. The static electricity that is generated by the human body is measurable.
k. Frustration is a normal part of the burden of insecure writers.
l. There has been a computer boom that has swept the country.
m. There are eight holes at 45-degree intervals around the outside surface of the rim, which are needed for the head bolts to fit through.

6. Improve the economy and directness of the following sentences by replacing nouns with verbs, changing negatives to positives, and clearing out clutter words, needless prefaces, and needless qualifiers.

a. We request the formation of a committee of students for the review of grading discrepancies.
b. I am not unappreciative of your help.
c. I tend to disagree with you.
d. It seems that I've made a mistake in your order.
e. Bill made the suggestion that we hire an additional salesperson.
f. Her quick wit is an extremely impressive aspect of her personality.
g. I find Boris to be an industrious and competent employee.
h. Actually, I am very definitely interested in this position.
i. Igor doesn't have any friends at the office.
j. In my opinion, George is a responsible employee.

7. Use two ways of combining each group of sentences below into a single, fluent sentence.

a. Sarah Fields is a structural engineer.
b. She has been assigned to the Northern Ontario Hydroelectric Project.
c. She is completing an inspection of the Winisk River dam site.

a. Calvin and Calvin Associates are an innovative architectural firm.
b. They have designed outstanding homes.
c. They sometimes do work for the city.
d. They lost the urban-renewal contract.

a. Boats with excessive horsepower should be banned from our lakes and ponds.
b. This would be one way to decrease noise and water pollution.
c. It also would save fuel.

a. I was employed by the Food Mart supermarket.
b. I held the position of service clerk.
c. It was my job to operate a cash register.
d. Also, I priced items and stocked shelves.

8. Rewrite the following paragraph to make it more concise and fluent.

There are two methods that may be used in glazing pottery. The best method is to use underglazes and glazes. Three coats of each are applied. The underglazes designate the color; the glazes give the pottery a shiny finish and a semitransparent color. The underglazes are put on first. They are usually put on with a fine paint brush. The glazes are put on next. A larger brush is usually used for this. The glazes are patted on, whereas the underglazes are brushed on. When glazes and underglazes are used, the pottery must be fired again. This method produces a shiny effect.

9. Rewrite the following statements in plain English.

a. This writer desires to be considered for a position with your company.
b. At this point in time we cannot agree to your terms.
c. Please refund our full purchase expenditure in view of the fact that the microscope is defective.
d. No decisions will be made until next week as far as the contract is concerned.
e. There are several banks that can be contacted in terms of obtaining a business loan.
f. I can wish you no better luck than that you find this job as enjoyable as I have.
g. Prior to this time we have had no such equipment failure.
h. In relation to your job, I would like to say that we can no longer offer you employment.
i. This report is useless as far as I am concerned.
j. Further interviews appear to be a necessity before we can identify the best qualified candidate.

k. I suggest that you might want to consider shipping your lobsters by air transport.

l. I suggest that you reduce the amount of cigarettes that you consume daily.

m. A good writer is cognizant of how to utilize grammar in a correct fashion.

n. This office is hopeful that, after a full transynchronization of effort, a more viable solution to our problem will eventuate.

o. Within the copier, a magnetic-reed switch is utilized as a mode of replacement for the conventional micro-switches which were in use on previous models.

10. Revise the following sentences by eliminating the overstatements and euphemisms.

a. I was less than candid.

b. This employee is poorly motivated.

c. It was decided to terminate your employment.

d. She expropriated company funds.

f. When the grenade exploded, his arm was traumatically amputated.

g. Quitting your job at this stage would be an act of self-destruction.

h. Cigarette smoking destroys your health.

i. You're the world's best boss!

j. The business world follows the law of survival of the fittest.

k. Today's students are nothing but illiterates.

11. Revise the following sentences to eliminate triteness and needless jargon.

a. The rising prime rate is sending shock waves through the business world.

b. The preparation of this report has been facilitated by the assistance of Professor Jones.

c. We will be appreciative of your input on this plan.

d. Timewise, this construction schedule is not viable.

e. In order to optimize our financial return, we should prioritize our investments.

f. There's never a dull moment on this job.

12. Revise the following sentences to make them precise.

a. Our outlet does more business than San Francisco.

b. Low-fat foods are healthy.

c. Due to hours of studying, I received an "A" on the exam.

d. Anaerobic fermentation is used in this report.

e. Prices hope to be held down by building a smaller engine.

f. The diesel can meet antipollution standards by installing exotic hardware.

g. When a plug is pushed in the outlet, it connects the cord to the power supply.

h. The diesel is not without faults and skepticism.

13. Revise the following sentences to make them more concrete and specific.

Example

A storm damaged the building.

Revised: A tornado tore the roof off our number 2 warehouse.

a. He received an excellent job offer.

b. The group presented its demands.

c. She repaired the machine quickly.

d. This thing bothers me.

e. His performance was awful.

f. The crew damaged a piece of furniture.

g. My new employee is disappointing.

h. They discussed the problem.

i. She claimed that he had never phoned her about the deal.

j. An animal injured a person at the construction site.

14. Eliminate the sexual bias or awkward expression in the following sentences.

a. When someone is elected senator, he receives a travel allowance.

b. The group's spokesman demanded improved working conditions.

c. When a nurse finishes her rounds, she must check in at the nurses' station.

d. The policeman didn't respond in time.

e. When a project engineer completes his assignment, he is transferred to another project.

f. The poetess read her most popular poems.

g. Should any lawyer arrive before the judge, direct him to the judge's chambers.

h. If a student needs a referral, s/he should be sent to Dean Simple.

i. The girls talked about world affairs, while the men took their time cooking dinner.

5

An Overview of the Writing Process

The Writing Process Defined

A Sample Writing Situation

Chapter Summary

Exercises

THE WRITING PROCESS DEFINED

Chapters 2 through 4 have covered some of the decisions writers make about their purpose, audience, content, organization, and style:

– What purpose do I wish to achieve with this document?
– Who are my readers, and what do they need to know?
– How can I find something worthwhile to say?
– How do I get organized?
– How will I know if the style is OK?

In what ways do writers combine these decisions to produce useful technical documents? What *process* do writers follow?

We've already seen that the actual drafting (putting words on the page) is only a small part of the writing process. Your biggest challenge lies in the other parts:

1. deciding on the exact meaning you wish to communicate
2. deciding on your intended audience and on how they will use your message

3. deciding on a plan for making your meaning clear to readers

4. discovering new meanings as you write, and deciding on ways of revising your original plan to accommodate these new meanings

In short, *writing is a process of making deliberate decisions in response to a specific situation.*

Every writer decides on answers to all these questions, but rarely in a neat, predictable order. Instead, each person approaches the writing process through a decision sequence that works best for *that* person. The writing process, in fact, is basically a *thinking* process. This chapter shows one typical writer's thinking in response to an actual situation.

A SAMPLE WRITING SITUATION

The company is Microbyte, maker of low-priced, dedicated word processors and portable microcomputers. The writer is Glenn Tarullo (B.S., Management; Minor: Computer Science). Glenn has been on the job two months as Assistant Training Manager for Microbyte's Marketing and Customer Service Division. For three years, Glenn's boss, Marvin Long, has periodically offered a training program for new managers. The program combines an introduction to the company with instruction in key management skills (time management, motivation, communication, etc.). Long seems pleased with his two-week program, but he has asked Glenn to evaluate it and write a report as part of a company move to upgrade personnel procedures.

Microbyte has a system for routing internal correspondence to division heads and decision makers. So Glenn knows his report will be read by Long's boss, George Hopkins (Assistant Vice-President, Personnel), and Charlotte Black (Vice-President, Marketing, the person who devised the upgrading plan). Copies also will go to other division heads, and to the division's chief executive: Brian Hull, Senior Vice-President for Human Resources.

Glenn spends two weeks (Monday, October 3 to Friday, October 14) monitoring and taking notes on Long's classes. On October 14, trainees are asked to evaluate the program. After reading these responses and reviewing his notes, Glenn concludes that the program was successful, but could stand improvement in such areas as course content and teaching techniques. His problem is to recommend improvements — without offending anyone (instructors, his boss, various executive guest speakers, and so on). Glenn is scheduled to present his report in conference with Long, Black, and Hopkins on Wednesday, October 19. Right after the final class (1 P.M., Friday, the 14th), Glenn gets to work on a first draft of his report. Let's trace his decisions.

Glenn decides to begin with a quick draft, off the top of his head — just to get warmed up. As a fairly new employee with an impossibly busy schedule (and now this memo deadline), Glenn spends half of Friday afternoon fretting over the many details of his situation, the many readers and other people involved, and the potential political problems. By 3 P.M., Glenn still hasn't written a word. Desperate, he decides to write whatever comes to mind. Here is his first draft.

Glenn's First Draft

Although the October Management Training Session was deemed quite successful, it would appear that several problems have emerged which require our immediate attention.

– There were too many of the instructors who had poor presentation skills. There were a few who never arrived on time. Mr. Thomas didn't stick to the topic but rambled incessantly. Mr. Jones and Ms. Wells seemed poorly prepared. Instructors in general seemed to lack any well-defined sense of clear objectives. Also, because too few visual aids were used, many presentations seemed colorless and apparently bored the trainees.

– The trainees (all new people) were not at all cognizant of how the company was organized or functioned. So the majority of them often couldn't relate to what the speakers were talking about.

– It is my impression that this was a weak session due to the fact that there were insufficient members (only five trainees). Such a small class makes the session a waste of time and money. For instance, Leslie Beck, Senior Vice President of Personnel, came down to spend over one hour addressing only a handful of trainees. Another factor is that the fewer trainees in a class, the less dialogue occurs, with people tending to just sit and get "talked at."

– Last but not least, executive speakers generally skirted the real issues, untruthful about what it was really like to work here. They never really explained how to survive politically (e.g., never criticize your superior; never complain about the hard work or long hours; never tell anyone what you *really* think; never observe how few women are in executive or managerial positions, or how disorganized things seem to be). New employees shouldn't have to learn these things the hard way.

In the final analysis, if these problems can be addressed with all due haste, it is my opinion we can look forward in the future to effectuating management training sessions of even higher quality than those we now have.

Glenn completes this first draft at 5:10 P.M. He's depressed over the results, so he decides to ask someone with experience for a quick critique. He chooses Ms. Blair Cordasco, a senior project manager, who has been with Microbyte fifteen years. She's already helped Glenn several times. Cordasco agrees to read Glenn's draft over the weekend; she suggests they have coffee Monday morning.

Glenn's Next Attempt

At 10:05 Monday morning, Cordasco goes over the document with Glenn. First, she points out obvious style problems: wordiness ("due to the fact that"), jargon ("effectuating"), triteness ("in the final analysis"), needless qualifiers ("it would appear that"), implied bias ("weak presentation," "untruthful"), among others. Can you identify other style problems in this draft?

Cordasco points out other problems as well: the piece is disorganized; the emphasis is much too critical (offensive to Glenn's boss, making him look bad to the V.P.'s); the views are too subjective (no one is interested in having Glenn "sound off" about what he thinks the company's political problems are); the report contains no useful advice; material about course follow-up and *positive* features of the program is missing. In this form, the report will only alienate important people and hurt Glenn's career.

Rethinking his situation, Glenn sees he needs more control. He decides to write a statement of purpose, outlining what he wants to accomplish before deciding *what* to say and *how* to say it. So he tries this statement:

> Statement of
> Purpose
> (rough draft)

> My purpose is to write a report that will cause positive change — without alienating anyone.

Glenn immediately sees that this statement is too indefinite to provide any real orientation. He realizes he needs to begin by *focusing* on his writing situation. So his thinking goes like this:

> Well . . . to begin, I'd better figure out *exactly* what my primary reader wants to know.
>
> Although Long requested the report, he did so because the Marketing V.P., C. Black, developed the scheme for division-wide improvements. So I really have two primary readers here. My boss and the big boss. (And, of course, George Hopkins, my boss's boss, will be reading this memo carefully, too.)
>
> My major question here is:
>
> Am I including enough details for all the bosses? And the answer to this question will require an answer to more specific questions:

> Anticipated Reader
> Questions

> What are we doing right, and how can we do it better?
>
> What are we doing wrong, and does it cost us money?
>
> Have we left anything out, and does it matter?
>
> How, specifically, can we improve the program, and how will those improvements help the company?

Since all readers have participated in many sessions (either as trainees, instructors, or guest speakers), they have the same background on this topic. So I don't have to worry about background explanations for any uninformed readers.

Now . . . back to readers' questions. To answer, I should begin by pointing out the *positive* features of the last training session. Then I can discuss the problems and make some recommendations. Maybe I can change the negative tone of my first draft by *suggesting improvements* instead of *criticizing weaknesses*. This way, I'm less likely to make anyone defensive. Also, I could briefly describe the *benefits* of following each suggestion.

Glenn realizes that if he wants successful programs in the future, he mustn't alienate the people involved. After all, he hopes to work with these people a long time.

OK . . . now this memo is shaping up. I guess I have a clear enough picture of my plan to write a better statement of purpose.

Statement of Purpose (revised)

My purpose is to provide my supervisor and interested executives with a detailed evaluation of the October training workshop by describing its strengths, suggesting improvements, and explaining the benefits of these changes.

Working from this statement, I should be able to revise my first draft into a decent report. But . . . that first draft is missing some important details I want to include. Maybe I should brainstorm here, to get *all* the details (including the *positive* ones) I want to cover.

Glenn's Brainstorming List

1. Better-prepared instructors and more visuals
2. On-the-job orientation *before* the training session
3. More members in training sessions
4. Executive speakers need to spell out qualities needed for success

                                            From Glenn's First Draft

5. Beneficial emphasis on interpersonal communication
6. Need follow-up evaluation (in six months?)
7. Four types of training evaluations:
   a. trainees' reactions
   b. testing of classroom learning
   c. transference of skills to the job
   d. impact of training on the organization (higher sales, more promotions, better written reports, etc.) *
8. Videotaping and giving critiques of trainee speeches worked well
9. Acknowledge the positive features of the session

    10. Ongoing improvement insures quality training
    11. Division of class topics into two areas was a good idea
    12. Additional trainees would increase classroom dialogue
    13. The more trainees in a session, the less time and money wasted
    14. Instructors shouldn't drift from the topic*
    15. On-the-job training to give a broad view of the division
    16. Clear course objectives to increase audience interest and to measure the program's success
    17. Marvin Long has done a great job with these sessions over the years*

By now it's 9:05 A.M.; the office is hectic. Glenn puts his list aside and spends the day on work that has piled up. Not until 4 P.M. does he return to his brainstorming list.

> Now what? I should delete items [marked by asterisks] my audience already knows or doesn't need to know: 7 can go (this audience needs no lecture in training theory); 14 is too negative and critical — besides, the same idea is stated more positively in 4; 17 is obvious brown-nosing, and I'm in no position to make such grand judgments.

> Now maybe I can unscramble this list by arranging items under the categories (strengths, suggested changes, and benefits) from my statement of purpose. (And within each category, I'll list items in the order I'll want to present them — and maybe I'll discover more items.)

Glenn's Brainstorming List Rearranged

Notice here how Glenn discovers other ideas about *content,* while he's deciding about *organization.*

> Strengths of the Workshop
> – division of class topics into two areas was useful
> – emphasis on interpersonal communication
> – videotaping of trainees' oral reports, followed by critiques

> Well, that's one category done. Maybe I should combine *suggested changes* with *benefits,* since I'll want to cover them together in the report, anyway.

> Suggested Changes/Benefits
> – more members per session would increase dialogue and use resources more efficiently
> – varied on-the-job experiences before the training sessions would give each member a broad view of the marketing division
> – executive speakers should spell out qualities required for success [and future sessions should cover professional behavior],[1] to provide trainees with a clear guide

---

[1] Brackets enclose the additional items that came to mind as Glenn worked.

   – follow-up evaluation in six months [by both supervisors and trainees would reveal the effectiveness of this training and provide suggestions for future improvements]

   – clear course objectives and more visual aids, to increase [instructor efficiency] and audience interest

Now that I have a fairly sensible arrangement, I can get this list into report form. (I'm sure to think of more material to add as I work.) Since this is an *internal* correspondence, I'll use a memo format.

Glenn's Second Draft

Sentences in Glenn's second draft are numbered for our later reference.

   – [1] In my opinion, the Management Training Session for the month of October was somewhat successful. [2] This success was evidenced when most participants rated their training as "very good." [3] But improvements still are needed.

   – [4] First and foremost, a number of innovative aspects in this October session proved especially useful. [5] Class topics were divided into two distinct areas. [6] These topics created a general-to-specific focus. [7] An emphasis on interpersonal communication skills was the most dramatic innovation. [8] This innovation helped class members develop a better attitude toward things in general. [9] Videotaping of trainees' oral reports, followed by critiques, helped clarify strengths and weaknesses.

   – [10] There is a detailed summary of the trainees' evaluations attached. [11] Based on these and on my past observations, I have several suggestions.

   – [12] All management training sessions should have a minimum of ten to fifteen members. [13] This would better utilize the large number of managers involved and the time expended in the implementation of the training. [14] The quality of class interaction with the speakers would also be improved with a larger group.

   – [15] There should be several brief on-the-job training experiences in different sales and service areas. [16] These should be developed *prior* to the training session. [17] This would provide each member with a broad view of the duties and responsibilities in all areas of the marketing division.

   – [18] Executive speakers should take a few minutes to spell out the personal and professional qualities essential for success with our company. [19] This would provide trainees a concrete guide to both general company and individual supervisors' expectations. [20] Additionally, by the next training session we should develop a presentation dealing with the attitudes, manners, and behavior appropriate in the business environment.

— [21] Do a six-month follow-up. [22] Get feedback from supervisors as well as trainees. [23] Ask for any new recommendations. [24] This would provide a clear assessment of the long-range impact of this training on an individual's job performance.

— [25] We need to demand clearer course objectives. [26] Instructors should be required to use more visual aids and improve their course structure based on these objectives. [27] This would increase instructor quality and audience interest.

— [28] These changes are bound to help. [29] Please contact me if you have further questions.

Although now better developed and sensibly organized, this version still is some way from the finished product on page 87. Glenn has to make further decisions about style, as well as content, arrangement, audience, and purpose. Let's follow Glenn's thinking at 8:15 Tuesday morning.

Well, this draft certainly beats my first, but it needs more work . . . maybe a sentence-by-sentence analysis. Here goes.

Sentence 1 begins with a needless qualifier, has a redundant phrase, and sounds vaguely insulting ("somewhat successful"). Maybe "generally successful" would be more precise and appropriate. Sentence 2 should be in the passive voice, to emphasize the training — not the participants. Also, 1 and 2 are choppy and repetitive, so I'll combine them.

Original In my opinion, the Management Training Session for the month of October was somewhat successful. This success was evidenced when most participants rated their training as "very good." (28 words)

Revision The October Management Training Session was generally successful, with training rated by most participants as "very good." (17 words)[2]

Sentence 3 is much too blunt and negative. I need an orienting sentence here that forecasts content more diplomatically.

Original But improvements still are needed.

Revision A few changes — beyond the recent innovations — should result in even greater training efficiency.

In sentence 4, "First and foremost" is trite, "aspects" is a clutter word, and word order needs changing to improve the emphasis (on innovations) and to lead into the subsequent examples.

---

[2] Notice throughout how careful revision sharpens the writer's meaning while cutting needless words.

Original   First and foremost, a number of innovative aspects in
this October session proved especially useful.

Revision   Especially useful in this session were several program
innovations.

Sentences 5 and 6 need combining, and content in 5 needs beefing up:
name the two areas. If 7 is labeled the "most dramatic innovation," it ought
to come last, for emphasis. In 8, "things in general" is too indefinite. And
7 and 8 need combining. Sentence 9 seems ok. All three examples would be
more readable in a *list*, and expressed in grammatically parallel phrasing
(maybe starting each with an "ing" phrase to signal the "action" verb,
as in 9).

Original   Class topics were divided into two distinct areas.
This created a general-to-specific focus. An emphasis
on interpersonal communication was the most dramatic innovation. This helped class members develop
an enthusiastic and relaxed attitude toward things
in general. Videotaping of trainees' oral reports, followed by critiques, helped clarify strengths and weaknesses.

Revision   – Dividing class topics into two distinct areas
helped create a general-to-specific focus: [the first
week's coverage of company structure and functions
created a clear context for the second week's coverage
of specific management skills].[3]
– Videotaping and critiquing trainees' oral reports
helped clarify strengths and weaknesses.
– Emphasizing interpersonal communication skills
[listening, showing empathy, and reading nonverbal
feedback] helped trainees [feel enthusiastic and re-
laxed] about [the group, their training, and the
company.]

And while I'm thinking about format, a couple of clear headings ("Workshop
Strengths," "Suggested Changes/Benefits") would segment the text, improving readability and appearance.

What else? Sentence 10 has a "there" opener that hurts the emphasis.
Also, 10 could be ambiguous (Were the trainees evaluated or did they evaluate?) Sentences 10 and 11 are supposed to be a transition from the innovations to the suggested changes. But these sentences are too indefinite. And
"past observation" in 11 is redundant.

---

[3] Notice that this revision has more words, but also much more concrete and specific detail
(in brackets). Completeness of information *always* takes priority over word count.

Original   There is a detailed summary of the trainees' evaluations attached. Based on these and on my past observations, I have several suggestions.

Revision   Innovations like these ensure quality training. And each future session can provide other ideas for change. Based on trainees' evaluations (summary attached) of the October session and my own observations, I recommend these additional changes.

Glenn follows this same editing and revising process through the rest of his memo. (Question 5 in the Exercises asks you to identify these remaining changes.) Wednesday morning, after much revising, Glenn hits the PRINT command on his word processor for the *final* draft shown below.

Glenn's Final Draft

October 19, 1984

TO: Marvin Long
FROM: Glenn Tarullo
RE: October Management Training Program: Evaluation and Recommendations

The October Management Training Session was generally successful, with training rated by most participants as "very good." A few changes, beyond the recent innovations, should result in even greater training efficiency.

Workshop Strengths
Especially useful in this session were several program innovations:

– Dividing class topics into two areas helped create a general-to-specific focus: the first week's coverage of company structure and functions created a context for the second week's coverage of management skills.

– Videotaping and giving critiques of trainees' oral reports helped clarify their speaking strengths and weaknesses.

– Emphasizing interpersonal communication skills (listening, showing empathy, and reading nonverbal feedback) helped trainees feel enthusiastic and relaxed about the group, their training, and the company.

Innovations like these insure quality training. And each future session ought to provide other ideas for change.

Suggested Changes/Benefits
Based on the trainees' evaluation of the October session (summary attached), and my observations, I recommend these additional changes:

- We should develop several brief (one-day) on-the-job rotations in different sales and service areas *prior* to the training session. This would give each member a "real-life" view of duties and responsibilities throughout the company.

- All training sessions should have at least ten to fifteen members. Larger classes would make more efficient use of resources and improve class/speaker interaction.

- We should ask instructors to follow a standard format (based on definite course objectives) for their presentation, and to use visuals liberally. A predictable structure, enriched by visuals, would increase instructor efficiency and audience interest.

- Executive speakers should take a few minutes to spell out personal and professional qualities essential in our company. Such advice would give trainees a concrete guide to both general company and individual supervisor expectations. Also, by the next training session, we should develop a presentation dealing with the attitudes, manners, and behavior appropriate in business.

- We should do a six-month follow-up of trainees (with feedback from supervisors as well as ex-trainees) to gain long-term insights, to measure the impact of this training on job performance, and to help design more advanced, follow-up management training.

Neither costly nor difficult to implement, these changes would yield immeasurable benefits in training efficiency and improved profits.

Copies: B. Hull, C. Hopkins, G. Black, J. Capilona, P. Maxwell, R. Sanders, L. Hunter

This is a useful report. But it did not leap magically from the writer's head onto the page, all in one giant step. Glenn made deliberate decisions about purpose, audience, content, organization, and style. More important, he *revised* the piece more than once.[4]

To create this report, Glenn engaged in a complex process. It was hard work, but worthwhile. Glenn's memo can have an important effect on his career. His boss (and his boss's boss), will evaluate Glenn by his ability to produce clear, helpful recommendations.

Although we can recognize some sequence in the writing process, we should not become slaves to any formula approach. For any writing, we generally begin with a purpose and a sense of our audience. And we do follow some sort of progression from generating content, to organizing, to improving style. But once we begin writing, we discover new ideas; we may redefine our purpose; we decide that sentences need shuffling and reorganizing; and we de-

[4] A special thanks to Glenn Tarullo for his perseverance. I made his task doubly difficult by having him explain each of his decisions during this writing process.

cide on clearer and more precise phrasing. In the process of discovering what we mean, and finding ways to express that meaning, we find ourselves revising. Although decisions about *content, arrangement* and *style* are treated separately in our discussions, in the writing process itself (the planning, drafting, and revising), these decisions often are indistinguishable.

Writers work on documents — and refine their drafts — in different ways. Some begin by brainstorming. Some come up with a statement of purpose immediately. Some begin with an outline. Others hate outlining, and simply write and rewrite, using scissors and tape to organize. Some even write a quick draft before thinking through their writing situation. Introductions and titles are more often written last. No one "step" in the process is ever complete until *the whole* is complete. (Notice, for instance, how Glenn sharpens his content *and* his style while he organizes.) *All* the decisions have to be made, but rarely are they made in the same order from day to day, even by a single writer.

No matter what the sequence, *revision* is a fact of life for anyone who writes. It is the one *constant* in the writing process. When you've finished a draft of your report, you have, in a sense, only begun. Sometimes you'll have more time to compose your report than Glenn did, sometimes less. Whatever your deadline, *always* leave time to revise. Revision checklists at the end of subsequent chapters will help you sharpen your editing and revising skills.

CHAPTER SUMMARY

Writers' decisions fall into five categories: purpose, audience, content, organization, and style. Because writing is a complex process, these decisions often are indistinguishable. But to focus your writing early, begin with a definite purpose and a clear sense of your audience. Then, during the writing process, you'll discover new ideas as you organize and rearrange to sharpen details and phrasing. No matter what sequence you follow, revise until your meaning is clear.

EXERCISES

1. In a numbered sequence, trace the order of decisions Glenn made in the process of composing his report (e.g., (1) wrote a rough draft, (2) identified his audience and their needs, (3) anticipated audience questions,

etc.). Could he have followed a more efficient sequence? (Say, not doing the first draft *before* brainstorming.) In his place, what sequence would you have followed?

2. Think of a course you've taken that had both definite strengths and weaknesses. Assume the instructor has asked you to evaluate the course and recommend improvements. Your secondary audience will be your instructor's chairperson and the Dean of Faculty. The instructor is up for tenure and is someone you like very much. You need to be candid in your report, but you don't want to make the instructor look bad. Write the report, being as specific as possible. (Note: use a fictional name for the instructor.)

3. Assume you are a training manager for XYZ Corporation. After completing this first section of the text and the course, what advice about the writing process would you have for a beginning writer who will need to write frequent reports on the job? In a one-page (single-spaced) memo to new employees, explain the writing process briefly, and give a list of general suggestions these beginning writers can follow.

4. Why did Glenn discuss the strengths of the workshop in that particular order? Explain your rationale in a brief memo to your instructor or in class discussion.

5. Compare Glenn's second draft (pages 84–85) with his final draft (pages 87–88). Identify all improvements in content, arrangement, and style besides those discussed on pages 85–87.

6. Assume you work in a large electronics company for a division manager named Bentley. You and division co-workers would like to institute an in-house day care center since, among you, you have eighteen preschool children. All of you presently have babysitters or take your children to centers or nursery schools scattered throughout the city. With an in-house center, employees would have less scurrying in the morning, thereby cutting back on lateness resulting from traffic, bad weather, etc. In addition, employees sometimes have to leave work early to pick up their children. Employees would pay for the two supervisors and do all the necessary paperwork to insure that the center meets state regulations. (The center would actually be cheaper than hiring individual sitters as you presently do.) The unused lounge on the second floor would be an ideal location.

An in-house center would have many benefits for employees, supervisors, and your employer. For instance, some supervisors might want to include their children. Brainstorm for ideas and benefits.

Since Bentley would have to give initial approval for the center, decide what kinds of questions she would ask. Draw up a list of ten questions you might have to answer.

Write your memo to Bentley. Submit all materials to your instructor.

STRATEGIES
FOR TECHNICAL
REPORTING

6

Summarizing Information

SUMMARIES DEFINED

A summary (or informative abstract) is a short version of a longer message. The summary provides readers with a concise and accurate view of the entire original. An economical way to communicate, a summary saves time, space, and energy.

PURPOSE OF SUMMARIES

Every effective statement conveys a message; it makes one or more points. Any attentive listener or reader can extract these major points without memorizing the original. For instance, your college notes do not include the lecturer's every word; instead, they summarize the main points in a way you will understand later. In studying a textbook for an exam, you extract the main ideas, deciding which are most important. Accordingly, good students

usually are those who take effective notes and who know what to study —
those who summarize well.

Out of school, you summarize daily, whether relating an anecdote or describing a magazine article to a friend. When you apply for jobs, your letter
and résumé summarize your qualities and qualifications. On the job, you have
to write concisely about your work. Perhaps you will record the minutes of a
meeting; summarize a lecture, news article, or report; or describe your progress
on a project. You might write proposals for new projects, bids for contracts,
or summaries of your research. Also, you will include summaries and abstracts
with your long reports. A routine assignment for a new employee in many
organizations is to provide superiors (decision makers) with summaries of the
latest literature in their field.

Whether you summarize someone else's information or your own, your job
is to communicate the *essential message* — to represent the full scope and detail
of the important material accurately in the fewest words. The principle is simple: include what your readers need and omit what they don't.

The essential message in any well-written piece is easy enough to identify,
as in this example:

> The lack of technical knowledge among owners of television sets leads
> to their suspicions about the honesty of TV repair technicians. Although TV
> owners might be fairly knowledgeable about most repairs made to their
> automobiles, they rarely understand the nature and extent of specialized
> electronic TV repairs. For example, the function and importance of the
> automatic transmission in an automobile are generally well known; however,
> the average TV owner knows nothing about the function of the flyback
> transformer in a TV set. The repair charge for a flyback transformer failure
> is roughly $150 — a large amount to a consumer who lacks even a simple
> understanding of what the repairs accomplished. In contrast, a $450 repair
> charge for the transmission on the family car, though distressing, is more
> readily understood and accepted.

Three significant ideas comprise the essential message here: (1) TV owners
lack technical knowledge and are suspicious of repair technicians. (2) An
owner usually understands even the most expensive automobile repairs. (3)
Owners do not understand or accept expenses for repairs and specialized parts
needed for their TV sets. A summary of the paragraph might read like this:

> Because TV owners lack technical knowledge of their sets, they are often
> suspicious of repair technicians. Although consumers may understand ex-
> pensive automobile repairs, they rarely understand or accept repair and parts
> expenses for their TVs.

This summary is almost 30 percent of the original length because the original
itself is short. With a longer original, a summary might be as short as 5 percent

or less. Length, however, is secondary to the need to represent the original faithfully.

Summaries are vital whenever people have no time to read in detail everything that crosses their desks.[1] Of course, only by reducing length without distorting the original message can a summary be effective.

ELEMENTS OF A GOOD SUMMARY

The following 235-word summary (or informative abstract) of a 5000-word report (pages 476–481) is a good distillation. Although less than 5 percent of the original, the summary covers the full scope of the report, emphasizing the significant points.

SUMMARY: SURVIVAL PROBLEMS OF TELEVISION SERVICE BUSINESSES

The high rate of business failure among qualified independent TV repair technicians (second only to service station failures) is rooted in a tradition of unsound business management and inadequate communication with customers. Soon after World War II, many radio-mechanic veterans opened radio shops, which were to form the basis of today's TV repair shops. The ensuing rapid technological progress saw these veterans and the newer technicians swept up in a frenzy of too much work and too little time to learn effective business methods. Moreover, the repair technicians' somewhat secretive ways of dealing with customers, coupled with the advent of television and its phenomenal growth, led to customer suspicions that helped form the "TV repairman syndrome." Even the best of today's technicians have difficulty in shaking this image. Accounts of dishonesty and ineptitude, prevalent in the early years, linger. Expensive repairs to electronic equipment are difficult to explain to the nontechnical consumer who feels that he is the victim of a supertechnology that requires elaborate servicing. Today's technician or shop owner is still hesitant to adopt sound management and collection methods, as well as to inform his beleaguered customers about the services he offers. Technicians' chances for survival will increase only when they learn improved business methods and sponsor collective advertising to improve their image and inform consumers.

The elements that make this summary effective are discussed on the next page.

---

[1] A recent U.S. president reportedly required all significant world events for the last twenty-four hours to be compressed into one typed page and placed on his desk the first thing each morning. Another president had a writer who summarized articles from more than two dozen major magazines.

Essential Message

A good summary answers the reader's implied question: "What does the original say?" We have just seen that the essential message is the minimum needed for the reader to understand the shorter version. It is the sum of the significant points — and *only* the significant points — in the original. Significant points include controlling ideas (thesis statements and topic sentences); major findings and interpretations; important names, dates, statistics, and measurements; and major conclusions or recommendations. Significant points do not include background discussions; the author's personal comments, digressions, or conjectures; introductions, explanations, lengthy examples, visuals, long definitions, or data of questionable accuracy. (These distinctions are illustrated on pages 99–103.)

Nontechnical Style

When your audience is large, expect more people to read the summary than any other part of your report. Write at the lowest level of technicality. Translate technical terms and complex data into plain English; for example, if the original states: "For twenty-four hours, the patient's serum glucose measured a consistent 240 mg%," you might rephrase: "For twenty-four hours, the patient's blood-sugar level remained critically high." When you do know specifically the people who will read the report, keep these people in mind. If they are expert or informed, you won't need to simplify as much (as discussed in Chapter 2). It is safer, however, to risk oversimplifying than to risk confusing your reader.

Independent Meaning

In meaning, as well as style, your summary should stand alone as a self-contained message. Readers should have to read the original only for a closer view — not to make sense of your message.

No New Data

Your job is to represent the original faithfully. Avoid personal comments or judgments ("This interesting report . . ." or "I strongly agree with this last point," etc.). Add nothing to the original.

Introduction-Body-Conclusion Structure

Most good writing has an introduction, a body, and a conclusion; so should your summary.

1. Begin with a clear statement of the controlling idea.
2. Present the supporting details in the same order as in the original.
3. Close with the original's conclusions and recommendations.

To improve coherence, use transitional words ("however," "in addition," "while," "therefore," "although," "in contrast," etc.).

Conciseness

A summary, above all, is concise. Because messages differ greatly, however, we cannot set a rule for summary length. Your best bet is to know your readers and their exact needs. Unless a length is specified as part of the job, all we can say is that the summary must be short enough to be economical and long enough to be clear and comprehensive. A long, clear summary is always better than a short, foggy one.

WRITING THE SUMMARY

Summarize your own work only after completing the original. Follow these instructions for paring down any piece — your own or another's.

1. *Read the entire original.* When summarizing another's work, read the whole thing before writing a word. Get a complete picture. You have to understand the original fully before you can summarize it effectively.

2. *Reread and underline.* Reread the original two or three times, underlining significant points (usually found in the topic sentences of individual paragraphs). If the piece is in a book, journal, or magazine that belongs to someone else, write the material on a sheet of paper instead.

Identify the key (thesis) sentence, which states the controlling idea. Omit all minor details such as introductions, explanations, examples, and definitions.

3. *Edit the underlined data.* Reread the underlined material and cross out needless words. Leave only key phrases you can later rewrite as sentences.

4. *Rewrite in your own words.* Rewrite the edited, underlined material in your own words, following the original order. Include all important data in

the first draft, even if it's too long; you can always trim later. Avoid judgments "The author is correct in assuming . . .") and add nothing.

5. *Edit your own version.* When you have everything readers need, edit for conciseness.

a. Cross out all your own needless words without harming clarity or good grammar. (See Appendix A.) Do *not* delete "a," "an," or "the" from any writing. Use complete sentences.

> The summer internship program in journalism gives the ~~journalism~~ student ~~first-hand~~ experience ~~at what goes~~ on ~~within~~ a ~~real~~ newspaper staff.

b. Cross out needless prefaces such as "The writer argues . . ." or "The researchers discovered . . ." or "Also discussed is. . . ."

c. Use numerals for numbers, except when beginning a sentence. (See Appendix A.)

d. Combine related ideas within longer sentences through subordination. (See Appendix A.)

Choppy Sentences

The occupational outlook for journalists is good. There was a 53 percent increase in journalism jobs between 1947 and 1972. The national job increase was only 41 percent. Indications point to a continuation of this trend. This is partly due to an increase in weekly newspapers.

Revised

The occupational outlook for journalists is good, as evidenced by the 53 percent increase in journalism jobs between 1947 and 1972, in contrast with a national job increase of only 41 percent. Indications point to a continuation of this trend, which is partly due to an increase in weekly newspapers.

Five short sentences are combined into two longer ones, emphasizing relationships only implied in the original.

6. *Check your version against the original.* When your own version is refined, verify that you have preserved the essential message, followed the original order, and added no extraneous comments.

7. *Rewrite your edited version.* Rewrite, following an introduction-body-conclusion structure. Add transitions (Appendix A) to reinforce the logical connection between related ideas ("*X* therefore *Y*" shows that *X* is related to *Y* in a cause-and-effect relationship, and so on).

8. *Document your source.* If summarizing another's work, identify the source in a bibliographical note immediately following the summary, and place directly quoted statements within quotation marks. (See Chapter 15 for various documentation formats.)

When summarizing your own work, eliminate steps 1 and 8. Otherwise the procedure is identical. Although the abstract is written last by the writer, it is read *first* by the reader. Though you may be tired, take the time to do a good job.

APPLYING THE STEPS

Let's apply the above steps to an actual summarizing process. Steps 1, 2, and 3 have been completed on the following article:

BRIGHTER PROSPECTS FOR WOMEN IN ENGINEERING

One of the major deterrents to women considering engineering as a career is the all-male image. This barrier is rapidly disappearing as the engineering image changes from that of a hard-hat roustabout at a construction site to that of a thoughtful, logical individual who is genuinely interested in solving the engineering and social problems which face us today. True, she may still show up at a construction site in her hard hat, but her time is more apt to be spent at a desk working on new solutions. A female engineer — unlikely? Not quite.

Although women make up an unimpressive 1 percent of the engineering population, their ranks have been growing. The latest Society of Women Engineers survey of schools accredited by the Engineering Council for Professional Development shows that female engineering enrollment increased from 1,035 during 1959–60 to 3,905 during the 1972–73 school year. This increase may not be as large as it appears on the surface. Only 128 schools replied to the 1959–60 survey. But with the advent of the Civil Rights Act of 1964 and, more recently, the implementation of the federal affirmative action program, as well as an increased awareness on the part of the schools, 201 responded last year. However, since the number of female engineering students per school has increased, even as the number of males enrolled at these schools has decreased, there is little doubt that the percentage of women enrolled in engineering undergraduate programs is growing.

Combine as key sentence (thesis).

Delete example.

Include significant statistic.

Delete data of questionable accuracy.

Include significant finding.

Delete explanation repeating above finding.

Jobs Come Fast, Promotions Slowly

What happens when the newly minted female engineer tries to enter the field? Initially, she is sought after by almost every employer in sight. Once she is on the job, however, things change. On the average, promotions do not come as rapidly for women as they do for men.

Include significant finding.

Discrimination can be a double-edged sword, however, for unlike her male counterpart, the female engineer is highly visible, and if she does an outstanding job, she may very well be rewarded faster than a man would be. If her performance is average or slightly below average, she may be judged in terms of a number of myths. Perhaps chief among them is the notion that men (and women) don't like to work for women. In my personal experience, I have found that people who enjoy their work get it done without any thought to whether their supervisor is a man or a woman.

Delete author's personal comment.

Include significant finding.

Delete author's personal comment.

Some echoes of other misconceptions about women are still heard among engineers, and undoubtedly contribute to the lag in promoting women to top management ranks. Examples of these myths are: (a) a company's public image will suffer if a woman takes over a top management position, because men have traditionally been the corporate leaders; (b) a woman won't travel on sales trips, to plant inspections, to professional conferences and so forth; (c) women don't want to accept responsibility; (d) a woman's family will always take precedence over her career. (One must ask, why shouldn't it take precedence over a man's as well?)

Include factual interpretation and continuation of above finding.

Delete rhetorical question.

It has also been argued that promotion policies don't favor women because companies prefer long tenure for those elevated to executive positions, and they believe that turnover rates are greater for women. But, not only do government figures show that professional women have working careers comparable in length to those of men, it is also clear that promotions generally accrue to men regardless of age and experience. Over 20 percent of all male engineers are in management, as opposed to an estimated 3 percent of female engineers.

Include continuation of finding.

Include significant finding.

Admittedly, because the number of women in the profession is small, the above figure is open to sampling error. Indeed, as many as 40 percent of the women surveyed in 1972 by the Society of Women

Delete statistics of questionable accuracy, along with the related explanation.

Engineers stated that they supervised groups which ranged in size from teams to major organizations. It must be noted, however, that members of SWE (and engineering societies in general) are probably among the more qualified and professionally active engineers.

Attitudes Vary

A questionnaire on discrimination was included in the <u>1972</u> SWE survey. In a classic case of "which-came-first-the-chicken-or-the-egg?" the results showed that those <u>women</u> who were <u>very successful</u> in terms of salary, responsibility, and years of experience <u>felt they had not encountered any discrimination.</u> Those women who were <u>moderately successful indicated</u> that there was <u>no discrimination</u> encountered <u>from</u> their <u>immediate superiors.</u> They felt, however, that people in the <u>upper</u> levels of <u>management</u> hierarchy <u>did discriminate</u> and that there was some evidence of discrimination by <u>coworkers.</u>

   <u>Those</u> women <u>on</u> the <u>low side</u> of the average in terms of salary and responsibility <u>indicated</u> that they had encountered <u>discrimination at all levels.</u> It can be argued that these women have less ability than their male cohorts, and use "discrimination" as an excuse for their lack of advancement.

Salaries

<u>All women encounter discrimination</u>, perhaps not intentional or even conscious, from their male colleagues. This contention is borne out by the results of the SWE salary survey, compared with the results of the survey of Engineers Joint Council for the profession as a whole. For engineers <u>with 11 years' experience</u> (the median for women), <u>the median salary for female engineers is $14,200 per year, while that for all engineers with 11 years' experience is $16,700</u>, according to the EJC. Both surveys were completed <u>in 1972.</u> The disparity may be even greater because again, SWE members are more professionally active than all engineers taken as a group.

   Of course, the engineering profession is not alone in this disparity in salaries. In the federal civil service, men average $14,328 per year, and women only $8,578. This is not because there are separate pay scales for women, but rather because women em-

Annotations in margin:

Include significant date.

Include significant finding.

Include continuation of finding.

Delete obvious explanation.

Include factual interpretation.

Delete background information.

Include significant statistics and date.

Delete author's conjecture.

Delete author's digression.

ployees are heavily concentrated in lower-grade jobs.

All is not bleak, however. In 1973, the average
starting salary offered to women engineering graduates at the bachelor's degree level was $936 per
month — $15 a month more than the average for
men, according to the College Placement Council.
This represents a closing of the gap when compared
with 1971, when women were offered $8 a month
less than men. Engineering — the profession offering
the highest starting pay for those with bachelor's degrees, remains the only profession where salary offers
are higher for women than for men.

> Include significant finding, supporting statistics, and date.

> Include significant conclusion.

If one considers salary offers from private industry only, the salary gap between male and female
engineers was even greater than the averages indicate, and favored women. But the federal government, which offered significantly lower salaries to
entry-level female engineers than to males, dragged
the overall averages closer together.

> Delete explanation.

The Future?

The current energy crisis and materials shortage indicate that this country is fast moving from a state
of have to have not. The only way to maintain our
current standard of living is through technology,
which means that engineers will continue to be in
great demand. It also means that the image of
engineering will continue to change as attention
is focused on sociological-technological problems.
Consequently, we can expect women to enter the
engineering profession in greater numbers.[2]

> Include significant conclusion and restatement of thesis.

With the reading, rereading and underlining, and editing completed, we can
move on to step 4: rewriting. Here is how the first rewrite of the underlined
material might read; ideas are simply listed in order, without concern for
length or subordination.

FIRST REWRITE

The all-male image has deterred women from engineering careers. This
image of a hard-hat worker is giving way to that of a thoughtful individual
working on today's engineering and social problems. Women comprise only
1 percent of the engineering population, but their ranks are growing. Female
engineering students have increased in number while males have decreased.

[2] Naomi J. McAfee, "Brighter Prospects for Women in Engineering," *Engineering Education* 64, no. 7 (April 1974): 502–504. Reprinted with permission from
Engineering Education © 1974 The American Society for Engineering Education.

The new female engineer easily finds work but is not rapidly promoted. Because of high visibility, she may be promoted faster than a man if her performance is outstanding. If performance is average or below, she may be judged in terms of several myths: that people don't like to work for a woman; that a woman in top management harms a company's public image; that a woman won't travel on business; that women won't accept responsibility; that her family takes precedence over her career; and that turnover rates are higher for women. But careers of professional women are as long as those of men. Men usually receive the promotions, regardless of age or experience.

In 1972, highly successful women engineers reported no discrimination. Moderately successful women sensed no discrimination from immediate superiors, but felt that higher management and coworkers did discriminate. Those with minimal success felt discrimination at all levels. All women do encounter salary discrimination. In 1972, the median salary for female engineers with 11 years' experience was $14,200 yearly, as opposed to $16,700 for all engineers with equal experience. However, in 1973, the average starting salary for women engineers with bachelor's degrees was $936 per month — $15 more than for men. This figure contrasts with 1971 figures when women were offered $8 a month less than men. Engineering is the only profession where salary offers are higher for women than for men.

We now have a shortage of energy and materials. We can only maintain our living standard through technology. Demand for engineers will continue to grow. Their image will continue to change with new emphasis on sociological-technological problems. Thus, women are expected to enter the engineering field in greater numbers.

This version has all significant information, without regard for coherence. In the final draft, transitional terms and punctuation signals (see below) will be added, and related ideas combined to tighten the whole structure. Also, the length (roughly 330 words or 25 percent of the 1400-word original) will be further reduced. Word length, however, is less important than accurate emphasis and faithful representation. This version preserves the original emphasis by recording the brighter prospects as well as the continuing problems. Factual statements are faithfully represented because they are *fully* expressed. Imagine the distortion if, to save space, the statement, "The new female engineer easily finds work but is not rapidly promoted," were only partially expressed as, "The new female engineer easily finds work."

Here is the final draft with steps 5, 6, 7, and 8 completed. Transitional terms and connecting devices, including punctuation signals, are underlined:

SUMMARY OF "BRIGHTER PROSPECTS FOR WOMEN IN ENGINEERING"

The all-male, hard-hat image, <u>which</u> has deterred women from engineering, is changing to that of a thoughtful individual working on today's engineering and social problems. <u>Although</u> women comprise only 1 percent of engineers, their growing ranks <u>are evidenced by</u> an increase in female engineering students, <u>contrasted with</u> a decrease in male students. The graduating female easily finds work <u>but</u> no rapid promotion unless her performance is outstanding. An average or below-average performance may be judged in terms of several myths: that people dislike working for women; that a woman in top management harms a company's image; that she won't travel or accept responsibility; that her family takes precedence over her career; and that female turnover rates are higher. <u>In fact,</u> women's professional careers are as long as men's, <u>but</u> men usually receive the promotions.

In 1972, highly successful women engineers reported no discrimination. Moderately successful women sensed none from immediate superiors, <u>but</u> felt that higher management and coworkers did discriminate. Marginally successful women claimed discrimination at all levels. Women <u>do encounter</u> salary discrimination: the 1972 median salary for female engineers with 11 years' experience was $14,200 yearly, <u>as opposed to</u> $16,700 for all equally experienced engineers. <u>However,</u> the 1973 average starting salary for women graduates was $936 monthly — $15 more than for men. <u>By contrast,</u> in 1971, women received $8 less than men. <u>Thus,</u> only in engineering are women receiving higher offers.

Current energy and materials shortages increase our reliance on technology to maintain living standards; <u>consequently,</u> the demand for engineers with a sociological-technological commitment should attract more women. (Naomi J. McAfee, "Brighter Prospects for Women in Engineering," *Engineering Education* 64 (April 1974): 502–504.)

This version is trimmed, edited, and tightened: word count is reduced to roughly 20 percent of the original. A summary this long serves well in many situations, but other audiences might want a briefer and more compressed summary — say, 125 to 150 words, or about 10 percent of the original:

The all-male image of engineering is changing, and although women comprise only 1 percent of engineers, their ranks are growing. Female students are increasing while males decrease. Although female graduates easily find work, only the outstanding are rapidly promoted, with average or lower performance often judged according to conventional myths about women in "male" professions.

In 1972 no discrimination was reported by the highly successful women engineers; selective discrimination, by the moderately successful; and general discrimination, by the marginal. Women *do* encounter salary discrimination: the 1972 median salary for experienced females was $14,200, compared with $16,700 for all equally experienced engineers. However, in 1973 women graduates commanded starting salaries of $15 more monthly than men, whereas in 1971 they had received $8 less. Only in engineering are women receiving higher offers.

As resource shortages increase reliance on technology to maintain living standards, the demand for "sociological-technological" engineers should attract more women.

Notice that the essential message is still intact; related ideas are again combined and fewer supporting details are included. Clearly, length is adjustable according to your audience and purpose.

WRITING THE DESCRIPTIVE ABSTRACT

A summary (or informative abstract) reflects *what the original contains,* whereas a descriptive abstract (or abstract) reflects *what the original is about.* These differences can be clarified by an analogy. Imagine you are describing your summer travels to a friend; you have two options: (1) You might simply mention the places you visited, in chronological order. This catalogue of major areas would convey the basic nature of your trip. (2) In addition to describing your itinerary, you might describe the significant experiences you had in each place. Option 1 is like a roadmap, an overview of the areas traveled. This option is analogous to a descriptive abstract. Option 2, on the other hand, is expanded to include the significant experiences. This second option is analogous to a summary (or informative abstract), which gives the major facts from the original.

A descriptive abstract, then, presents the broadest view, and offers no facts from the original. Whereas the summary contains the meat of the original, the descriptive abstract contains only its skeletal structure; a descriptive abstract is a kind of "summary of a summary," as shown below:

ABSTRACT OF "BRIGHTER PROSPECTS FOR WOMEN IN ENGINEERING"

The changing status of women in engineering is examined in terms of women's population increase, job opportunities, promotions, salaries, and long-term prospects.

Because it merely previews the original, a descriptive abstract is always brief — usually no longer than a short paragraph. One- or two-sentence abstracts often follow article titles in journal or magazine tables of contents; they give a bird's eye view.

PLACING SUMMARIES IN REPORTS

If your reader asks for a descriptive abstract of your report, place it in front, on a separate page, right after your table of contents (or on the title page). It is usually single-spaced. With a descriptive abstract in front, your summary (or informative abstract) will go in the conclusion of your report. Sometimes you will place your summary in front, instead of writing a descriptive abstract. Practice can vary from company to company.

CHAPTER SUMMARY

A summary (or informative abstract) — like this one — is an economical way to communicate because it compresses a longer message into essentials. An effective summary extracts *only* the major points from the original and (usually) presents them in a nontechnical style. The summary stands independently as a complete message and adds nothing to the original.

The key word in summary writing is *conciseness;* that is, the summary must be brief but also clear and comprehensive. Better to make it a bit long than to omit some key point and distort the original.

You can summarize your own work only after you have written the complete version.

Follow these steps in writing your summary:

1. Read the entire original.
2. Reread and underline (or copy) the major points.
3. Edit the underlined or copied data to cut out needless words.
4. Rewrite the material in your own words.
5. Edit your version by crossing out needless words and prefaces, using digits for numbers, and combining related ideas through subordination.
6. Check your version against the original for accuracy.
7. Rewrite in an introduction-body-conclusion structure with transitions.
8. Document your source if you have summarized another's work.

Whereas a summary (or informative abstract) reflects what the original contains, a descriptive abstract reflects what the original is about (a summary of a summary). Descriptive abstracts are short and simply give a bird's-eye view.

A descriptive abstract always belongs in front of a report (or on the title page), but the summary (or informative abstract) may be in the conclusion or in front, as readers request.

REVISION CHECKLIST FOR SUMMARIES

Use this checklist as a guide to refining your summaries.

Content

1. Does the summary contain only the essential message (controlling idea; major findings and interpretations; important names, dates, statistics, measurements; conclusions and recommendations)?
2. Does it make sense as an independent piece?
3. Is the summary accurate in scope and detail (checked against the original)?
4. Is it free from personal comments or other additions to the original?
5. Is it free from needless details?
6. Is it short enough to be economical and long enough to be clear and comprehensive?
7. Is the source documented?
8. Does the descriptive abstract express what the original is about?

Arrangement

1. Is the summary coherent?
2. Are there enough transitions for readers to follow the line of thought?
3. Does it have an introduction-body-conclusion structure?
4. Does it follow the order of the original?
5. Have you placed the summary (informative abstract) or descriptive abstract at the proper location in your report?

Style

1. Is it at the best level of technicality for *all* anticipated readers?
2. Is it free from needless words?
3. Are all sentences clear, concise, and fluent (pages 43–57)?
4. Is it written in correct English (Appendix A)?

Now list those features of your summary that need improvement.

EXERCISES

1. In one paragraph, describe the differences between a summary and an abstract.
2. *In class:* Organize into groups of four or five and choose a topic for group discussion: a social problem, a political issue, a campus problem, plans

for an event, suggestions for individual energy conservation, etc. Discuss the topic for one full class period, taking notes of significant points. Afterward, organize and edit your notes in line with the directions for "Writing the Summary." Next, write a summary of the group discussion in no more than 200 words. Finally, as a group, compare your individual summaries for accuracy, emphasis, conciseness, and clarity.

3. In one or two paragraphs, discuss the kinds of summary-writing you expect in your occupation. Will your reading audience be mainly colleagues, superiors, customers, clients, or other general readers? (If you don't know, ask someone in the field.)

4. Read each of the following paragraphs, and list the significant ideas comprising each essential message. Write a summary of each paragraph.

In recent years, ski-binding manufacturers, in line with consumer demand, have redesigned their bindings several times in an effort to achieve a noncompromising synthesis between performance and safety. Such a synthesis depends on what appear to be divergent goals: Performance, in essence, is a function of the binding's ability to hold the boot firmly to the ski, thus enabling the skier to change rapidly the position of his skis without being hampered by a loose or wobbling connection. Safety, on the other hand, is a function of the binding's ability both to release the boot when the skier falls, and to retain the boot when subjected to the normal shocks of skiing. If achieved, this synthesis of performance and safety will greatly increase skiing pleasure while decreasing accidents.

Contrary to public belief, sewage-treatment plants do not fully purify sewage. The product that leaves the plant to be dumped into the leaching (sievelike drainage) fields is secondary sewage containing toxic contaminants such as phosphates, nitrates, chloride, and heavy metals. As the secondary sewage filters into the ground, this conglomeration is carried along. Under the leaching area develops a contaminated mound through which ground water flows, spreading the waste products over great distances. If this leachate reaches the outer limits of a well's drawing radius, the water supply becomes polluted. Furthermore, because all water flows essentially toward the sea, more pollution is added to the coastal regions by this secondary sewage.

5. In 500 words or less, summarize your reasons for applying for a certain job or to a certain school.

6. Attend a campus lecture on a topic of interest and take notes of the significant points. Write a summary of the lecture's essential message.

7. Use your own modified technique of the steps in "Writing the Summary" as a study aid in preparing for an examination in one of your courses. After taking the exam, write one or two paragraphs evaluating this technique.

8. Find three examples of abstracts or summaries from journals and magazines and bring them to class. As a group, analyze selected examples on an overhead projector for clarity and meaning.

9. Find an article about your major field or area of interest and write both an abstract and a summary of the article.

10. Select a long paper you have written for one of your courses; write an abstract and a summary of the paper.

11. Read the following article and write both a descriptive abstract and a summary (or informative abstract) of it, using the steps under "Writing the Summary" as a guide. Define a specific audience and use for your material. Bring your summary to class and exchange it with a classmate for editing according to the revision checklist. Revise your edited copy before submitting it to your instructor.

EPA SETS STANDARDS FOR NEW MOTORCYCLES AND MOTORCYCLE REPLACEMENT EXHAUST SYSTEMS

The U.S. Environmental Protection Agency has issued standards which limit the noise from newly manufactured motorcycles and motorcycle replacement exhaust systems.

The standards will be phased in over a two- to five-year period beginning in 1983. Mopeds are considered motorcycles by EPA and will be covered, but will have only one standard to meet — also imposed in 1983. No existing motorcycles, or any built before 1983, will be affected.

Some 93 million people are daily affected by traffic noise. Motorcycles are an integral and important part of the traffic stream.

The motorcycle manufacturing industry has been greatly concerned about potential restrictions on commerce as a result of being required to produce new motorcycles that will comply with a multiplicity of differing state and local noise standards. This regulation will preempt state and local noise standards for newly manufactured motorcycles, thereby providing national uniformity of treatment.

Motorcycles are the source of more annoyance and adverse community response than any other single traffic noise source. EPA realizes that much of this negative response comes about because of excessively loud, exhaust-modified motorcycles. Because of this, the Agency believes that both the noise from newly manufactured motorcycles and from modified motorcycles must be controlled if the public health and welfare benefits Congress expected when it passed the Noise Control Act are to be realized. To control the noise from exhaust-modified motorcycles, the combined efforts of the federal government and state and local governments are essential. In addition to providing the labeling and antitampering provisions of the regulation, EPA will assist state and local governments in establishing complementary noise control programs including ordinances that will prohibit the use of noisy exhaust systems.

EPA has set 80 decibels (dB) as the most stringent noise standard for street motorcycles and small off-road motorcycles. Although the standards are less stringent than those that were proposed, EPA anticipates that these standards will, on the average, reduce the noise from new street motorcycles by 5 dB and by 2 to 7 dB on new offroad motorcycles by 1986. The exhaust system regulation and the "antitampering" and labeling provisions of the motorcycle regulation, in combination with strong complementary state and local programs, should help reduce exhaust-modified motorcycles to between one-half and one-fourth their current numbers.

These reductions are expected to result in a 55 to 75 percent decrease in interferences with human activities (such as sleeping, conversation), depending on the extent to which state and local governments are able to contribute to reducing the numbers of exhaust-modified motorcycles. Likewise, these reductions are expected to result in a 7 to 11 percent decrease in the severity and extent of overall traffic noise impact, again depending on in-use enforcement.

The standards and effective dates applicable to new motorcycles and to new motorcycle replacement exhaust systems are as follows:

1. Street motorcycles: 83 dB, January 1, 1983; 80 dB, January 1, 1986.
2. Moped type street motorcycles: 70 dB, January 1, 1983.
3. Off-road motorcycles (displacement 170 cc and below): 83 dB, January 1, 1983; 80 dB, January 1, 1986; (displacement more than 170 cc): 86 dB, January 1, 1983; 82 dB, January 1, 1986.

EPA expects the costs of compliance to be reflected in increased purchase prices for motorcycles and exhaust systems. For street motorcycles, increases will average approximately 2 percent (or $36). The estimated purchase price increase for off-road motorcycles will average 2 percent (or $21). For replacement exhaust systems, the estimated purchase price increase will average 25 percent (or $30).

Although higher retail prices could result in some initial lost sales, total industry sales (in terms of both units and dollars) are projected to significantly expand in the next decade. Furthermore, because all mopeds that the Agency has tested, which are being sold in the United States, already comply with the 70 dB level set for these vehicles, EPA foresees no impact on moped prices and consequently on moped sales.

This is the fourth noise control regulation EPA has issued to limit traffic noise. Regulations have been issued for interstate motor carriers (October 19, 1974), newly manufactured medium and heavy trucks (April 13, 1976), and newly manufactured garbage trucks (October 1, 1979). On the same day that the motorcycle noise standard was issued, EPA proposed an amendment to the testing requirement of the regulations.

The proposed amendment would require manufacturers to take one additional step in their testing program over and above what is now required of them as a result of the final regulations. Specifically, under the proposed amendment, manufacturers would be required to remove all easily removable components from their exhaust systems before conducting the tests necessary to show compliance with applicable standards. The Agency believes that this amendment will enhance the effectiveness of the regulation since the control of motorcycle noise is dependent on exhaust systems retaining their noise suppression performance beyond the time of sale. These amendments are expected to encourage manufacturers to design exhaust systems in ways which will reduce the incidence of tampering by consumers.[3]

[3] U.S. Environmental Protection Agency, EPA Environment News, Boston, February 1981, pp. 12–13.

7

Defining
Your Terms

DEFINITION

To define a term is to give the precise meaning you, the writer, intend when you use it. Assuming a writer has something worthwhile to say, *clarity* becomes the vital feature of any document. Clear writing begins with clear thinking; clear thinking begins with a definite understanding of what all the terms mean. So, clear writing depends on definitions that both reader and writer understand. Unless you are sure the reader knows the exact meaning you intend, always define something before you discuss it.

Too often, we use words that seem to express our meaning, without realizing that our particular choice could have a vastly different meaning for readers. For instance, it's easy enough for a computer salesperson to claim that automated offices will increase *white-collar productivity* — but what exactly does that term mean? Whereas blue-collar productivity can be measured concretely by the amount and quality of goods produced, white-collar productivity is much harder to measure or define. We know that a word processor, for instance, will help writers revise, but will it help writers think faster or come up with better ideas? Before they can sensibly discuss white-collar productivity, both writer and reader have to settle on the term's precise meaning for them.

PURPOSE OF DEFINITIONS

Virtually every specialty has its own technical "language." Engineers, architects, and programmers talk about "prestressed concrete," "tolerances," or "microprocessors"; psychologists, social workers, counselors, and police officers use terms like "manic-depressive psychosis," "sociopathic behavior," or "repression"; lawyers, real estate brokers, and investment counselors discuss "easements," "liens," "amortization," or "escrow accounts" — and so on. Whenever such terms are unfamiliar to a nonspecialized audience, they need to be defined.

When writing to colleagues, you rarely have to define specialized terms in your field (unless the term is new), but reports often are written for the layperson — the client or some other general reader. When you write for nonspecialists, think about their needs. Don't force readers to a dictionary or encyclopedia to make sense of your message. Assume your readers know less than you — general readers, much less! Make your meaning clear with good definitions.

Most of the specialized terms mentioned above are concrete and specific. Once a term such as "microprocessor" has been defined in enough detail to suit the reader's needs, its meaning will not differ appreciably in another context. And when a term is highly technical, a writer easily can figure out that it should be defined for certain readers. Any nonspecialist knows that he or she has no idea what "prestressed concrete" or "diffraction" means. However, readers are less likely to be aware that more familiar terms like "disability," "guarantee," "tenant," "lease," or "mortgage" acquire very specialized meanings in specialized contexts. This is where definition (by all parties) becomes crucial. What "guarantee" means in one situation is not necessarily what it will mean in another. That's why a contract is a detailed (and legal) definition of the subject of the contract.

Let's assume you're shopping for disability insurance to provide a steady income in case injury or illness should cause you to lose your job. Besides comparing prices of various policies, you will want each company to define "physical disability." Although Company A offers the least expensive policy, it might define physical disability as your inability to work at any job what-soever. Therefore, should a neurological disease prevent you from continuing your work as a designer of delicate, transistorized electronic devices, without disabling you for work as a salesperson or clerk, you might not qualify as "disabled," according to Company *A*'s definition. In contrast, Company *B*'s policy, which is more expensive, might define physical disability as your in-ability to work at your specific job. Thus, although all companies use the term "physical disability," they don't mean the same thing.

Similar problems in definition arise with certain purchases. If you're buying a condominium, obtain the developer's full definition of "condominium." Other-wise, your failure to read the "fine print" might be disastrous. The same is true for the terms of "warranty" on a new automobile. Because you are legally responsible for all documents bearing your signature, you must understand the importance and technique of clear definition.

USING DEFINITIONS SELECTIVELY

Growth in technology and specialization will continue to make definitions vital to communication. Use definitions, however, only when the audience needs them. Know for whom you're writing, and why. Reports in *Psychology Today* (with a general readership), for instance, define many terms not defined in reports to psychologists. Expert or informed readers naturally require fewer definitions. If unable to pinpoint your audience, assume a general readership and define generously.

Depending on your subject, purpose, and audience, definitions vary greatly in length. Often you will need only a *parenthetical definition* — a few words or a synonym in parentheses after the term. Sometimes your definition will require one or more complete sentences. Certain terms require a definition that extends to hundreds of words — an *expanded definition*. (Each type is discussed later.)

The choice of parenthetical, sentence, or expanded definition depends on what information readers need, and that, in turn, depends on why they need it. For instance, "carburetor" could be defined in a single sentence (as shown on page 116), telling interested readers what it is and how, in general, it works. This definition, however, should be greatly expanded for the student mechanic who needs to know where the word *carburetor* comes from, how the device was developed and perfected, what it looks like, how it is used, and exactly how its parts work.

In most cases, abstract and general terms ("loan," "partnership," etc.) will need expanded definitions.

ELEMENTS OF A USEFUL DEFINITION

For all definitions, use these guidelines:

Plain English

Your purpose is to clarify meaning, not muddy it. Use language your readers will understand.

Incorrect   A tumor is a neoplasm.

Correct   A tumor is a growth of cells that occurs in the body, grows independently of surrounding tissue, and serves no useful function.

Incorrect   A solenoid is an inductance coil that serves as a tractive electromagnet. (*This definition might be appropriate for an electrical engineering manual, but is too specialized for the general reader.*)

Correct   A solenoid is an electrically energized coil that converts electrical energy to magnetic energy capable of performing various mechanical functions.

Basic Properties

Any single thing has characteristics that make it different from all others. Its definition should express clearly these basic properties. Thus, a thermometer can be defined in terms of its singular function: it measures temperature; this is the essential information a reader needs. All other data about a thermometer, such as types, special uses, materials used in construction, and cost are secondary. On the other hand, a book cannot primarily be defined in terms of its function, because books can have several functions. A book can be used to write in or to display pictures (if the pages are blank), to record financial transactions, to read (if the pages are printed or written), and so on. Also, other items — individual sheets of paper, posters, newspapers, picture frames, etc. — serve the same functions. The basic property of a book is physical: it is a bound volume of pages. To be called a book, an item must have that feature; it is what your reader would have to know *first*, to understand what a book is. Comments about types of books, uses, sizes, contents, etc. provide only secondary information.

Objectivity

Make your definition objective. Tell readers what the item is, not what you think of it. Personal comments and interpretations should come only *after* your data, and only then at the specific request of your reader. "Bomb," for instance, is defined properly as "an explosive weapon detonated by impact, proximity to an object, a timing mechanism, or other predetermined means." If instead you define a bomb as "a weapon devised and perfected by hawkish idiots to eventually destroy themselves and the world," you are editorializing; furthermore, you are not presenting a bomb's basic property.

Likewise, in defining something like "diesel engine," simply tell your reader what it is and how it works. You might think that diesels are too noisy and sluggish for automobiles, but reserve these judgments until *after* your definition — that is, if your reader has asked for them at all (as in a comparison of minibuses for your town's public transportation system). Otherwise, let the data speak for themselves.

CHOOSING THE TYPE OF DEFINITION

After deciding to define a term in your report, choose the most appropriate type of definition: parenthetical, sentence, or expanded.

Parenthetical Definition

A parenthetical definition is the simplest type. It explains the term in a word or phrase and is often a synonym in parentheses following the term:

> The effervescent (bubbling) mixture was quickly discarded.
> The leaching field (sievelike drainage area) needs 15 inches of crushed stone.

Another option is to express your definition as a clarifying phrase:

> The trees on the site are mostly deciduous; that is, they shed their foliage at season's end.

Use parenthetical definitions to give readers a general understanding of specialized terms so they can easily follow the discussion where these terms are used. A parenthetical definition of "leaching field" might be adequate in a progress report to a client whose house you are building. A town report titled "Groundwater Contamination from Leaching Fields," however, would call for an expanded definition of the report's subject.

Sentence Definition

Often, a clear definition requires more than just a word or phrase in parentheses. A sentence definition (which may be stated in more than one sentence) follows a fixed structure: (1) the term to be defined, (2) the class (specific group) to which the term belongs, and (3) the features that differentiate the term from all others in its class.

Term	Class	Distinguishing Features
polygraph	a measuring instrument	that simultaneously records changes in pulse, blood pressure, and respiration, and is often used in lie detection
carburetor	a mixing device	in gasoline engines that blends air and fuel into a vaporized mixture for combustion within the cylinders
transit	a surveying instrument	that measures horizontal and vertical angles
diabetes	a metabolic disease	caused by a disorder of the pituitary gland or pancreas, and characterized by excessive urination, persistent thirst, and, often, an inability to metabolize sugar
liberalism	a political concept	based on belief in progress, the essential goodness of man, the autonomy of the individual, and standing for the protection of political and civil liberties
brief	a legal document	containing all the facts and points of law pertinent to a specific case, and filed by an attorney before arguing the case in court
stress	an applied force	that strains or deforms a body

In their presentation, these elements are combined into one or more complete sentences.

> Diabetes is a metabolic disease caused by a disorder of the pituitary gland or pancreas. This disease is often characterized by excessive urination, persistent thirst, and often an inability to metabolize sugar.

Sentence definition is especially useful if you need to stipulate the precise working definition of a term that has several possible meanings. In a construction, banking or real estate report, for example, a term such as "qualified buyer" could have different meanings for different readers. The same is true for "compact car" in a report comparing various brands of cars for use in the company fleet.

State your working definitions at the beginning of your report, as in the following example:

> Throughout this report, the term "disadvantaged student" is taken to mean. . . .

Classifying the Term

Be specific and precise in your classification. The term's class will reflect its similarities to all other items with common attributes (as discussed in Chapter 8). The narrower your class, the more specific your meaning. For example, "transit" is correctly classified as a "surveying instrument," not as a "thing" or simply as an "instrument." Likewise, "stress" is correctly classified as "an applied force"; to say that stress "is what . . ." or "takes place when . . ." or "is something that . . ." fails to reflect a specific classification. Also, select the most precise terms of classification: "Diabetes" is precisely classified as "a metabolic disease," not as "a medical term."

Differentiating the Term

Differentiate the term by separating the item it names from every other item in its class. If you can apply the distinguishing features to more than one item, your definition is imprecise. Make these features narrow enough to pinpoint the item's unique identity and meaning, yet broad enough to be inclusive. For example, a definition of "brief" as "a legal document introduced in a courtroom" is too broad because the definition doesn't differentiate "brief" from all other legal documents. Conversely, a differentiation of "carburetor" as "a mixing device used in automobile engines" is too narrow because it fails to acknowledge the carburetor's use in all other gasoline engines.

Also, avoid circular definitions (repeating, as part of the distinguishing features, the word you are defining). Thus "stress" should not be defined as "an applied force that places stress on a body." The class and distinguishing features must express the item's basic property (e.g., "an applied force that strains or deforms a body").

Expanded Definition

The sentence definition of "solenoid" on page 123 would be good for a general reader who simply needs to know what a solenoid is. An instruction manual for mechanics or mechanical engineers, however, would define this item in detail (as on pages 123–124); these readers need to know what a solenoid is, how it works, and how to use it. So the choice of length and detail in a definition of a concrete and specific term depends on the purpose of the definition and the needs of the audience.

The problem with defining an abstract and general word, such as "condominium" or "bodily injury," is different. "Condominium," for example, is a vaguer term than "solenoid" (solenoid *A* is pretty much like solenoid *B*) because the former refers to a wide range of ownership agreements; therefore, its meaning is much more variable and needs to be spelled out.

Concrete, specific terms such as "diabetes," "transit," and "solenoid" often can be defined in a sentence, and will require an expanded definition only for certain audiences. Terms such as "disability" and "condominium," however, will almost always require expanded definition. The more general or abstract the term, the more likely the need for an expanded definition.

An expanded definition may be a single paragraph (as in defining a simple tool) or may extend to many pages (as in defining a digital dosimeter — a device for measuring exposure to radiation); sometimes the definition itself *is* the whole report.

The following excerpt from an automobile insurance policy defines the coverage for "bodily injury to others." Its style and detail make this definition clear to general readers. Instead of the fine print "legalese" seen in many policies, this definition is written in plain English.

PART 1. BODILY INJURY TO OTHERS

Under this Part, we will pay damages to people injured or killed by your auto in Massachusetts accidents. Damages are the amounts an injured person is legally entitled to collect for bodily injury through a court judgment or settlement. We will pay only if you or someone else using your auto with your consent is legally responsible for the accident. The most we will pay for injuries to any one person as a result of any one accident is $5,000. The most we will pay for injuries to two or more people as a result of any one accident is a total of $10,000. This is the most we will pay as the result of a single accident no matter how many autos or premiums are shown on the Coverage Selections page.

We will *not* pay:

1. For injuries to guest occupants of your auto.
2. For accidents outside of Massachusetts or in places in Massachusetts where the public has no right of access.
3. For injuries to any employees of the legally responsible person if they are entitled to Massachusetts workers' compensation benefits.

The law provides a special protection for anyone entitled to damages under this Part. We must pay their claims even if false statements were made when applying for this policy or your auto registration. We must also pay even if you or the legally responsible person fails to cooperate with us after the accident. We will, however, be entitled to reimbursement from the person who did not cooperate or who made any false statements.

EXPANDING YOUR DEFINITION

The following strategies will help you expand definitions. Amplify each strategy by description and by synonyms or analogies whenever possible. Always begin an expanded definition with a formal sentence definition, then use only expansion strategies that serve your reader's needs.

Etymology

Often, the origin of a word (its development and changing meanings) helps clarify a definition. *Biological control* of insects, for instance, is derived from the Greek "bio," meaning *life* or *living organism,* and the Latin "contra," meaning *against* or *opposite.* Biological control, then, is the use of living organisms against insects. Standard college dictionaries contain etymological information, but your best bet is *The Oxford English Dictionary* and various encyclopedic dictionaries of science, technology, and business (partially listed on pages 284–285).

Modern technical terms often are derived from two or more traditional terms. *Transceiver,* for instance, is derived from "transmitter" and "receiver," and defined as "a module composed of a radio receiver and transmitter."

Sometimes the origin of a term can be colorful as well as instructional. *Bug* (jargon for any type of programming error), for instance, is said to derive from an early computer at Harvard that malfunctioned because of a dead bug blocking the contacts of an electrical relay. Since, like most of us, programmers hate to admit mistakes, the term became popular as a euphemism for *error.* Correspondingly, *debugging* is the act of correcting errors in a program.

History and Background

The definition of specialized terms like "radar," "bacteriophage," "laser," "silicon chips," or "X ray" can often be clarified through a background discussion: discovery of the item, development, method of production, applications, and possibilities for use in exploration, medicine, etc. Specialized encyclopedias are a good source of background information.

Negation

Readers can better grasp some meanings (especially of abstract terms) by understanding clearly what the term *doesn't* mean. For instance, the earlier insurance policy defines coverage for "Bodily Injury to Others" partly through negation: "We will *not* pay: 1. For injuries to guest occupants of your auto, etc."

Examples

A definition containing familiar examples is helpful. Tailor the example to your reader's level of specialization. Thus a definition of "economic inflation" for the general reader could use rising fuel prices to illustrate higher costs per unit volume. Likewise, a definition of "clothing fashion" could be clarified by the example of changes in skirt lengths. For a more specialized reader, a definition of the mineral "borax," for a student of ecology or chemistry, should mention its use as a cleaning agent and detergent softener.

Visuals

A well-labeled visual can clarify definitions of most objects. The figure should be introduced by an identifying sentence: Figure 7-1 illustrates the construction of a spark plug." If your visual is borrowed, credit your source at the bottom left. Further explanation, if needed, should *follow* the visual. Unless

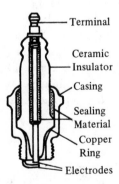

From *The McGraw-Hill Dictionary of Technical and Scientific Terms* (New York: The McGraw-Hill Book Company, 1974), p. 1388. Reprinted by permission.

FIGURE 7-1   A Spark Plug in Cross-section

a visual takes up one whole page, or more, don't place it on a separate page. Include it in the definition text.

Analysis of Parts

Many items or processes consist of several parts and are best defined through detailed explanation of each part.

> The standard frame of a pitched-roof wooden dwelling is composed of floor joists, wall studs, roof rafters, and collar ties.

> Psychoanalysis is an analytic and therapeutic technique consisting of four major parts: (1) free association, (2) dream interpretation, (3) analysis of repression and resistance, and (4) analysis of transference.

In discussing each part, of course, you would further define specialized terms like "floor joists" and "repression."

Comparison and Contrast

To compare is to identify similarities or likenesses; to contrast is to emphasize differences or dissimilarities. Whenever possible, compare the item with a more familiar one, or contrast it to various other models, sizes, etc.

Comparison	A cog railway, like a roller coaster, relies on a center cogwheel and a cogged center rail for transporting its cars up steep inclines.
Contrast	The X-55 fiberglass ski provides more edge control in icy conditions than its lighter but more durable aluminum counterpart, Model A-32.
Comparison and Contrast	Mediation, like arbitration, is a form of settling disputes; however, it differs from arbitration in that the decision of the mediator is not binding to the parties in the dispute.

Basic Operating Principle

Any mechanical item works according to a basic operating principle whose explanation should be part of your definition:

> Air-to-air solar heating involves the circulation of cool air, from inside the home, across a collector plate (heated by sunlight) on the roof. This warmed air is then circulated back into the home.

A clinical thermometer works on the principle of heat expansion: as the temperature of the bulb increases, the mercury inside expands and a thread rises into the hollow stem.

Even abstract items or processes can be explained this way:

Economic inflation functions according to the principle of supply and demand: if an item or service is in short supply, its price increases in proportion to its demand; this principle is evidenced by the effect of massive United States grain sales to foreign countries.

Special Materials or Conditions Required

Some items and processes are highly sensitive or volatile. These may require special materials, conditions, or handling. A detailed definition should include this important information.

Fermentation is the chemical reaction that splits complex compounds into simpler substances (as when yeast converts sugar to carbon dioxide and alcohol). This process has several special requirements: (1) a causative agent such as yeast, (2) a controlled pH (acid-base balance), (3) airtight and sterilized containers, and (4) a narrow temperature range.

To make good beer, home brewers need to know all these requirements. Abstract concepts may also be defined in terms of special conditions.

To be held guilty of libel, a person must have defamed someone's character through written or pictorial statements.

APPLYING THE STEPS

The following expanded definitions use various methods of amplification, as appropriate to the needs of their audiences. Specific expansion strategies are identified in the right margin. Notice that each definition, like a good essay, is unified and coherent: each paragraph is developed around a central idea and logically connected to other paragraphs. Visuals are incorporated into the text. Transitional words and phrases underscore the logical connection between related ideas. Each definition is written at a level of technicality that will connect with the intended audience.

AUDIENCE-AND-USE ANALYSIS

The intended readers below are beginning student mechanics. Before they can repair a solenoid, they will need to know where the term comes from, what a solenoid looks like, how it works, how its parts operate, and how it is used. This definition is designed as merely an *introduction*, so it offers only a general (but comprehensive) view of the mechanism.

Because the intended readers are not engineering students, they do *not* need details about electromagnetic or mechanical theory (e.g., equations or graphs illustrating voltage magnitudes, joules, lines of force, etc.).

By analyzing his audience carefully, our writer was able to decide answers to the question, "Just how much is enough?"

EXPANDED DEFINITION
OF "SOLENOID"

A solenoid is an electrically energized coil that forms an electromagnet capable of performing various mechanical functions. The term, "solenoid," is derived from the word, "sole," which, in reference to electrical equipment, means "a part of," or "contained inside, or with, other electrical equipment." The Greek word, *solenoides,* means "channel," or "shaped like a pipe."

Formal sentence definition

Etymology

A simple, plunger-type solenoid consists of a coil of wire attached to an electrical source, and an iron rod that passes in and out of the coil at right angles to the spiral. A spring holds the bar outside the coil when the current is deenergized, as shown in Figure 1.

Description and analysis of parts

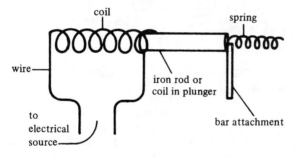

Visual

FIGURE 1   Lateral Diagram of a Plunger-Type Solenoid

When the coil receives electrical current, it becomes a magnet and thus draws the iron bar inside, along the length of its cylindrical center. With a lever attached to its end, the bar can transform electrical energy into mechanical force. The amount of me-

Special conditions and principle of operation

chanical force produced is determined by the product of the number of turns in the coil, the strength of the exciting current, and the magnetic conductivity of the iron rod.

The plunger-type solenoid, shown in Figure 1, is commonly used in the starter motor of an automobile engine. This type is 4½ inches long and 2 inches in diameter, with a steel casing attached to the casing of the starter motor. A linkage (pivoting lever) is attached at one end to the iron rod of the solenoid, and at the other end to the drive gear of the starter, as shown in Figure 2. When the ignition key is turned, current from the battery is supplied to the solenoid coil, and the iron rod is drawn inside the coil, thereby shifting the attached linkage. The linkage, in turn, engages the drive gear, activated by the starter motor, with the flywheel (the main rotating gear of the engine).

Example and analysis of parts

Explanation of illustration

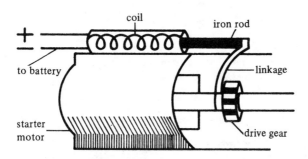

Visual

FIGURE 2   Lateral Diagram of Solenoid and Starter-Motor Assembly

Because of the many uses of the solenoid, its size varies according to the amount of work it must do. Therefore, a small solenoid will have a small wire coil, hence a weak magnetic field. The larger the coil, the stronger the magnetic field; in this case, the rod in the solenoid is capable of doing harder work. For example, an electronic lock for a standard door would require a much smaller solenoid than one for a large bank vault.

Comparison of sizes and applications

The audience for the following definition (an entire community) is too diverse to define precisely, so the writer wisely decides to write at the lowest level of technicality — to ensure that all readers will understand.

AUDIENCE-AND-USE ANALYSIS

The following definition is written for members of a community whose water supply (all obtained from wells, since the town has no reservoir) is doubly threatened: (1) by chemical seepage from a recently discovered toxic dump site, and (2) by a two-year drought that has severely depleted the water table. This definition forms part of a report analyzing the severity of the problems and exploring possible solutions.

To understand the problems, these readers first need to know what a water table is, how it's formed, what affects its level and quality, and how it figures into town planning decisions. The concepts of *recharge* and *permeability* are vital to reader understanding of the problem here, so these terms are defined parenthetically. These readers have no interest in geological or hydrological (study of water resources) theory. They simply need a broad picture.

WATER TABLE:
AN EXPANDED DEFINITION

The water table is the level below the earth's surface at which the ground is saturated with water. Figure 1 shows a typical water table which might be found in the East. Wells driven into such a formation will have a water level identical to that of the water table.

Formal sentence definition

Example

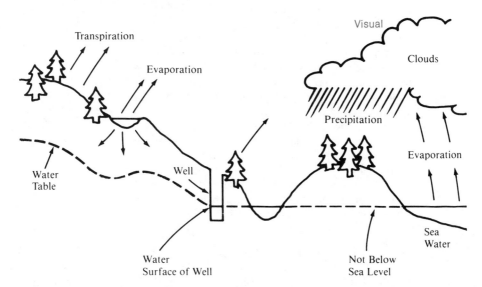

FIGURE 1   A Typical Water Table (Eastern USA)

The world's supply of fresh water comes almost
entirely as precipitation that begins with the evaporation of sea and lake water. This precipitation falls to
earth, and follows one of three courses: it may fall   Basic operating
directly onto bodies of water, such as rivers or lakes,   principle
where it is directly used by humans; it may fall onto
land, and either evaporate or run over the ground to
the rivers, etc.; or it may fall onto land, be contained,
and seep into the earth. This latter precipitation
makes up the water table.

Similar in contour to the earth's surface above, the   Comparison
water table generally has a level that reflects such
features as hills and valleys. In areas where the water
table intersects the ground surface, a stream or pond
results.

A water table's level, however, will vary, depend-
ing on the rate of recharge (replacement of water).   Basic operating
The recharge rate is affected by rainfall or soil per-   principle
meability (the ease with which water flows through
a soil). So a water table is never static; rather it is
the surface of a body of water striving to maintain
a balance between the forces which deplete it and
those which replenish it. In Florida and some west-   Example
ern states where the water table is depleted, for instance, the earth collapses, leaving sink holes.

The water table's depth below ground is vital in
water resources engineering and planning. For in-   Special conditions
stance, it determines an area's suitability for waste-   and examples
water disposal, or a building lot's ability to handle
sewage. A high water table could become contaminated by a septic system. Also, bacteria and chemicals seeping into a water table can pollute an entire
town's water supply. Another consideration in water
table depth is the cost of drilling wells. These concerns obviously affect an industry's or homeowner's
decision to locate in a certain area.

The rising and falling of the water table gives an
indication of the pumping rate's effect on a water   Special
supply (drawn from wells), and of the sufficiency of   conditions
the recharge rate in meeting demand. This kind of
information helps water resources planners decide
when new sources of water must be made available.

PLACING DEFINITIONS IN REPORTS

Poorly placed definitions interrupt the information flow. Eliminate this threat
to coherence by following these suggestions.

Parenthetical Definitions

If you have only a few informal definitions, place them in parentheses immediately after the terms. More than two or three definitions per page will be disruptive, and belong elsewhere. Rewrite them as sentence definitions and place them in a "Definitions" section of your report introduction, or in a glossary (see Chapter 12).

Sentence Definitions

If your sentence definitions are few, place them in a "Definitions" section of your report introduction. Otherwise, place them in a glossary. Any definitions of terms in the report's title belong in your report introduction.

Expanded Definitions

Place expanded definitions in one of three locations:

1. If the definition is essential to the reader's understanding of the *entire* report, place it in your report introduction. A report titled "The Effects of Aerosol Spray on the Earth's Ozone Shield," for instance, would require expanded definitions of "aerosol" and "ozone" early in the report.

2. When the definition clarifies a major part of your discussion, place it in the related section of your report. In a report titled "How Advertising Influences Consumer Habits," "operant conditioning" might form a major topic of the report; this term should then be defined early in the appropriate section. Too many expanded definitions *within* a report, however, can be disruptive.

3. If the definition is an aid to understanding, but serves as a *secondary* reference, it belongs in an appendix (see Chapter 12). An investigative report on fire safety in a public building, for example, might include an expanded definition of "smoke detectors" in an appendix.

CHAPTER SUMMARY

Precise meanings are vital to clarity. Every field has its own specialized language that has to be translated for nonspecialists. Also, many terms whose meanings seem obvious — like "guarantee" or "disability" — have different meanings in different contexts. So always specify your meaning.

For a parenthetical definition (the simplest type), place a synonym or explanatory phrase, usually in parentheses, right after the word. For a sentence definition, indicate the term, the class in which the item it names belongs, and the features that distinguish the item from all others in its class. For an expanded definition, discuss the term from just enough of the following approaches to specify its meaning for your particular audience:

1. etymology
2. history and background
3. negation
4. familiar examples
5. visuals
6. detailed explanation of various parts
7. comparison with similar items or contrast with dissimilar ones
8. the basic operating principle
9. special materials or conditions required

Place definitions where they are most useful: in parentheses after the word, in a "Definitions" section of your introduction, in a glossary, at appropriate points in your discussion, or in an appendix.

REVISION CHECKLIST FOR DEFINITIONS

Use this list to check the content, arrangement, and style of your definition.

Content

1. Have you chosen the type of definition (parenthetical, sentence, expanded) best suited to your subject, purpose, and reader's needs?

2. In defining a concrete and specific term (such as *transit* or *ophthalmoscope*), have you given all the details, and *only* those details, that will serve the reader's needs?

3. In defining an abstract and general term (such as *condominium* or *partnership*), have you specified its meaning in the context where you are using it?

4. Does the definition express the basic properties of the item (the unique features that make the item what it is)?

5. Is your definition objective (free from implied judgments)?

6. For an expanded definition, have you used all expansion techniques that will serve your specific reader?

7. Is the definition free from needless details?

8. Have you documented all data sources?

9. Have you used visuals adequately and appropriately?

Arrangement

1. Does your sentence definition follow the term-class-distinguishing-features structure?
2. Is your expanded definition logical and coherent (like an essay)?
3. Is there adequate transition between related ideas?
4. Have you placed your definition in the most appropriate report location?

Style

1. Is your definition in plain English?
2. Will its level of technicality connect with the intended audience?
3. Is it free from needless words?
4. Are all sentences clear, concise, and fluent?
5. Is it written in correct English (Appendix A)?

Now list those features of your definition that need improvement.

EXERCISES

1. In a few paragraphs, discuss the differences among parenthetical, sentence, and expanded definitions, citing specific examples where each would be used for a certain purpose and audience.

2. Sentence definitions require precise classification and differentiation. Is each of the following definitions adequate for a general reader? Rewrite those that seem inadequate. If necessary, consult dictionaries and specialized encyclopedias.

   a. A bicycle is a vehicle with two wheels.
   b. A transistor is a device used in transistorized electronic equipment.
   c. Surfing is when one rides a wave to shore while standing on a board specifically designed for buoyancy and balance.
   d. Bubonic plague is caused by an organism known as *pasteurella pestis*.
   e. Mace is a chemical aerosol spray used by the police.
   f. A Geiger counter measures radioactivity.
   g. A cactus is a succulent.
   h. In law, an indictment is a criminal charge against a defendant.
   i. A prune is a kind of plum.

   j. Friction is a force between two bodies.

   k. Luffing is what happens when one sails into the wind.

   l. A frame is an important part of a bicycle.

  m. Hypoglycemia is a medical term.

   n. An hourglass is a device used for measuring intervals of time.

   o. A computer is a machine that handles information with amazing speed.

   p. A Ferrari is the best car in the world.

   q. To meditate is to exercise mental faculties in thought.

3. Think of a situation where someone failed to define a term adequately (as in giving instructions or an assignment), and caused you to misunderstand the message. Describe the situation and its consequences in two or three paragraphs.

4. Both standard college dictionaries and specialized encyclopedias contain definitions. Standard dictionaries, however, define a word for the general reader, whereas specialized reference books provide definitions for the specialist. Choose an item in your field and copy the meaning as found (1) in a standard dictionary and (2) in a technical reference book. For the technical definition, label each expansion method carefully on a photocopy you have made. Finally, rewrite the specialized definition for a general reader.

5. Using reference books as necessary, write sentence definitions for the following terms or for terms from your field. To express the basic properties of each item clearly, narrow your classification and name the distinguishing features.

biological insect control	economic inflation	estuary
generator	anorexia nervosa	acid rain
dewpoint	low-impact camping	classical conditioning
microprocessor	hemodialysis	hypothermia
capitalism	gyroscope	thermistor
economic recession	floppy disk	aquaculture
marsh	oil shale	nuclear fission
chemotherapy	parachute	

6. Select an item or concept from the list in question 5 or from an area of interest. Identify the particular audience and its particular needs. Begin with a sentence definition of the term. Then write an expanded definition for a first-year student in that particular field. Next, write the same definition for a layperson (client, patient, or other interested party). Leave a three-inch margin on the left side of your page to list your expansion strategies (use at least four in each version). Submit, with your two versions, a brief explanation of your changes from the first to the second version.

7. The memo below, written for laypersons, contains an expanded definition of "epilepsy," along with instructions (omitted here) for dealing with epileptic seizures in class. Your assignment is twofold: (1) identify the expansion strategies used in the definition and (2) locate and copy an expanded definition of "epilepsy" written for technically informed readers.

(Consult *The Merck Manual* or medical reference books, *not* general encyclopedias.) Document your source. Discuss the specific differences (in content, arrangement, and style) between the two versions. Submit your
analysis and your copy of the second version to your instructor.

POWNAL COLLEGE HEALTH OFFICE

OCTOBER 20, 1981

TO: Faculty

FROM: Hester Pryor, Director of Health Services

SUBJECT: EPILEPSY AND THE ROLE OF FACULTY MEMBERS

Each year, a number of our students are stricken by seizures. A seizure
during a class period may disrupt not only the entire class but also the
future of the stricken student, *unless the faculty member knows how to
handle such incidents wisely.*

Epilepsy: A Definition

The term *epilepsy* is derived from the Greek word for "seizure." The concept of "being seized" (as though from the outside of oneself) was passed
down through the ages until the nineteenth century, when Hughlings Jackson, the great English neurologist, defined epileptic seizures as a state
produced by "a . . . sudden, violent, disorderly discharge of brain cells."
Jackson's definition, implying the discharge of excessive electrical (nervous)
energy, has been substantiated by brain wave studies made possible by the
development of the electroencephelograph.

Many seizures fall under the general classification of epilepsy, but for our
purposes, only the three predominant types of seizures will be discussed.

Petit Mal

The petit mal seizure is a brief interruption of consciousness characterized
by the appearance of daydreaming. The person generally has a blank look,
and some small muscular movements may occur in the face. Usually the
seizure lasts only a few seconds. Such people will have no awareness that
they have had a seizure.

Grand Mal

From the point of view of the faculty member and the other students, the
grand mal seizure is the most alarming type. The person may suddenly
slump over or fall to the floor. Muscles become rigid and then make con-
vulsive movements. Bladder control may be lost. The person may have difficulty breathing, and saliva will begin to collect and run from the mouth.
The attacks usually are brief, and after resting, the person sometimes can
continue with the class.

Psychomotor Seizures

Psychomotor seizures take many forms. In some cases, a student will experience strange sensations (e.g., unpleasant odors, or distorted perception). In others, the person will seem to be making purposeful movements, but these will bear no relation to the immediate situation. Some people will make lip-smacking or chewing movements.

It is important to remember that, with epileptics, the attitudes of those around them are of vital importance. The more normal a life they are able to lead, the better their response to medical treatment.

8

Dividing and Organizing

DEFINITIONS

Sometimes we divide a single thing into its parts to make sense of it. At other times we group an assortment of things into categories to sort them out. Partition and classification are techniques for sorting things out. The two related activities serve different purposes: *partition* identifies the parts of a single item; *classification* creates categories for sorting similar items.

USING PARTITION AND CLASSIFICATION

Whether you use partition or classification depends on your subject. If you must describe a golf club, for example, you have little choice but to begin by partitioning it into its major parts: grip, shaft, and head. But if someone has unexpectedly given you 228 record albums that you want to arrange in some way so you can easily locate the particular record you want, partition will not help. You will have to classify the records by dividing the pile into smaller

categories. You might want to classify them as classical, jazz, rock, and miscellaneous. Or you might classify them according to your likes and dislikes.

Close examination of any complex problem usually requires both partition
and classification. If, for example, you are designing a new supermarket, you
must first partition the whole market into its parts:

- display and shopping area
- receiving and storage area
- meat refrigeration and preparation area
- checkout area
- small office area

Next, you need to sort out your inventory by grouping the thousands of items
into smaller classes according to their similarities, for example:

- frozen foods
- dairy products
- meat, fish, and poultry
- pet foods
- fruits and vegetables
- paper products
- canned goods
- beverages
- baked goods
- cleaning products

In turn, you divide each of these sections further. You might divide "meat"
into three smaller groups.

- beef
- pork
- lamb

Under these headings you will group the various cuts of meat (steaks, ribs,
etc.) in each category. And you might carry the division further for certain
meat products, like types of ground beef:

- regular
- lean
- extra lean
- diet lean

This kind of dividing and grouping continues until you have enough categories
or classes to sort the hundreds of crates and cartons of inventory that sit in
your receiving and storage area. You have divided your store into its parts

and grouped your inventory into classes to sort out the items. You have used both partition and classification.

Whether you partition or classify, follow strict guidelines so your arrangement will make sense.

GUIDELINES FOR PARTITION

Apply Partition to a Single Item

Table 8-1 shows how a single item can be divided. Notice that the component nutrients add up to 100 percent; the total egg equals the sum of its parts.

Make Your Partition Complete and Exclusive

Include *all parts* of the item. If you have omitted one or more parts for good reason, say so in your title ("The Exterior Parts of a Typewriter"). On the other hand, be sure that each part belongs to the item. For instance, do not include "typewriter ribbon" in your partition of the exterior of a typewriter.

Make Your Division Consistent with Your Purpose

Most things can be divided in different ways for different purposes. You might divide an apple into the meat, skin, and core, but that division is useless to the nutritionist who wants to know the food value of an apple, and isn't interested in the fact that it has a core (unless it has nutritional value). You could divide a house according to at least three different bases: (1) the rooms it comprises, (2) the materials used in construction, or (3) the steps required to build it. The sum of the parts in each of these divisions equals 100 percent of the house. The basis you choose, however, will depend on your purpose and the reader's

TABLE 8-1   Approximate Composition of a Whole Goose Egg

Component	Percentage (by Weight)
Shell	14.0
Water	60.0
Protein	13.0
Fat	12.0
Ash	1.0

needs. To interest a prospective buyer, you might choose option 1; to provide cost or materials specifications, option 2; to give instructions to the do-it-your-selfer, option 3. Sometimes you might need two or more partitions of the same item.

Let your title promise what the partition will deliver: "The Jones House, Partitioned According to Materials for Construction." Under this title, you would include every item, from nails to plumbing fixtures. If your purpose is less ambitious, limit your title: "Concrete Materials," "Electrical Materials," or the like.

Subdivide as Far as Necessary

Subdivide as much as necessary to show what makes up the item. This text-book, for instance, is divided into chapters. In turn, each chapter is subdivided into major topics such as "Guidelines for Partition." Major topics are again di-vided into minor topics such as "Apply Partition to a Single Item," and so on.

Follow a Logical Sequence

Most items have a particular logic of organization that determines the order for listing their parts (based on the way they work or are put together, or on relative importance, etc.). The division of a basic house, for example, logically follows a *spatial sequence* — the foundation, the floor, the frame, the siding, and the roof — the sequence in which the parts are arranged. A progress report (a partition of your activity over a certain period) would follow a *chrono-logical sequence* — from earliest to latest. A partition of your monthly budget might follow a descending scale of the *order of importance* of each expendi-ture. A problem can be analyzed by division into its *causes and effects*. Other specific orders of development are discussed in Chapters 3 and 9.

Keep Parts of Equal Rank Parallel

All parts at any one level of division (major parts, minor parts, subparts, etc.) are considered equal in rank. So list them in equal, or parallel, grammatical form.

> Faulty
>
> A deed contains the following seven items:
>
> 1. The buyer and seller must be identified.
> 2. A granting clause.
> 3. Consideration, not necessarily money, must be mentioned.

4. An explanation of the rights being transferred.
5. A full-length description of the property.
6. Proper execution, signature, seals, and delivery.
7. It must be properly recorded in the county where the property lies.

Here, items 1, 3, and 7 are not parallel to the others; they are expressed as sentences, whereas the others are expressed as phrases. Revise 1, 3, and 7 to read as follows:

1. identification of buyer and seller
3. mention of consideration, not necessarily money
7. proper recording in the county where the property lies

Or, you could write 2, 4, 5, and 6 as sentences.

Keep Parts from Overlapping

The logic of division requires that each item exclude all others. Consider this faulty partition of the executive structure of a corporation:

Faulty

chairman of the board	vice-president
board of directors	administrative officers
president	

These parts overlap because each of the first four positions may be listed under the heading "administrative officers." A correct version would omit this last heading.

Use Precise Units of Measurement

If the parts are in percentages, pounds, feet, or the like, say so, as in Table 8-2.

Choose the Clearest Format

Select the best format for your partition: prose discussion, list (as in an outline), table, or chart. Assume, for instance, that you are researching federal budgets for the past decade, and you need to partition the budget for 1974 in two ways: by sources of income and types of expenditure. Here is how your prose version might read:

> The federal budget dollar for fiscal year 1974 can be divided into two broad categories: sources and expenditures. Income sources are broken down as follows: Individual income tax provided $0.42 of every dollar of federal income. Social insurance taxes and contributions provided $0.29, whereas corporation income taxes provided $0.14. The smaller income sources in-

TABLE 8-2   The Parts of a Selected Multivitamin

Ingredient	Quantity
Vitamin A	15 International Units
Vitamin E	15 International Units
Vitamin C	60 mg
Folic Acid	0.4 mg
Vitamin B$_1$	1.5 mg
Vitamin B$_2$	1.7 mg
Niacin	20 mg
Vitamin B$_{16}$	2 mg
Vitamin B$_{12}$	25 micrograms
Vitamin D	400 International Units
Iron	18 mg

cluded excise taxes ($0.06), borrowing ($0.05), and miscellaneous sources ($0.04).

The greatest federal expenditure was in human resources, which consumed $0.47 of every federal dollar. Next was national defense, requiring $0.30. Smaller expenditures were for physical resources ($0.10), interest ($0.07), and miscellaneous spending ($0.06).

With an enumerative list like this, a prose discussion can be difficult to follow. Table 8-3 shows the same division in another format.

TABLE 8-3   A Partition of the Federal Budget Dollar for Fiscal Year 1974

Where It Came From	Amount
Individual income taxes	$0.42
Social insurance taxes and contributions	0.29
Corporate income taxes	0.14
Excise taxes	0.06
Borrowing	0.05
Other	0.04
Where It Went	**Amount**
Human resources	$0.47
National defense	0.30
Physical resources	0.10
Interest	0.07
Other	0.06

Depending on the writer's purpose, any item in the table could be partitioned further. Notice that all items add up to 100 percent. Figure 8-1 shows

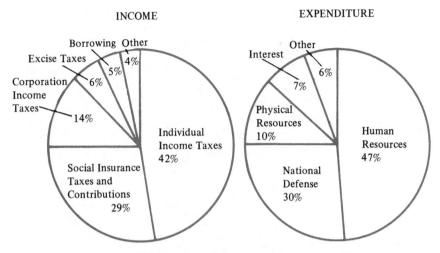

FIGURE 8–1   A Partition of the Federal Budget Dollar for Fiscal Year 1974

pie-chart versions of the same data. For this partition, the pie chart seems most effective because it dramatizes the vast differences both in sources of income and in types of expenditure. (See Chapter 13 for instructions on composing visual aids.)

Different subjects for partition lend themselves to different formats. A subject whose parts are ranked in order of occurrence might best be partitioned as a flowchart, while one with parts ranked in order of importance might call for an organizational chart, as in Figure 8-2.

GUIDELINES FOR CLASSIFICATION

Apply Classification to a Group of Items

Assume you are writing a report on vegetarian diets and have decided to divide foods into three classes: fat sources, starch sources, and protein sources. Using a table, you might classify the foods you've designated as protein sources (see Table 8-4). Notice that the classification is limited by the term "selected." Notice also that the whole egg partitioned in Table 8-1 is here just one item in the classification system.

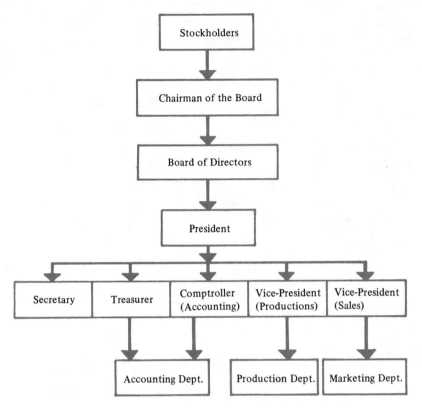

FIGURE 8-2   The Organizational Chart of a Basic Corporation

Make Your Classification Complete, Exclusive, and Inclusive

List *all items* that logically belong to a particular class.

> *Red Meats*
> − pork
> − lamb
> − beef

Also, divide the assortment into enough classes to contain every item you are sorting. A division of red meats into pork and beef, for example, would not be inclusive because lamb products (chops, legs, ribs, etc.) would have no place in which to be grouped. Finally, be sure that each item belongs in that particular class.

Red Meats
- pork
- lamb
- beef
- frozen meats

"Frozen meats" are a function of temperature, not a type of meat. This system also overlaps because all three types of meat might be included in the category "frozen."

Limit Your Classification to Suit Your Purpose

A classification becomes more specific and useful as you limit its focus. A résumé classification labeled "Work Experience," for instance, is more informative than one labeled "Experience." The classification "Vegetables" would require a list of every vegetable ever grown. Depending on your purpose, you might want to limit the classification to "Vegetables Sold by Our Food Co-op," "Canned Vegetables," "Green Vegetables," or another precise designation. Let your title promise what you will deliver.

Choose Bases of Comparison Consistent with Your Purpose

After limiting your classification, you may need to select one or more bases to compare or contrast the items in your list. For example, in planning a high-protein diet, you could choose this basis: "Green Vegetables Classified on the Basis of Protein Content." Or you might choose several bases: protein content, caloric content, chlorophyll content, etc. A classification that includes one or more bases is called a formal classification. Table 8-5 gives the same material as Table 8-4, but as a formal classification. Because only one basis of comparison is used here, the items are arranged in a specific order of presentation — in this case, a descending order.

TABLE 8-4   Selected Protein Sources

Brewer's yeast	Sunflower (seed)
Soybean (seed)	Whole egg
Groundnut (peanut)	Coconut
Cottonseed	Cow's milk (whole)
Sesame	Potato

Source: Adapted from Johnson and Peterson, *Encyclopedia of Food Technology,* p. 722. Used by permission of AVI Publishing Company.

TABLE 8-5   Selected Protein Sources Classified in Descending Order
of Protein Content

Source	Protein (gm/100)
Brewer's yeast	38.8
Soybean (seed)	38.0
Groundnut (peanut)	25.6
Cottonseed	20.2
Sesame	18.1
Sunflower (seed)	12.6
Whole egg	12.4
Coconut	6.6
Cow's milk (whole)	3.5
Potato	2.0

Source: Adapted from Johnson and Peterson, *Encyclopedia of Food Technology*,
p. 722. Used by permission of AVI Publishing Company.

If your purpose in classifying some types of food is to achieve a healthful
diet, you would arrange the items according to several bases of nutrient content — protein, fat, iron, vitamins, etc. — all in a combined table such as
Table 8-6. Here the bases are listed in the left vertical column so the table
can be contained within the width of one page.

Express All Bases in Precise and Objective Terms

Your bases must express specific similarities and differences; vague terms of
assessment are useless. Modifiers such as "long," "nice," "heavy," "good," and
"strong" are not specific enough. The classification "Good Vegetables" is meaningless unless "good" can be defined and measured — with units of nutritional
value expressed in grams, milligrams, international units, or the like.

Keep Categories of Equal Rank Parallel

Consider these class headings:

Frozen Foods   *Dried Foods*   *Foods That Are Smoked*

This system is not parallel unless the last heading is revised to read "Smoked
Foods."

TABLE 8–6   Nutrient Content of Red Meats, Poultry, and Fish

	Chicken	Nonfatty Fish	Herring	Beef	Lamb	Pork
			Per Ounce (28.35 gm) Raw Meat			
Protein (gm)	5.9	4.5	4.5	4.2	3.7	3.4
Fat (gm)	1.9	0.1	4.0	8.0	8.8	11.4
Calories (kcal)	41	19	54	89	94	116
Calcium (mg)	3	1	28	3	3	3
Iron (mg)	0.4	0.3	0.4	1.1	0.6	0.3
Vitamin D (mg)	—	—	6.38	—	—	—
Vitamin A (mg)	—	—	13	—	—	—
Vitamin B_1 (mg)	0.01	0.02	0.01	0.02	0.04	0.28
Vitamin B_2 (mg)	0.05	0.03	0.09	0.06	0.07	0.06
Niacin (mg)	1.7	0.8	1.0	1.4	1.4	1.4
Pantothenic acid (mg)	0.19	0.06	0.28	0.11	0.14	0.17
Vitamin B_6 (mg)	0.28	0.06	0.13	0.08	0.09	0.14
Biotin (mg)	2.83	2.83	x	0.85	0.85	1.1
Folic acid (mg)	0.85	14.1	x	2.83	0.85	0.85
Vitamin B_{12} (mg)	x	0.28	2.83	0.56	0.56	0.56
Vitamin E (mg)	0.06	x	x	0.17	0.23	0.19

Source: Johnson and Peterson, *Encyclopedia of Food Technology*, p. 626. Used by permission of AVI Publishing Company.
"—" indicates nutrient not present.
"x" indicates content not yet determined.

Keep Categories from Overlapping

Consider these headings:

> *Pork   Beef   Ham   Lamb*

The headings overlap because ham is not an exclusive category; it is a pork product.

Choose the Clearest Format

Remember you have two purposes in classifying: (1) to help you organize your material and (2) to save readers time. So select the clearest format for your classification: either a prose discussion, a list (as in an outline), a table, or a graph. (See page 262 for a classification graph.)

Sometimes, the bases in a classification cannot be expressed in simple units of measurement or by single words. Because the following classification (from an environmental pamphlet for general readers in agriculture) contains definitions of various pest-control methods, it is cast in a prose format.

To illustrate the importance that audience analysis plays in a writer's decision about "How much is enough?", the example below, like many throughout this text, is preceded by an audience-and-use analysis.

AUDIENCE-AND-USE ANALYSIS

The intended readers below are expected to have little or no awareness of organic pest-control methods. So the classification is designed merely as an *introduction* to substitutes for synthetic chemicals. Specific alternatives are merely identified and described in the most general terms. Readers here need an answer to this question only: What are the categories of possible alternatives? They do *not* need detailed explanations or instructions. Readers who are interested might later consult specific manuals for details. But the purpose here merely is to provide an overview.

Terms such as *tillage, field sanitation, crop rotation,* and *legumes* are not defined here because these readers have agricultural backgrounds.

Notice that the methods are classified in a logical and clear order: on the basis of methods requiring increasing human involvement (from easiest to most difficult).

ORGANIC PEST CONTROL: A SENSIBLE ALTERNATIVE

Organic pest control is a method for decreasing weeds, insects, and plant disease by the use of natural substances and nondisruptive practices. Unlike the synthetic chemical pesticides in current use, organic control does not harm the environment. Organic control includes a variety of techniques that can be classified under the following headings: spontaneous control, cultural control, biological control, mechanical control, and natural chemical control.

Spontaneous control is anything normally occurring in the environment that will control pests. It occurs without human interference. For instance, climatic factors such as light, moisture, temperature, and topography are important in pest control. Another form of spontaneous control includes natural pest enemies, such as predators and parasites, that have not been introduced by humans.

Cultural control is achieved by normal farming practices designed to reduce pest populations. Examples include the use of resistant plant species, certain tillage practices, field sanitation, and crop rotation.

Biological control occurs when natural pest enemies are introduced by humans. Insects can be used to control weeds and other insects. A single lady beetle, for instance, consumes up to 50 harmful insects per day. Bacteria and other organisms also can be used to limit insect populations.

In certain cases, mechanical or physical control can be effective. Examples include using logs or fences as insect or pest barriers, setting trapping devices for insects, or manipulating air supply, temperature, and moisture in silos and produce bins.

Certain plants contain natural substances with insecticidal properties. Two of these naturally occurring insecticides are pyrethrum (derived from chrysanthemums) and rotenone (derived from various legumes). These substances are less harmful than the synthetic chemical insecticides because they have low toxicity to humans and wildlife, and they degrade rapidly.

These organic control methods are increasingly being included in pest-management programs to limit the use of dangerous synthetic pesticides.

CHAPTER SUMMARY

To divide a single thing into its parts, we use partition. To group an assortment of things into specific categories, we use classification. Partition is always applied to a single object. Its purpose is to separate systematically that whole object into its parts, pieces, or sections. Classification is always applied to an assortment of objects that have some similarities. Its purpose is to group these objects in a systematic way.

Whether you choose to apply partition or classification will depend on your subject and your purpose. In analyzing complex problems, you will often have to use both techniques.

Follow these guidelines for dividing a single item into its parts:

– Apply partition to a single item.
– Include all parts, and be sure each part belongs.
– Divide in a way consistent with your purpose.
– Subdivide as far as needed to show what makes up the item.
– Make sure your sequence of division follows the item's logic of organization (spatial, chronological, or the like).
– Make parts of equal rank parallel in grammatical form.
– Make sure that parts do not overlap.
– Choose the clearest format (prose, list, or visual).

Follow these guidelines for dividing an assortment of things into classes:

– Apply classification to a group of items.
– Include all items that belong to a particular class; be sure that each item belongs in that class; and divide the assortment into enough classes to contain every item you are sorting.
– Limit your classification to a specific focus.

– Choose useful bases, expressed in precise and objective terms.
– Make equal categories parallel, and make sure categories do not overlap.
– Choose the clearest format.

REVISION CHECKLIST FOR PARTITION AND CLASSIFICATION

Use this list to refine your system of partition or classification.

Content

1. Have you given a clear, limiting, and accurate title?
2. Have you documented all data sources?
3. Have you included all items that belong, but no extraneous ones?
4. Have you avoided overlapping?
5. Have you given all units of measurement (grams, pounds, etc.)?
6. Have you subdivided all items in your partition as far as necessary?

Arrangement

1. Have you chosen the clearest format for your purpose (table, chart, graph, prose discussion)?
2. Do you integrate the partition or classification into the surrounding text (introduce and discuss)?
3. Does the partition follow a logical sequence (spatial, chronological, etc.)?
4. Do you divide all items in the partition according to a basis that suits your purpose and audience?
5. Do you arrange all items in the classification according to a basis that suits your purpose and audience?
6. If you use a chart, graph, or table, does it follow the criteria for visuals, discussed in Chapter 13?

Style

1. Are all items in parallel grammatical form?
2. Are the bases objective and precise?
3. Is the prose version in correct English (Appendix A)?

Now list features of the partition or classification that need improvement.

EXERCISES

1. In a short essay, discuss the types of partition and classification systems you expect to use on the job. Be as specific as possible in identifying typical subjects, your intended audience, and uses of your data.

2. Is the following classification effective? Use the revision checklist as a guide for evaluation, and write out your suggestions for improvement.

1984 Beer Production
in the Top Six National Breweries

Langdon Brewing Co.	14,678,400
Kastel Inc.	21,739,200
Flagstaff Brewing Co.	7,635,800
King Inc.	10,478,300
Rothberg Brewing Co.	23,547,700
Case Brewing Co.	9,658,400

3. Discuss briefly the difference between classification and partition. Name three items that can be classified and three that can be partitioned.

4. Jot down each of your activities on any day. When your list is complete, group related activities under specific headings: ("Social Activities," "Schoolwork," etc.). Next, choose a basis of comparison for arranging items within a given class in a specific order (e.g., descending order of time spent on each activity, order of desirability, etc.). Finally, choose one class of activities — "Schoolwork," for example — and present your formal classification as a bar graph, a table, and a prose discussion. Which form of presentation seems most "readable" here?

5. Assume you are planning a week-long camping trip into a wilderness area. Make a list of all the items you will need. Next, group related items under specific headings in order to organize your inventory. Finally, take one item from the list (e.g., "tent"), and partition it into its various parts (pegs, poles, canvas, guy lines, mosquito netting, etc.).

As an alternative assignment, assume you are planning a one-week spring vacation in Bermuda, and perform the same tasks.

6. Some items listed in the following classification systems are not logically related. Identify the specific error in each group — faulty parallelism, overlapping, inconsistent general meaning, or incompleteness — and correct it.

Classification of selected
alcoholic beverages

    beer
    wine
    whiskey
    bourbon

Classification of
winter sports

    hockey
    curling
    handball
    skiing
    figure skating
    sledding

Classification of automobiles
on the basis of body type

    sedans
    family cars
    roadsters
    coupes
    hardtops
    station wagons

Classification of common
technical writing tasks

    proposals
    sending memos
    specifications
    work orders
    progress reports
    instructions
    writing letters
    reports that are formal
    product and mechanism
      descriptions

7. Find an example of a formal classification in a magazine, such as *Consumer Report*, or in a textbook. Is the classification effective? Using the criteria for effectiveness discussed in this chapter, compose a one-paragraph answer. Do the same for a partition.

8. Choose an item with a singular meaning (bicycle, digital calculator, clock, retail store, stereo system, etc.). Partition the item, for a specified audience and purpose, on as many levels as necessary.

(1st level)

Major parts of a bicycle
    frame
    wheels
    drivetrain
    attachments

(2nd level)

		Major parts	
Parts of a frame	*Major parts of a wheel*	*of a drivetrain*	*Major attachments*
	hub		
	spokes		
	rim		

(3rd level)

Etc.	*Parts of a hub*	Etc.	Etc.
	etc.		

Notice how this partition (composed for the repair-it-yourself bicycle owner) moves from division to subdivision, and so on. Hence, the partition is complete at each level. Complete the entire partition by filling in the required items, or construct one of your own, using a subject that you know well.

9. Assume your company's vice-president has decided it's time to purchase a new line of tools or equipment (chain saws, microscopes, drafting tables, cash registers, company cars, or the like). You've been asked to compare various brands of the item and to submit your findings to the vice-president. Identify one or more bases that will best serve your reader's needs. Do the research and compose a classification report. Use visuals where applicable.

10. Read the following essay, and answer the questions below.

COMPUTER SCIENTISTS:
A CLASSIFICATION

Computer scientists deal with the software (programming) end of computer technology. They hold B.S. degrees from accredited schools, with majors in software engineering or computer science. The curriculum focuses on the logic and language of programming, and, briefly, on the hardware of the machine. (Programming is the process of instructing computers to perform various applications. A program defines in complete and minute detail just what the computer is to do in particular circumstances.)

Computer scientists play major roles in science, government, and business. Their programs might be used for guiding satellites, developing federal and state budgets, controlling inventories, making and confirming reservations, or grading examinations — among countless other uses. This broad demand for programs creates a diversity of job opportunities. The major specialties within computer science can be grouped into three categories: systems programming, applications programming, and systems analysis.

Systems programmers write programs that run the computer equipment itself. These programs act as an interface (connector) between the machines and the users' programs. Examples of systems programs include those that control the computer's operating system, monitor, printer, and interpreter. Much systems programming is done in machine language (using a binary-number system: 0's and 1's).

The job of the applications programmer overlaps with that of the systems programmer in that both entail the same type of programming logic, but for different purposes and in different codes (languages). Applications programs put the computer to work on specific jobs such as keeping track of accounts in banks and insurance comanies. These programs act as an interface between the systems program and the user. Applications programs usually are written in high-level languages (COBOL, Pascal, BASIC, FORTRAN, etc.).

Systems analysts are the organizers. They monitor the systems and applications programs, and are responsible for eliminating any bugs (program failures). Instead of actually writing the programs, systems analysts update those written by systems programmers, as required. They may also analyze an organization's particular needs, and write specifications for a computer system. These specifications are then given to a programmer. Systems analysts require several years of experience as programmers.

In any role, computer scientists are responsible for analyzing a problem, and then reducing it to the sequence of small, deliberate steps the computer can use to solve the problem.

The essay above was written for entering freshmen, who will be reading it as part of a university's *Career Handbook* while they are choosing specific majors. Using your intuition, characterize in detail the intended readers of the above essay. (Use the analysis on page 144 as a model.) How much are these readers likely to know about computer science? How much do they need to learn? Why has the writer not been more specific? Why are definitions (of programming, machine language, etc.) included? How did this writer order her classification — according to what basis?

Now, compose a classification of the jobs commonly available to graduates in your major. Write for an audience of entering freshmen, and include a written audience-and-use analysis with your essay. Choose a definite basis for ordering your material.

11. Our world is full of alternatives, of better ways of doing many things: say, of storing or disposing of hazardous wastes, of reducing environmental damage from acid rain, of reducing our energy consumption, or of finding alternatives to nuclear energy. Using the classification essay on pages 144–145 as a model, identify something that could be better done in other ways. For instance, you might explain to members of your community (in a cold climate) economical alternatives to central heating (kerosene heaters, passive solar heating, wood stoves, quartz heaters), arranged according to cost, safety, convenience, or some other basis. Choose a subject from the list below, or make up one of your own, and write an essay defining each alternative briefly but clearly. Select a definite basis for ordering your material, and include an audience-and-use analysis with your essay. Some possible topics include:

– alternatives to meat as protein sources (soy, wheat, eggs, legumes, etc.), on the basis of cost or nutritional value
– alternatives to weight lifting for body building (gymnastics, calisthenics, martial arts, Nautilus, isometrics, etc.), on the basis of rapid results, long-term benefits, or convenience
– alternatives to conventional frame housing (geodesic domes, earth houses, cement-block or stone houses, envelope houses, etc.), on the basis of energy efficiency
– alternatives to single-family dwellings (condominiums, duplexes/triplexes, co-op housing, communal housing, etc.), on the basis of cost.

12. Assume you have decided to purchase your own personal computer as an aid in school and professional work. Identify one or more bases that will best serve *your* needs (for word or data processing, statistical analysis, mathematical modeling and simulation, spreadsheet projections, visual design, or the like) at the lowest possible cost. Using your bases to compare various brands of computers (and available software), do the research and compose a classification report.

9

Charting
Your Course:
The Outline

OUTLINES DEFINED

Outlines come in many shapes and sizes: a few key words or phrases on a page, a grocery list, a daily reminder calendar. Each reminds us of things to be done. The size and complexity of your own outline will vary with the writing task. For most short reports, a list of words or phrases will be enough of a guide; a long, complex report or proposal may require a systematic arrangement of topics, with a formal numbering and lettering system. In any case, never write the final draft of your report — long or short — without following some sort of outline.

PURPOSE OF OUTLINING

Why does a writing assignment fall easily into place for one writer but not another? *Because the successful writer spends more time thinking and planning than writing.* Good writing calls for deliberate decisions: about what to say, how to say it, how to organize it, and how to revise it to make the whole thing work. In Chapter 5, we saw that these decisions rarely occur in a predictable order, but they must *all* be made at some point during the writing process. And a major part of any writer's challenge is getting organized.

Last-minute changes in building a house are made more easily on the blueprints than by tearing down actual walls or moving a bedroom wing. Likewise, an outline is easier to modify than a written report. Like the blueprint, the outline is not a set of commandments, but simply a tool — an orienting device. For a research report, for example, you can sketch an informal outline long before writing the actual report. This list of topics will provide a direction for gathering data. As your investigation proceeds, you can revise the outline. Then you might compose a formal outline before writing the first draft of the report, or *after* the first draft, to evaluate the organization of your message.

CHOOSING A TYPE OF OUTLINE

The Informal Outline

An informal outline is a simple list, probably all you need for a short report. First, write a statement of purpose. Next, brainstorm (Chapter 2) for ideas. After selecting ideas (expressed as phrases), arrange them in a sensible sequence. Here is a sample statement of purpose, with its outline:

> STATEMENT OF PURPOSE
>
> This report describes the building and plumbing inspector's evaluation to determine the waste-disposal system needed for lakefront building site #52.
>
> TOPICS
>
> 1. Location of the Site
> 2. Description of the Site
> 3. Type of Water Supply
> 4. Instructions for Constructing the Gray-Water System (sink drain)
> 5. Instructions for Constructing the Privy (Outhouse) Pit
> 6. Instructions for Applying for a Gravel Waiver

For this sequence, the writer observed the principle that anything to be discussed must first be described. With this list, the writer gets oriented and organized. The report can now be composed easily.

If you plan a longer report, an informal outline is still handy. In this case, it is a tentative (or working) outline, because it keeps you on track without excluding possible changes as you work.

Here is an informal outline for a report titled "An Analysis of the Advisability of Converting Our Office Building from Oil to Gas Heat." The writer begins by formulating her statement of purpose.

> The president of Abco Engineering Consultants has requested a report on the advisability of converting our office building from oil to gas heat.

As the writer brainstorms, she produces the following list of major topics for research and discussion.

- Estimation of Gas Heating Costs
- Removal of the Oil Burner and Tank
- Installation of a Gas Burner
- Description of Our Present Heating System
- Installation of a Gas Pipe from the Street to the Building

She then shuffles these topics until she finds the most sensible sequence — in this case, a chronological sequence.

1. Description of Our Present Heating System
2. Removal of the Oil Burner and Tank
3. Installation of a Gas Pipe from the Street to the Building
4. Installation of a Gas Burner
5. Estimation of Gas Heating Costs

Notice that the topics follow the sequence of steps in the actual conversion process, beginning, of course, with a description of the present system. The writer now has a general plan for gathering data. She can expand this outline and make it more specific by adding subtopics from her brainstorming list, as shown below. When her data gathering is complete, the writer composes her first draft. Sometime before writing her final draft, she will develop a formal outline, to check the organization of her report.

The Formal Topic Outline

The formal topic outline is a more detailed and systematic arrangement, using formal notation (numbers, letters, and other symbols marking logical divisions). You may choose between two common systems of notation: the roman numeral–letter–arabic numeral system or the decimal system.

Roman Numeral–Letter–Arabic Numeral Notation

Here are the five topics from the informal outline developed into a formal outline, using roman numeral–letter–arabic numeral notation:

II. REPORT BODY (or COLLECTED DATA)
    A. Description of Our Present Heating System
       1. Physical condition

        2. Required yearly maintenance
        3. Fuel supply problems
          a. Overworked distributor
          b. Varying local supply
        4. Cost of operation
    B. Removal of the Oil Burner and Tank
        1. Data from the oil company
        2. Data from the salvage company
          a. Procedure
          b. Cost
        3. Possibility of private sale
    C. Installation of a Gas Pipe from the Street to the Building
        1. Procedure
        2. Cost of installation
        3. Cost of landscaping
    D. Installation of a Gas Burner
        1. Procedure
        2. Cost of plumber's labor and materials
    E. Estimation of Gas Heating Costs
        1. Rate determination
        2. Required yearly maintenance
        3. Cost data from neighboring facility
        4. Overall cost of operation
          a. Cost of conversion
          b. Cost of maintenance
          c. Cost of gas supply

The writer now adds the introduction and conclusion.

   I. INTRODUCTION
    A. Background
    B. Purpose of the Report
    C. Intended Audience
    D. Information Sources
    E. Limitations of the Report
    F. Scope
        1. Description of our present heating system
        2. Removal of the oil burner and tank
        3. Installation of a gas pipe from the street to the building
        4. Installation of a gas burner
        5. Estimation of gas heating costs

  II. BODY (as shown earlier)

 III. CONCLUSION
    A. Summary of Findings
    B. Comprehensive Interpretation of Findings
    C. Recommendations

In a short report, the introduction and conclusion are still included, but usually shortened to one or two sentences apiece. The entire outline easily converts into a table of contents for your finished report, as shown in Chapter 12.

Decimal Notation

Here is a partial version of the same outline in decimal notation:

2.0 Collected Data
    2.1 Description of Our Present Heating System
        2.1.1 Physical condition
        2.1.2 Required yearly maintenance
        2.1.3 Fuel supply problems
            2.1.3.1 Overworked distributor
            2.1.3.2 Varying local supply
        2.1.4 Cost of operation
    2.2 Removal of the Oil Burner and Tank
        2.2.1 Data from the oil company
        2.2.2 Data from the salvage company
            2.2.2.1 Procedure
            2.2.2.2 Cost
        2.2.3 Possibility of private sale

The decimal outline makes it easier to refer readers to various sections. Both systems, however, achieve the same organizing objective. Unless readers express a preference, use the system that you prefer.

The Formal Sentence Outline

The above outline is called a *topic outline* because each division is expressed as a topic phrase. Although a topic outline is most popular, it may be expanded one step further, into a *sentence outline*.

  II. COLLECTED DATA
    A. Our present heating needs are supplied by circulating hot air generated by a Model A-12, electrically fired, Zippo oil burner fed by a 275-gallon fuel tank. Both are twelve years old.
      1. Both burner and tank are in good working order and physical condition, as they have been carefully maintained.
      2. The oil burner and associated components require cleaning once yearly at a service charge of $18. The air filter, costing $2.25, is replaced three times yearly at a total cost of $6.75.
      3. Three times in the past two years, our system has run out of fuel for twelve to twenty-four hours for one of two reasons:

      a. Our town has only one fuel oil distributor who lacks the man-
        power and machinery to provide immediate service to all cus-
        tomers during the peak heating season.
      b. Etc.

Each sentence in turn serves as a topic sentence for a pargraph in the report.
Sentence outlines are used mainly in large projects where various team members write individual sections of a report.

ELEMENTS OF A FORMAL OUTLINE

Full Coverage

Make the list of major topics broad enough to encompass your subject. For
example, the outline for the heating report would not be adequate without
"Description of Our Present Heating System." Readers would have no basis
for comparing the two forms of heating.

    Besides making your outline inclusive, make it specific enough so you can
discuss each topic in detail. Thus, the sample outline partitions "Estimation
of Gas Heating Costs" into four subtopics. The fourth subtopic, "Overall cost
of operation," is further partitioned into its three constituent sub-subtopics.
This finite breakdown helps the reader understand each step in the cost
determination.

Successive Partitioning

Each division must have at least two parts. Also, keep parts or subparts of
equal rank at the same level:

    Faulty
        C. Installation of a Gas Pipe from the Street to the Building
          1. Procedure
          2. Cost of installation
        D. Cost of Landscaping

These topics are not correctly partitioned: *D* is not equal to *C*, but only a
subtopic of *C*, since the lawn has to be dug up and repaired as part of the
installation procedure.

    Each successive level is a division of the immediately preceding level.
Therefore, the subdivisions must add up to the immediately preceding item
at the next higher level. Thus, "Procedure," "Cost of installation," and "Cost
of Landscaping" add up to "Installation of a Gas Pipe. . . ."

Logical Notation and Consistent Format

We have defined notation as the system of numbers, letters, and other symbols marking the logical divisions of your outline. Format, on the other hand, is the arrangement of your material on the page (the layout). Proper notation and format show the subordination of some parts of your topic to others. Because your formal outline contains both major and minor parts, be sure that all sections and subsections are ordered, capitalized, lettered, numbered, punctuated, and indented to show how each part relates to other parts, and to the overall discussion.

The general pattern of notation, then, is as follows:

        I.
          A.
            1.
            2. *
          B.
            1.
            2.
                a.
                b.
                   (1)†
                   (2)
          C.
        II. etc.

Here is the same general pattern of notation in decimal form:

        1.0
          1.1
            1.1.1
            1.1.2
          1.2
            1.2.1
            1.2.2
                1.2.2.1
                1.2.2.2
                    1.2.2.2.1
                    1.2.2.2.2
          1.3 etc.

---

* Any division must yield at least two subparts. For example, you could not logically divide "Types of Strip Mining' into "1. Contour Mining," without other subparts. If you can't divide your major topic into at least two subtopics, change your original heading.

† Further subdivisions can be carried as far as needed, as long as the notation for each level of division is individualized and consistent.

Indent consistently from category to category and from level to level. The same applies for line spacing. If, for instance, you indent your first A notation five spaces from the margin, indent all other uppercase letter notations identically. If you leave a triple space between notations I and II, be sure to triple space between II and III. Likewise, if you single space your first set of arabic numerals — 1, 2, and 3 — single space all other sets of arabic numerals at this level of division.

Express topics at particular levels in consistent letter case: all BLOCK LETTERS; First Letter of Each Word in Caps (except articles, conjunctions, and prepositions); or First letter capitalized.

Parallel Construction for Parallel Levels

Make all items of equal importance parallel, or equal, in grammatical form. Then your outline will emphasize the logical connections among related ideas.

Faulty

    E. Estimation of Gas Heating Costs
        1. Rate determination
        2. The system requires yearly maintenance
        3. Cost data were obtained from a neighboring facility
        4. Overall cost of operation

Each item at this level of division is presented as equal in importance to the other items that comprise the cost estimation, but 1 and 4 are phrases, whereas 2 and 3 are sentences.

Correct

    E. Estimation of Gas Heating Costs
        1. Rate determination
        2. Required yearly maintenance
        3. Cost data from neighboring facility
        4. Overall cost of operation

Conversely, each item could be expressed as a complete sentence. See Appendix A for a full discussion of parallelism.

Clear and Informative Headings

As you compose the outline, be sure topic headings contain *specific* information. Choose informative words. Under "Description of Our Present Heating System," for example, a heading titled "Fuel" is less informative than "Fuel supply problems."

Also, avoid repetitions that add no information.

Faulty

C. Environmental Effects of Strip Mining
   1. Effects on land   3. Effects on water
   2. Effects on erosion   4. Effects on flooding

Correct

C. Environmental Effects of Strip Mining
   1. Permanent land scarring   3. Water pollution
   2. Increased erosion   4. Increased flood hazards

In the correct version, each heading has a key phrase that summarizes the message.

Parts in Logical Sequence

Which details does your reader need — and in what order? Which item comes first? Last? Does your subject have any special traits that might determine the sequence? Answer these questions as you plan your outline. Here are some possible sequences for the body section.

Chronological Sequence

In a chronological sequence, follow the time sequence of your subject (for example, the sequence of steps in a set of instructions). Also, follow this sequence to explain the order in which the parts of a mechanism operate (for example, how the heart works to pump blood). Begin with the first step, and end with the last.

Spatial Sequence

In a spatial sequence, follow the physical arrangement of parts (left to right, top to bottom, front to rear, etc.), as when you describe how an office will be remodeled to accommodate automated equipment.

Reasons For and Against

In giving reasons for and against something, follow the sequence in which both sides of an issue are argued — first one side, then the other — as in an analysis of the value and danger of a proposed flu-vaccination program.

Problem-Causes-Solution

Follow the sequence of the problem-solving process, from a description of the problem, through diagnosis, to a solution. An analysis of the rising rate of business failures in your area would require this sequence.

Cause and Effect

In a cause-and-effect sequence, follow actions to their results. Use this sequence in analyzing the therapeutic benefits of transcendental meditation, for example.

Comparison-Contrast

In evaluating two items, first discuss their similarities and then their differences. An example would be the item-by-item comparison of two proposed sites for dumping low-level nuclear wastes.

Simple to Complex

Explain a complex subject by beginning with its most familiar or simplest parts. An explanation of satellite transmission, for example, follows the logic of the learning process by describing what we see on our TV screens *before* discussing the complex mechanism that transmits the signal.

Sequence of Priorities

Sometimes, you will place items in sequence according to their relative importance, as in a proposal for increasing the school budget in your town.

These sequences are illustrated in the sample reports throughout this text, and especially in Chapter 3 sample paragraphs. Many reports involve more than one sequence. The heating-conversion outline, for example, fuses the chronological sequence with comparison-contrast and cause-and-effect. A paragraph generally follows *one* sequence, as shown on pages 34–39.

Items Relevant to Purpose

Be sure that all items in your outline add up to your statement of purpose. Omit irrelevant topics. A sixth topic titled "Temperature Forecast for Next Winter," for example, would be irrelevant to the stated purpose of the report outlined earlier.

COMPOSING THE FORMAL OUTLINE

Preliminary Steps

Complete the following steps *before* beginning work on your formal outline:

1. *Write out a full statement of purpose.* Decide specifically what you want to accomplish in this report and write your intention in one or two sentences. You will then have a point of reference as you work.

2. *Brainstorm your subject.* Do some hard thinking to identify ideas and topic divisions. The more time you spend on this step, the more details you will have. See Chapter 2 for brainstorming.

3. *Construct your informal (or working) outline.* Organize the relevant major topics from your brainstorming list in the most logical sequence for your stated purpose.

4. *Collect your data.* Using your rough outline as a guide, collect all the information you need. If your data suggest additional topics or subtopics, revise your working outline.

5. Compose a rough draft of the report. Refine the working outline into a final plan (your formal outline). When you have answered the first essential question — *What is my purpose?* — your formal outline will help you answer the second — *How will I achieve my purpose in a way that is worthwhile and clear to my reader?* In addition to the major topics from your brainstorming list, the formal outline should identify all subtopics.

The following model can be adapted to most reports directed toward reaching a decision. (For information-type reports, the CONCLUSION becomes simply a Summary.)

GENERAL OUTLINE MODEL

I. INTRODUCTION
   A. Definition, Description, and History (and significance of the subject)
   B. Statement of Purpose
   C. Target Audience (including assumptions about the reader's prior knowledge)
   D. Information Sources (including research methods and materials)
   E. Working Definitions
   F. Limitations of the Report
   G. Scope of Coverage (major topics from your brainstorming list in the sequence in which you will discuss them)

II. BODY (the parts of your subject, divided into subparts, as necessary)
   A. First Major Topic
      1. First subtopic of A

2. Second subtopic of **A**
   a. First subtopic of 2
   b. Second subtopic of 2
      etc. (subdivision carried as far as necessary to isolate the important points in your topic)
B. Second Major Topic
   etc.

III. CONCLUSION (where everything is tied together)
   A. Summary of Information in II
   B. Comprehensive Interpretation of Information in II
   C. Recommendations and Proposals Based on Information in II

Suggestions for developing each section follow.

Introduction

Many introductions contain the following parts:

1. *Definition, description, and history.* Before discussing the subject, give readers a clear picture of its background and significance.

2. *Statement of purpose.* A statement of purpose is like your thesis statement in an essay. Why are you writing this report? What do you plan to achieve?

3. *Target audience.* Identify the audience. If you can, explain how the reader will use your information. How much prior knowledge does your reader need in order to understand the report?

4. *Information sources.* If your report includes data from outside sources, identify them briefly here (you will identify them in detail in your footnotes). Outside sources include interviews, questionnaires, library research, company brochures, government pamphlets, personal observation, etc.

5. *Working definitions.* Do you need to define any technical terms, such as "autoanalyzer," or general terms with special meanings, such as "liability"? If you have ten or more terms, save them for a glossary at the end.

6. *Limitations of the report.* State the reasons for any incomplete information. For instance, perhaps you were unable to locate a key book or interview a key person. Or perhaps your study can only be titled "preliminary," instead of "definitive" (the final word on the subject), because new facts have yet to be made public. Or perhaps your report discusses only *one side* of an issue, as in a study of the *negative* effects of strip mining in your county.

7. *Scope of coverage.* In your final subsection, preview the scope of your report by listing all major topics discussed in section II, the body.

Not all reports require each of these subsections in the introduction. *Sources, Definitions,* and *Limitations* might be optional.

Body

The body is the heart of your report. Here you develop major topics and sub-topics. Whether you are describing something, analyzing something, or giving instructions, the data in this section support and clarify your statement of purpose, your conclusions, and any recommendations. "Show me!" is the implied demand any reader will make of your report. Your body section should deliver a step-by-step view of the process by which you move from introduction to conclusion. Any interpretations or recommendations will be only as credible as the evidence that supports them.

Whenever possible, give your body section an explicit title, to reflect the specific purpose of your report. For example, if your report is a physical description, you might title your body section "Description and Function of Parts." The same section in a set of instructions might be titled "Instructions for Performance," or "Collected Data" in a report that analyzes a problem or answers a question. (See section titles in sample reports in this book.)

Conclusion

Your conclusion contains no new data. Instead it reviews, interprets, or otherwise clarifies the body. The following subsections are most often used, but the subsections in your conclusion will vary with different reports. In a report describing a mechanism, for example, your conclusion might simply review the major parts of the mechanism, then briefly describe one complete operating cycle.

1. *Summary of information in the body.* When your discussion is several pages long, summarize it.

2. *Comprehensive interpretation of information in the body.* Tie your report together with an overall interpretation of data and conclusions based on facts.

3. *Recommendations and proposals based on information in the body.* Base recommendations or proposals directly on your conclusions.

Although a good beginning, middle, and ending are indispensable, feel free to modify, expand, or delete any subsections as you see fit. In fact, no single model should be followed slavishly by any writer. *The organization of any report ultimately is determined by what your readers need.*

THE REPORT DESIGN WORKSHEET

The report design worksheet in Figure 9–1 works as a supplement to an outline, to help you zero in on your audience and purpose. This version is based

REPORT DESIGN WORKSHEET

<u>Preliminary Information</u>

What is to be done?   *A report on the feasibility of converting our home office from oil to gas heat*

Whom is it to be presented to, and when?   *Charles Jones, company president; April 1*

	Primary Reader(s)	Secondary Reader(s)
<u>Audience Analysis</u>		
Position and title:	*President, Abco Engineering Consultants*	*Company officers engineering staff*
Relationship to author or organization:	*Employer*	*supervisors, colleagues, junior members*
Technical expertise:	*nontechnical (for this subject)*	*nontechnical*
Personal characteristics:	*highly efficient; demands quality and economy*	*all serious-minded professionals*
Attitude toward author or organization:	*is considering me for promotion to assistant V.P.*	*friendly and respectful; officers will vote on my promotion*
Attitude toward subject:	*highly interested because of last winter's inconvenience*	*interested*
Effect of report on readers or organization:	*will be read closely and acted upon*	*will be read and discussed at our next staff meeting*

<u>Reader's Purpose</u>

Why has reader requested it?   *wants to make a practical decision*

What does reader plan to do with it?   *use the data to make the best choice*   *confer with the president about the choice*

What should reader know beforehand to understand it as written?   *nothing special; history of problem is reviewed in report*   *same*

What does reader already know?   *remembers last winter's problems*   *same*

What amount and kinds of detail will reader find significant?   *brief description of conversion procedures and detailed cost analysis*   *same*

What should reader know and/or be able to do after reading it?   *make an educated decision*   *advise the president about his decision*

<u>Writer's Purpose</u>

Why am I writing it?   *to communicate my research findings clearly*

What effect(s) do I wish to achieve?   *to have my readers conclude that conversion is not economically feasible; to persuade them to accept my recommendation of an alternative to conversion.*

FIGURE 9-1   Completed Report Design Worksheet

Design Specifications

Sources of data: *gas company, our oil company representative; Tubo Plumbing Corp., Jumbo Salvage Co., Watt Electronics, Inc.*

Tone: *semiformal*

Point of view: *first- and second-person*

Needed visuals and supplements: *title page, letter of transmittal, table of contents, informative abstract, data sheet appendix reviewing the procedure for cost analysis*

Appropriate format (letter, memo, etc.): *formal report format with full heading system*

Rhetorical mode (description, definition, classification, etc.--or some combination): *primary mode: analysis; secondary modes: description, process narration*

Basic organization (problem-causes-solution, intro-instructions-summary, etc.): *questions - answers - conclusions and recommendations*

Main points in introduction: *Background*
Purpose
Intended Audience
Data Sources
Limitations Scope

Main points in body: *Description of Present System*
Removal of Oil Burner and Tank
Installation of Gas Pipe
Installation of Gas Burner
Estimation of Gas Heating Costs

Main points in conclusion: *Summary of Findings*
Interpretation of Findings
Recommendation

Other Considerations *no frills; these readers are all engineers interested in hard facts.*

FIGURE 9-1 (*Continued*)

on a worksheet model developed by Professor John S. Harris of Brigham Young University. It is useful for any report or letter. Figure 9–1 has been completed for the heating-conversion report outlined earlier.

CHAPTER SUMMARY

An outline is a plan for anything you write. It partitions a subject into its parts and classifies them on the basis of their similarities. To stay in control, the successful writer spends more time planning than writing.

Choose the best type of outline, depending on your purpose and on the length and complexity of your report:

1. *An informal outline:* a simple list of words or phrases, especially useful for a short report.

2. *A formal topic outline:* a systematic arrangement of topics, using a formal notation system.

3. *A formal sentence outline:* a further development of the topic outline. Each topic is expressed as a sentence.

Make your formal outline broad enough to encompass your subject and specific enough so you can discuss each topic in detail. Make each division yield at least two subparts, and place all subparts of equal rank at the same level. Use a logical system of notation and a consistent format, expressing all divisions in parallel form. In a formal topic outline use clear and explicit headings and choose the most logical sequence for arranging the parts. Be sure all items are relevant to your statement of purpose.

Follow these steps in constructing your formal outline:

1. Before writing the actual outline, complete all preliminary steps.
    a. Formulate your statement of purpose.
    b. Brainstorm your subject for specific ideas and topic breakdowns.
    c. Write your informal outline.
    d. Collect your data.
    e. Write a rough draft.

2. In your introduction, map out your background, which begins with a definition or description of your subject and its history. Then state your purpose, and describe your audience. Next, identify outside sources of data, and explain any limitations of your report. Place working definitions in the following section, and end your introduction with a list of major topics in the body.

3. In your body, divide major topics and subtopics in logical sequence, presenting all relevant evidence. Give your body section an explicit title.

4. In your conclusion, summarize the main points in your body, give an overall interpretation of these points, and draw conclusions based on facts.

Base recommendations or proposals on your conclusions. Vary these subsections according to the report.

Modify any subsections in this three-section structure as needed.

REVISION CHECKLIST FOR OUTLINES

Use this list to check the quality of your outline.

1. Is this the best type of outline for your purpose (informal, formal topic, formal sentence)?
2. Is your outline broad enough to encompass the full topic?
3. Is it specific enough in its divisions to show all major and minor points?
4. Does each division yield at least two subparts?
5. Are parts or subparts of equal rank placed at the same level?
6. Do subparts at any given level add up to the immediately preceding items at the next higher level?
7. Should some minor points be major points, or vice versa?
8. Is the system of notation logical (roman numerals for major areas; capital letters for major topics; arabic numerals for subtopics; lowercase letters for further division; "(1)" for even further division)?
9. Is your format consistent (uniform indentation, spacing, and letter case)?
10. Are items of parallel importance expressed in parallel grammatical form?
11. Are all headings clear and explicit?
12. Is the subject arranged in the most logical sequence?
13. Is every item in the best possible location?
14. Are all topics and subtopics directly relevant to your stated purpose?
15. Are all necessary topics and subtopics included?
16. Is the introduction-body-conclusion structure fully developed?
17. Does the total of all parts in the body add up to the statement of purpose?
18. Is the outline clear and easy to follow (does it make sense)?
19. Are all topics consistently stated either as phrases or sentences?

Now list those elements of your outline that need improvement.

EXERCISES

1. In one paragraph, explain the difference between an informal outline and a formal outline. What is the major function of each?
2. Locate a short article (1,000 words maximum) from a journal in your field and make a topic or sentence outline of the article. Does the article

conform to the outlining principles discussed in this chapter? If not, how-could the article be improved? Discuss your conclusions in class or in one or two paragraphs.

3. For each of the following report topics, indicate the most appropriate sequence for organizing the subject. (For example, a proposal for a micro-computer lab at your college would follow a spatial sequence.)

– a set of instructions for operating a power tool
– a campaign report describing your progress in gaining support for your favorite political candidate
– a report analyzing the weakest parts in a piece of industrial machinery
– a report analyzing the desirability of a proposed nuclear power plant in your area
– a detailed breakdown of your monthly budget to trim excess spending
– a report investigating the reasons for student apathy on your campus
– a report investigating the effects of the ban on DDT in insect control
– a report on any highly technical subject, written for a general reader
– a report investigating the success of a no-grade policy at other colleges
– a proposal for a no-grade policy at your college

4. *In class:* Organize into groups of four or five. Choose *one* of the following topics and, *after* you have formulated a clear statement of purpose, brainstorm to identify its parts. Extend your partition into as many parts and subparts as possible. Rearrange the parts in logical order and in a consistent format, as you would in the body section of an outline. When each group completes this outlining process, one representative can write the final draft on the board for class revision.

– job opportunities in your career field
– a physical description of the ideal classroom
– how to organize an effective job search
– how the quality of your higher educational experience can be improved
– arguments for and against a formal grading system
– arguments for and against a college-wide computer-literacy requirement

5. Use the checklist at the end of this chapter to evaluate the formal topic outline on pages 153–154. Suggest revisions. Do the same in class for the outlines by groups doing exercise 4.

6. Assume you are preparing a report titled "The Negative Effects of Strip Mining on the Cumberland Plateau Region of Kentucky." After brainstorming and researching your subject, you settle on the four following major topics:

– economic and social effects of strip mining
– description of the strip-mining process
– environmental effects of strip mining
– description of the Cumberland Plateau

Arrange these topics in the most sensible sequence.

When your topics are arranged, assume that subsequent research and further brainstorming produce the following list of subtopics:

- method of strip mining used in the Cumberland Plateau region
- location of the region
- permanent land damage
- water pollution
- lack of educational progress
- geological formation of the region
- open-pit mining
- unemployment
- increased erosion
- auger mining
- natural resources of the region
- types of strip mining
- increased flood hazards
- depopulation
- contour mining

Arrange each subtopic (and perhaps some sub-subtopics) under its appropriate topic headings. Use an effective system of notation and a good format to create the body section of a formal outline.

Hint: Assume your thesis sentence is the following: "The federal government needs to acknowledge that decades of strip mining (without reclamation) in the Cumberland Plateau have devastated this region's environment, economy, and social structure."

10

Describing Objects and Mechanisms

DESCRIPTION DEFINED

To describe is to create a picture with words (and diagrams). Descriptions serve a specific purpose: they convey appropriate information about an item to someone who will use it, buy it, operate it, or assemble it, or to someone who has to know more about it for good reason. Any item can be described in countless ways. Therefore, *how* you describe — your plan of attack — depends on your purpose and on the needs of your audience.

PURPOSE OF DESCRIPTION

Description is part of all writing. Manufacturers, for example, use descriptions to stimulate interest in products; banks require detailed descriptions of any business venture before approving a loan; architects and engineers describe and perfect their plans on paper before actual construction begins; medical personnel maintain periodic descriptions of a patient's condition to ensure effective treatment.

The questionnaire in Figure 10–1 is used by police to obtain a precise

DESCRIPTION QUESTIONNAIRE

Case No: _____

Interviewer: _____   Witness: _____
Place of Interview: _____   Address: _____
Date: _____   Phone #: _____

Description of Suspect

Sex ____ Nationality _____ Age ____ Height _____ Weight _____
Build _____ Who does this person look like? _____
In what way? _____
Hair: Color _____ Long ____ Short ____ Bald ____ Curly ____
      Straight ____ Other ____
Face: Round ____ Oval ____ Square ____ Other ____
Race_____ Color Skin _____
Complexion: Light _____ Dark _____ Ruddy _____
Unusual Facial: Scars _____ Pockmarks _____ Dimples _____
                Other ____
Eyes: Color _____ Shape _____ Brows: Color ____ Bushy _____
                                       Thin _____ Average ____
Nose: Large _____ Small _____ Wide _____ Flat _____
      Pronounced _____ Nostril Shape _____
Mouth: Lip Shape _____ Large _____ Small _____ Wide ____
       Thin _____ Color _____
Teeth: Large _____ Pronounced _____ Crooked _____
       Missing _____ Stained _____
Speech: Manner of Talking _____
Words Spoken: _____
_____

Chin: Pronounced _____ Recessed _____ Wide _____ Narrow _____
      Dimple _____
Mustache: _____ Beard _____ Sideburns _____ Color _____
          Shape _____
Cheeks: Pronounced _____ Recessed _____ Flat _____ Color ____
Ears: Large ____ Small ____ Flattened ____ Protruding _____
Neck: Large _____ Thin _____ Long _____ Short ____
Shoulders: Wide ____ Narrow ____ Chest: Broad____ Flat_____
                                       Other _____
Hand: Which used _____ Shaking _____ Calm _____
      Gloves Worn _____ Tattoos _____ Watch _____
      Unclean _____ Scars _____ Other _____

FIGURE 10–1 Part of a Descriptive Questionnaire Used by Police Investigators (The questionnaire continues with *fingers, fingernails, clothing,* etc.)

description of a suspect. When a witness provides the details, a police artist converts the word picture into a sketch. Clearly, such a description is more useful than "The suspect is tall and dark, with a medium build," or "The suspect is ugly and evil-looking."

Another application can be seen in job descriptions that outline duties, responsibilities, and requirements. Your own job description will spell out what the organization expects of you. If you become a manager, you in turn might write job descriptions for other positions. Figure 10–2 shows a job description for the director of a college computer center. The details serve as guidelines for evaluating the employee's performance.

No matter what the subject, readers need the answers to some or all of these questions:

1. What is it?
2. What does it do?
3. What does it look like?
4. What is it made of?
5. How does it work?
6. How has it been put together?

These are the questions that description answers. The police questionnaire, for example, answers "What does it (the person) look like?" The job description answers "What is it?" — a question which usually requires an answer to some of the other questions. In this case, we need a description of the parts that make up the whole job. The purpose of description, then, is to answer as many of these questions as are applicable.

MAKING DESCRIPTIONS PRECISE

Descriptions are mainly *subjective* or *objective* — that is, based either on feelings or fact. Subjective description emphasizes the perceiver's attitude toward the thing, whereas objective description emphasizes the thing itself. Because they show an *impartial* view, objective descriptions are the most precise.

Subjective Description

Subjective description is dictated by feelings or opinions. Essays describing "My Biggest Complaint," "An Unforgettable Person," or "A Beautiful Moment" are expressions of opinion; they are written from a personal point of view. Subjective description aims at expressing feelings, attitudes, and moods. You create an *impression* of your subject ("The weather was miserable"), more

DIRECTOR OF THE COMPUTER CENTER

The Director of the Computer Center is responsible for the administrative direction of the College's Computer Center.

The Director of the Computer Center is responsible to the President of the College for the proper operation of the Computer Center as a teaching resource of the College, and particularly for its data processing programs. He is also responsible for the execution of certain administrative requests for computer information. He:

Advises the President regarding the policies and procedures needed for the effective use of computer facilities.

Serves the College's Advisory Committee on Data Processing Program and serves as consultant for data processing programs.

Advises the College Committee on Administrative Data Processing Procedures and the College Committee on use of Computing Facilities regarding programming time allocations, new equipment and other matters related to the instructional and administrative uses of the computer.

Supervises the data processing of college computerized records.

Oversees the development of all computer programs.

Advises Divisional Chairmen, faculty and other professional staff regarding the possible application of the computer to their instructional or administrative duties.

Submits budget requests to the Dean of Administration.

Participates in local, state and national professional associations in the field of data processing.

Serves on the Administrative Advisory Council.

FIGURE 10–2   A Typical Job Description

than communicating factual information about it ("All day, we had freezing rain and gale-force winds").

Objective Description

Objective description virtually is not influenced by emotions or preferences. If you have ever been in an auto accident, you probably completed a form similar to the one in Figure 10–3, which forces the writer to be objective. In it, you are asked to describe the area and give a *factual* account of the accident. The objective details — if accurately recorded — should speak for themselves.

Objective description is not only concerned with concrete physical details, however. A psychologist's description of a patient's mental condition is objective, insofar as it is informed opinion based on demonstrable fact.

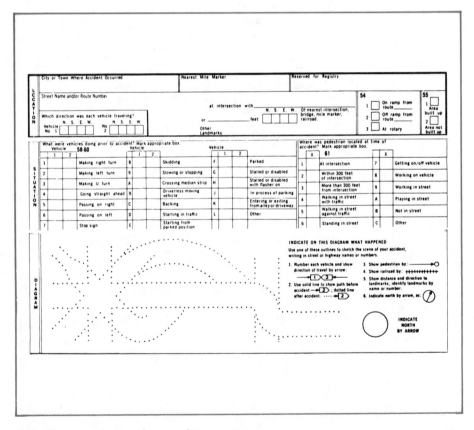

FIGURE 10–3   Portion of an Accident Report Form

With the exception of promotional writing, most of your descriptions on the job will be objective. Your employer is less interested in your feelings about an item than about its factual details. The six questions on page 172 require objective answers about an item's appearance or function — or both. Notice that the question "What is your opinion of the item?" is not on that list. If you *are* asked to give an opinion or an interpretation, do so only *after* giving all the objective details. Objective description records exactly what you see, which should be what any other person would see. Here are some guidelines for maintaining objectivity.

Be Totally Familiar with the Item

Objective description relies on precise details; study the item to learn every detail before you write.

Record Observable Details Faithfully

To identify observable details, ask yourself these questions: What characteristics could any observer recognize? What details would a camera record?

> Subjective   His office has an *awful* view, *terrible* furniture, and a *depressing* atmosphere.

The italicized words only *tell;* they do not *show.*

> Objective   His office has broken windows looking out on a brick wall, a rug with a 6-inch hole in the center, chairs with bottoms falling out, missing floor boards, and a ceiling with plaster missing in three or four places.

Use Precise and Factual Language

Use high-information words that create pictures. Name specific parts without calling them "things," "gadgets," "watchamacallits," or "doohickeys." Avoid judgmental words ("good," "super," "impressive," "poor"), unless your judgment is specifically requested and can be supported by facts. The same holds true for words like "large," "small," "long," and "near"; substitute exact measurements, weights, dimensions, and ingredients.

Use words that specify location and spatial relationships: "above," "oblique," "behind," "tangential," "adjacent," "interlocking," "abutting," and "overlapping." Use position words: "horizontal," "vertical," "lateral," "longitudinal," "in cross-section," "parallel."

Indefinite	Precise
at high speed	eighty miles per hour
a small office	an eight-by-twelve-foot office
an adequate internal memory	a 128K memory
a late-model car	a 1984 Ford Thunderbird two-door hardtop
a high salary	$50,000 per year
long hours	sixty hours weekly
an inside view	a cross-sectional, cutaway, or exploded view
a tall mountain	a vertical rise of 2,500 feet from sea level
impressive gas mileage	forty miles per gallon, city; fifty, highway
a poorly made tool	a tool with brittle plastic fittings
right next to the foundation	adjacent to the right side
partially exposed	overlapping
a small red thing	a red activator button with a 1-inch diameter and a concave surface

Don't confuse precise language with needlessly complicated technical terms, however. Don't say "phlebotomy specimen" instead of "blood" in describing a microscopic blood analysis. The clearest writing uses precise — but simple — language. Nontechnical language is preferable to specialized terminology, as long as the simpler terms do the job. Always think about your specific readers' needs.

The following description of a stethoscope is indefinite:

> The standard stethoscope is an *ordinary looking thing* which is *small* and *light* in weight. This *well-made gadget* consists of *several* parts that work together to perform an *important* function.

If you knew nothing about a stethoscope, this description wouldn't help. In fact, the subject could be just about anything with parts that do something. The italicized words tell us nothing.

Here is a version with precise details about appearance and function.

> The stethoscope is roughly twenty-four inches long and weighs about five ounces. The instrument consists of a sensitive, sound-detecting and amplifying device whose flat surface is pressed against a bodily area. This device, in turn, is attached to rubber and metal tubing that transmits the body sound to a listening device inserted in the ear.
>
> Seven interlocking pieces contribute to the stethoscope's Y-shaped appearance: (1) diaphragm contact piece, (2) lower tubing, (3) Y-shaped metal piece, (4) upper tubing, (5) U-shaped metal strip, (6) curved metal tubing, and (7) hollow ear plugs. These parts are assembled into a continuous unit.

ELEMENTS OF USEFUL DESCRIPTION

Clear and Limiting Title

Limit your topic by promising exactly what you will deliver — no more and no less. A title like "A Physical Description of a Typical Ten-Speed Racing Bicycle" promises a description of the entire item, down to the smallest bolt or cotter pin. If your intention, however, is to describe the bicycle's braking mechanism only, be sure that your title so indicates: "A Physical Description of a Center-Pull Caliper Braking Mechanism."

Overall Appearance and Component Parts

Allow readers to visualize the item as a whole before you describe each part. First, describe the general features (as in the stethoscope description on page 176).

Function of Each Part

Explain the role of each part. In the stethoscope description, the function of the diaphragm contact piece can be explained this way:

> The diaphragm contact piece is caused to vibrate by body sounds. This part is the "heart" of the stethoscope, as it receives, amplifies, and transmits the auditory impulse.

Comparison with More Familiar Items

The following comparison helps readers visualize the item:

> The diaphragm contact piece is a shallow metal and plastic bowl roughly the shape and size and twice the thickness of a silver dollar.

Introduction-Body-Conclusion Structure

Follow the structure of all good communication: (1) introduce the subject, (2) discuss it, and (3) summarize the main points.

Visuals

Use drawings, diagrams, or photographs generously (see Chapter 13). Our overall description of the stethoscope is greatly clarified by Figure 10–4.

Appropriate Details

Your choice of the kinds and amount of details is crucial. Provide enough details to give a clear picture. But do not burden your readers with needless details. Identify your readers and their reasons for reading your description.

Assume you are about to describe (to an uninformed person) a certain brand and model of bicycle. The picture you create will depend on the details you select. How will your reader use this description? What is his or her level of technical understanding? Is the reader a customer likely to be interested in what the bike looks like — its flashy looks and racy style? Are you writing for a repair technician who needs to know the order in which the parts operate? Or is your reader a helper in your bicycle shop who needs to know how to assemble this particular kind of bicycle? Because anything could be described in dozens of ways, you need to identify your specific purpose and audience.

If you can't identify your reader, write for general readers, and observe these guidelines: (1) details that are less technical are more widely understood and (2) too many details are no better than too few.

Consider, for instance, a description of a brand of dishwasher you would write for the average consumer. You can assume readers know what a dishwasher looks like. And you do not want to bury them in details they cannot

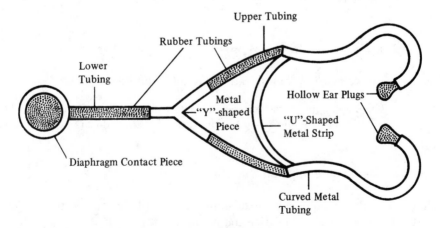

FIGURE 10–4 Stethoscope with Diaphragm Contact Piece

use. So, depending on your purpose, you might describe only the parts used in operating the appliance. Or, you might describe the interior chamber so readers will know how many dishes the machine will clean. In a repair manual for service technicians, however, your description would include many technical details.

COMPOSING A DESCRIPTION

Before writing a word, answer these questions:

1. Why am I interested in this item?
2. Who is my reader, and why is he or she interested?
3. How much does he or she already know?
4. Is my reader interested in the item's function or appearance — or both?
5. What details does this particular reader require, and in what order?

Your purpose in describing anything will be to tell readers what it looks like, how it has been put together, or how it works. Sometimes your purpose will be a combination of these. Once you have identified that purpose, you will need to devise a plan. Which descriptive sequence is best for helping readers picture this item? Where do you begin? What comes next?

Clearest Descriptive Sequence

Virtually any item has its own logic of organization, based on (1) the way it appears as a static object, (2) the way its parts operate in order, or (3) the way its parts are assembled. We can describe these relationships, respectively, through a spatial, functional, or chronological sequence.

Spatial Sequence

Part of all physical descriptions, a spatial sequence answers these questions: What is it? What does it do? What does it look like? What parts and material is it made of? Use this sequence when you want your reader to visualize the item as a static object or mechanism at rest (a house interior, a document, the Statue of Liberty, a plot of land, a chainsaw, or a computer keyboard). Will your reader best understand the item from outside to inside, front to rear, left to right, top to bottom? (What logical path do the parts create?) A retractable pen, for example, would logically be viewed from outside to inside; a hammer, from top to bottom.

If you want to emphasize certain parts of the item, place those parts either first or last. You might describe a newly designed condominium high-rise in a top-to-bottom sequence to emphasize the more dramatic penthouse suites. Instead, to emphasize the lobby, lounge, pool, and playroom on the first floor, you might follow a bottom-to-top sequence. When striving for emphasis, base the sequence on the angle of vision you wish to create for your reader.

Functional Sequence

Like the spatial sequence, the functional sequence answers the questions about identification, appearance, and parts and material; however, it also answers the question, How does it work? It is best used in describing a mechanism in action, such as a 35-millimeter camera, a nuclear warhead, a smoke detector, or a car's cruise-control system. The logic of the item is reflected by the order in which its parts function. A mechanism usually has only one functional sequence. For instance, the stethoscope description on page 176 follows the sequence of parts through which sound is transmitted.

In describing a solar home-heating system, you would logically begin with the heat collectors on the roof, moving on through the pipes, pumping system, and tanks for the heated water, to the heating vents in the floors and walls — in short, from the source to the outlet. After describing the heating system according to the functional sequence of operating parts, you could describe each part according to a particular spatial sequence.

Chronological Sequence

In addition to answering the questions about identification, function, appearance, and parts and material, a chronological sequence answers the question, How has it been put together? The chronology is determined by the sequence in which the parts are assembled.

Use the chronological sequence for an item that is best understood in terms of its assembly (such as a piece of furniture, an umbrella tent, or a prehung window or door unit). Architects might find a spatial sequence best for describing a proposed beachhouse to clients; however, they would use a chronological sequence (of blueprints) for describing the same house to the builder.

Combined Sequences

The description of a trout-activated feeder mechanism in Figure 10–5 employs all three sequences; first, a spatial sequence (top-to-bottom) for describing the overall mechanism at rest; next, a chronological sequence for explaining the order in which the parts are assembled; finally, a functional sequence for describing the order in which the parts operate.

AUDIENCE-AND-USE ANALYSIS

The audience for the description in Figure 10–5 (owners/operators of trout hatcheries) are technically informed. They already know about other types of feeder mechanisms. Therefore, they will need no background explanation about the function of feeder mechanisms in general. Nor will they need definitions of technical terms such as *surface feeders* and *gravity feed*. Instead, they are interested in the mechanism's *exact* dimensions and materials, the order in which the parts are assembled, and the way this mechanism works. Each section of the description can be greatly clarified by the insertion of visuals, including an enlarged drawing of the paper-clip attachment (the most crucial part of the mechanism at rest).

The Outline

Description of a complex mechanism almost inevitably calls for an outline. The following model can be adapted to virtually any description.

I. INTRODUCTION: GENERAL DESCRIPTION [1]
   A. Definition, Function, and Background of the Item
   B. Purpose of the Description, and Specific Audience
   C. Overall Description of the Item (with general visuals, if applicable)
   D. Principle of Operation (if applicable)
   E. List of Major Parts

II. DESCRIPTION AND FUNCTION OF PARTS
   A. Part One in Your Descriptive Sequence
      1. Definition
      2. Shape, Dimensions, Material (with specific visuals)
      3. Subparts (if applicable)
      4. Function
      5. Relation to adjoining parts
      6. Manner of attachment (if applicable)
   B. Part Two in Your Descriptive Sequence
      etc.

III. SUMMARY AND OPERATING DESCRIPTION
   A. Summary (used only in a long, complex description)
   B. Interrelation of Parts
   C. One Complete Operating Cycle

---

[1] In most descriptions, the subdivisions in the introduction can be combined and need not appear as individual headings in the report.

A DESCRIPTION OF AN

INEXPENSIVE TROUT-ACTIVATED

FEEDER MECHANISM

The Overall Mechanism

   Our trout-activated feeder mechanism is designed so
trout can obtain food at will.   The purpose of this description is to explain to trout hatchery owners and operators the
appearance, assembly, and operation of a simple and inexpensive feeder mechanism.

   The feeder mechanism is made of a 1-pint plastic funnel, a 5 x 1-1/4-inch wood strip, a 12-inch galvanized metal
rod (1/4-inch diameter), a wood disk (1/8-inch thick) with a
1/4-inch diameter hole in the center, and a 2-inch galvanized
paper clip (Figure 1).   The mechanism is a cone-shaped device
with a rod passing through its two openings.   The large
opening of the cone is on top.   Centered on the rod under
the smaller opening is a wooden disk of slightly larger
diameter than the opening that is fastened to the rod by a
paper clip.

FIGURE 10–5   A Description Using Combined Sequences

2

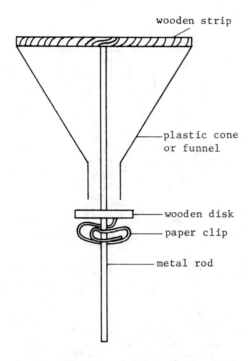

Figure 1.   An Overall Lateral View of the Feeder Mechanism

Order of Assembly

The plastic funnel is large enough to hold more food
than fifteen two-pound trout will need in one day. Placed
across the upper, large funnel opening, the wood strip holds
the metal rod that runs down through the cone. This rod then
passes through the bottom, smaller opening (and the disk)

FIGURE 10–5   (*Continued*)

3

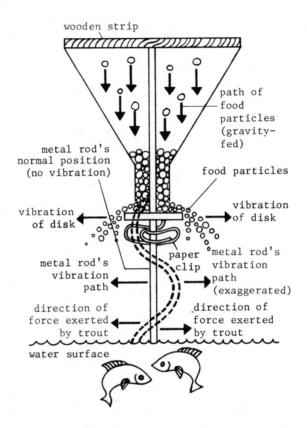

Figure 2.   A Lateral View of the Feeder Mechanism's
            Assembly and Operation

until it touches the surface of the water. The wooden disk

is attached to the rod just under the lower opening by a

FIGURE 10–5 (*Continued*)

4

paper clip (Figure 2). Bent twice, the paper clip forms a

notch where the metal rod is inserted. One bend is made in

the inner section of the clip, slightly behind the 180-degree

arc. The metal rod is inserted (and centered, crosslike)

between the inner and outer sections of the clip, and the

180-degree arc of the inner section is bent around the rod

at a 90-degree angle to the flat plane of the clip. At the

open end of the clip is a tip, bent 60 degrees straight

upward from the clip's edge, and inserted into the wood disk

(Figure 3).

Mechanism in Operation

As the trout surface, they strike the tip of the metal

rod, which is touching the water. The vibration travels up

the rod and shakes the disk, which is covered with food from

the funnel's gravity feed. As the disk vibrates, food

resting on it falls to the water surface (Figure 2).

FIGURE 10–5 (*Continued*)

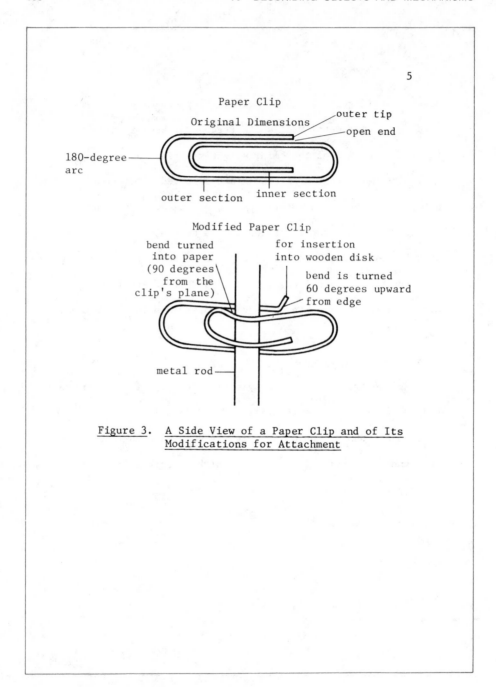

5

Paper Clip

Original Dimensions

Figure 3.   A Side View of a Paper Clip and of Its
            Modifications for Attachment

FIGURE 10–5   (*Continued*)

Introduction: General Description

The introduction is the background section. First, define the item, explain its function, and review its history. Next, explain the purpose of your description, and identify your audience.

Definition and Function

The stethoscope is a listening device that amplifies and transmits body sounds to aid in detecting physical abnormalities. Its function is to assist doctors and paramedics in diagnosing diseases.

History and Background

This instrument has evolved from the original wooden, funnel-shaped instrument invented by a French physician, R. T. Lennaec, in 1819. Because of the modesty of his female patients, he found it necessary to develop a device, other than his ear, for auscultation (listening to body sounds).

Purpose of the Description, and Intended Audience

This report explains to the beginning paramedical or nursing student the structure and operating principle of the stethoscope.

Finally, give a brief, overall description of the item, discuss its principle of operation, and list its major parts. The second version of the overall stethoscope description on page 176 follows a functional sequence.

Body: Description and Function of Parts

The body describes each major part. After arranging the parts in an overall sequence, follow the individual logic of each part. Begin your description of an individual part with a definition, followed by descriptions of the part's size, shape, and material. Also, discuss any important subparts. Finally, explain the part's function in the whole, its relation to adjoining parts, and its manner of attachment.

The intended readers of this description will use a stethoscope daily. Therefore, they will require intimate knowledge of its parts and their function.

Diaphragm Contact Piece

Definition, Size, Shape, and Material

The diaphragm contact piece is a shallow metal and plastic bowl, about the size and shape of a silver dollar (and twice its thickness), which is caused to vibrate by body sounds.

Subparts

Three separate parts make up the piece: hollow metal bowl, plastic diaphragm, and metal frame, as shown in Figure 10–6.

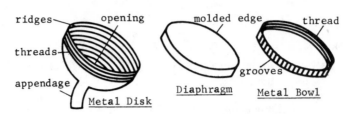

ridges   opening   molded edge   thread

threads

appendage   grooves

Metal Disk   Diaphragm   Metal Bowl

FIGURE 10–6   Frontal-Superior View of the Three Parts of a Diaphragm Contact
Piece

The metal bowl is stainless steel. Its inside surface is concave, with con-
centric ridges that funnel sound toward an opening in the tapered base,
then out through a hollow appendage. Lateral threads ring the outer circumference of the bowl to accommodate the interlocking metal frame. A fitted
diaphragm covers the bowl's upper opening.

The diaphragm is a plastic disk, 2 millimeters thick, 4 inches in circumference, with a molded lip around the edge. It fits over the metal bowl, and
vibrates sound toward the ridges. The diaphragm is held in place by a metal
frame that screws onto the bowl.

The stainless steel frame is 4 inches in circumference around its outer edge,
and 3½ inches around its inner edge. A ½-inch ridge between the inner and
outer edge accommodates threads for screwing the frame onto the bowl. The
outside circumference of the frame is covered with perpendicular grooves —
like those of the edge of a dime — to provide a gripping surface.

Function and Relation to Adjoining Parts

The diaphragm contact piece is the "heart" of the stethoscope as it receives,
amplifies, and transmits the auditory impulse through the system of attached
tubing.

Manner of Attachment

The diaphragm contact piece is attached to the lower tubing by an appendage on its upper end, which fits inside the tubing.

                    ✿ ✿ ✿ ✿ ✿ ✿ ✿ ✿ ✿ ✿

Each part of the stethoscope, in turn, is described according to its own logic
of organization.

Conclusion: Summary and Operating Description

In the conclusion, review your description briefly (if it is complex) and
explain how the parts combine to make the item function as a whole.

Summary

I've described the stethoscope according to the sequence of its functioning
parts, in enough detail to acquaint you with its physical characteristics and
operating principle.

Interrelation of Parts
The seven major parts of the stethoscope provide support for the instrument, flexibility of movement for the operator, and ease in auscultation.

One Complete Operating Cycle
In an operating cycle, the diaphragm contact piece, placed against the skin, picks up sound impulses from the body surface. These impulses cause the plastic diaphragm to vibrate. The amplified vibrations, in turn, are carried through a single tube to a dividing point. From here, the amplified sound is carried through two separate but identical series of tubes to hollow ear plugs.

Add well-labeled visuals whenever they clarify a description. Use transitional sentences to provide logical bridges between sections. The sentence below leads from the diaphragm contact piece to the lower tubing:

> The diaphragm contact piece is attached to the lower tubing by an appendage on its upper end, which fits inside the tubing.

As long as you follow an introduction-body-conclusion structure, you can modify the outline and development of individual sections. Depending on your subject, purpose, and reader, you might delete or combine some parts.

APPLYING THE STEPS

The following description of an automobile jack, written for a general reading audience, follows our outline model.

AUDIENCE-AND-USE ANALYSIS

Some readers of this description (written for an owner's manual) will have no mechanical background. Therefore, before they can follow instructions for *using* the jack safely, they will have to learn what it is, what it looks like, what its parts are, and how, generally, it works. They will *not* need precise dimensions (e.g., "The rectangular base is 8 inches long and 6½ inches wide, sloping upward 1½ inches from the front outer edge to form a secondary platform 1 inch high and 3 inches square"). The engineer who designed the jack might include such data in a description written for the manufacturer. For laypersons, however, only those dimensions are included that will help them recognize specific parts.

Also, this audience will need only the broadest explanation of how the leverage mechanism operates. While complex mechanical principles (*torque, fulcrum,* etc.) would interest design engineers, they would be of no use to readers who simply need to operate the jack safely.

A DESCRIPTION OF THE
STANDARD FORD BUMPER JACK

INTRODUCTION — GENERAL DESCRIPTION

Definition, Audience, and Purpose

The standard Ford bumper jack is a portable mechanism for raising the front or rear of a car through force applied with a lever. It allows even a frail person to lift one corner of a 2-ton automobile.

This description is for the general reader and car owner who will have to raise a car to change a flat tire.

Appearance, Function, and Major Parts

The jack consists of a molded steel base supporting a free-standing, perpendicular notched shaft (Figure 1). Attached to the shaft are a leverage mechanism, a bumper catch, and a cylinder for insertion of the jack handle. Except for the main shaft and leverage mechanism, the jack is made to be dismantled and to fit neatly in the car's trunk.

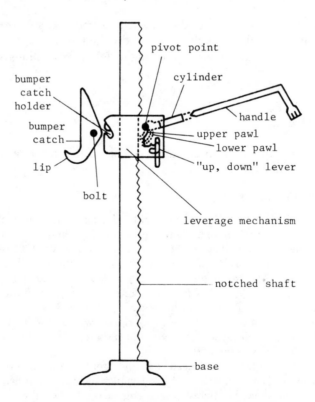

FIGURE 1. A Side View of the Standard Ford Bumper Jack

The jack operates on a leverage principle, with the human hand traveling 18 inches and the car only ⅜ of an inch during a normal jacking stroke. Such a device requires many strokes to raise the car off the ground, but may prove a lifesaver to a motorist on some deserted road.

Five main parts make up the jack: base, notched shaft, leverage mechanism, bumper catch, and handle.

DESCRIPTION OF PARTS AND THEIR FUNCTION

The Base

The rectangular base is a molded steel plate that provides support and a point of insertion for the shaft (Figure 2). The base slopes upward to form a platform containing a 1-inch depression that provides a stabilizing well for the shaft. Stability is increased by a 1-inch cuff around the well. As the base rests on its flat surface, the bottom end of the shaft is inserted into its stabilizing well.

The Shaft

The notched shaft is a steel bar (32 inches long) that provides a vertical track for the leverage mechanism. The notches, which hold the mechanism in its position on the shaft, face the operator.

The shaft vertically supports the raised automobile, and attached to it is the leverage mechanism, which rests on individual notches.

The Leverage Mechanism

The leverage mechanism provides the mechanical advantage needed for the operator to raise the car. It is made to slide up and down the notched shaft. The main body of this molded steel mechanism contains two units: one for transferring the leverage and one for holding the bumper catch.

The leverage unit has four major parts: the cylinder, connecting the handle

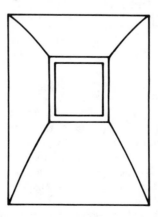

FIGURE 2.   A Top View of the Jack Base

and a pivot point; a lower pawl (a device that fits into the notches to allow forward and prevent backward motion), connected directly to the cylinder; an upper pawl, connected at the pivot point; and an "up-down" lever, which applies or releases pressure on the upper pawl by means of a spring (Figure 1). Moving the cylinder up and down with the handle causes the alternate release of the pawls, and thus movement up or down the shaft — depending on the setting of the "up-down" lever. The movement is transferred by the metal body of the unit to the bumper-catch holder.

The holder consists of a downsloping groove, partially blocked by a wire spring (Figure 1). The spring is mounted in such a way as to keep the bumper catch in place during operation.

The Bumper Catch

The bumper catch is a steel device that provides the attachment between the leverage mechanism and the automobile bumper. This 9-inch molded plate is bent to conform to the shape of the bumper. Its outer ½ inch is bent up into a lip (Figure 1), which hooks behind the bumper to hold the catch in place. The two sides of the plate are bent back 90 degrees to leave a 2-inch bumper contact surface, and a bolt is riveted between them. The bolt slips into the groove in the leverage mechanism and provides the point of attachment between the leverage unit and the car.

The Jack Handle

The jack handle is a steel bar that serves both as lever and lug-bolt remover. This round bar is 22 inches long, ⅝ inch in diameter, and is bent 135 degrees roughly 5 inches from its outer end. Its outer end is a wrench made to fit the wheel's lug bolts. Its inner end is flattened for removing the wheel covers and for insertion into the cylinder on the leverage mechanism that connects to the pivot point.

SUMMARY AND OPERATING DESCRIPTION

Interrelation of Parts

The jack's five main parts (base, notched shaft, leverage mechanism, bumper catch, and handle) make up an efficient, lightweight device for raising a car to change a tire.

One Complete Operating Cycle

The jack is quickly assembled by inserting the bottom of the notched shaft into the stabilizing well in the base, the bumper catch into the groove on the leverage mechanism, and the flat end of the jack handle into the cylinder.

The bumper catch is then attached to the bumper, with the lever set in the "up" position. As the operator exerts an up-down pumping motion on the jack handle, the leverage mechanism gradually climbs the vertical, notched shaft until the wheel is raised above the ground. Conversely, with the lever in the "down" position, the same pumping motion causes the leverage mechanism to descend the shaft.

CHAPTER SUMMARY

In describing anything, identify it, explain its function, portray its appearance, and list its parts and materials. Also, explain how a mechanical item works and, if appropriate, how it has been put together.

Subjective description emphasizes your attitude toward the subject, whereas objective description emphasizes the subject itself. To be objective, know your subject, stick to observable details, and use precise terms.

Begin with a forecasting title. First, describe the subject in terms of its overall characteristics; next, in terms of its individual parts and their functions. Whenever possible, use visuals and familiar comparisons. Make the details appropriate to the subject and the reader's needs.

Follow these steps in planning and writing your description:

1. After defining your purpose, identify the descriptive sequence or sequences that follow the logic of your subject.

    a. Use a spatial sequence for static objects or mechanisms at rest.

    b. Use a functional sequence for mechanisms in action.

    c. Use a chronological sequence for describing the order in which an item — either static or mechanical — has been assembled.

2. Make a detailed outline, and develop your report from it.

    a. In the introduction, first define the item, and explain its purpose and background. Next, discuss your aim, and specify your audience. Give an overall description of the item, and explain its operating principle. Finally, list the major parts described in the body.

    b. In the body, describe each major part, in order, giving its definition, shape, dimensions, material, subparts, function, relation to adjoining parts, and manner of attachment.

    c. In the conclusion, summarize the body, discuss the interrelation of parts, and describe one complete operating cycle. Modify this structure to suit your subject, purpose, and reader's needs.

REVISION CHECKLIST FOR DESCRIPTIONS

Use this list to refine the content, arrangement, and style of your description.

Content

1. Does the description have a title that promises exactly what is delivered?
2. Have you described the item's overall features, as well as each part?

3. Have you defined each part before discussing it?

4. Have you explained the function of each part?

5. Have you used comparisons with more familiar items whenever possible?

6. Have you used visuals whenever they provide clarification?

7. Will intended readers be able to visualize the item from your details?

8. Are any details unnecessary or confusing, given your stated audience?

9. Is your description keyed to your stated purpose? (Will it do what it is supposed to do for your audience?)

10. Does the description answer all the questions on page 172 that are applicable for this audience?

Arrangement

1. Does your description follow an introduction-body-conclusion structure?

2. Does your description follow your outline faithfully?

3. Does your description follow the clearest possible sequence or combination of sequences (spatial, functional, chronological)?

4. Are your headings appropriate and adequate?

5. Are there enough transitions between related ideas?

Style

1. Is your description objective, with all relevant, observable details recorded in precise, factual language?

2. Will the level of technicality connect with the intended audience?

3. Is your description written in plain English?

4. Is each sentence clear, concise, and fluent?

5. Is the description in correct English (Appendix A)?

Now list those elements of your description that need improvement.

EXERCISES

1. In two or three paragraphs, discuss the kinds of descriptive writing you expect to do in your field. Be as specific as possible in identifying the subjects, the situations, and the readers. After giving examples, illustrate with a scenario that answers the questions about purpose on page 179.

2. Rewrite these subjective statements, making them more objective:

    a. This classroom is (attractive, unattractive).
    b. The weather today is (beautiful, awful, mediocre).
    c. (He, she) is the best-dressed person in this class.
    d. My textbook is in (good, poor) condition.
    e. This room is (too large, too small, just the right size) for our class.

3. Write a three-paragraph subjective description titled "My Best Friend" in which you express your feelings. Next, use the sample descriptive questionnaire in Figure 10–1 as a guide for writing a factual description of the same person so readers could recognize this person in a crowd. Which of these descriptions is easier to write, and why?

4. After consulting library sources, faculty members, or workers in your field, write a one-page job description of the position you hope to hold in ten years. Or write a job description for the job you now hold. Using the sample description in Figure 10–2 as a model, include a description of function, duties, responsibilities, and qualifications.

5. Select three of your most valuable possessions (choose simple objects that can easily be described). Write a one-paragraph physical description of each in enough detail that a police officer could recognize the items if they were lost or stolen.

6. The following statements are indefinite. Choose one statement and expand it into a precise description in one paragraph:

    a. Come to my place for dinner and I will cook a meal you will never forget.
    b. Writing is done more efficiently on a word processor.
    c. The job opportunities in my major are numerous.
    d. I plan to be highly successful in my career.
    e. Attending a small college has (advantages, disadvantages).
    f. Regular access to a computer makes my work much easier.

7. Television commercials are, supposedly, a source of information about various products. But do we actually learn anything from modifiers like "super," "new," "improved," "jumbo," "nutritious," "elegant," and "exciting"? Often, the purpose of this indefinite language is to conceal the product's similarity to other products.

In class, analyze recordings or transcripts of commercials. What do you really learn from them? Try to isolate objective descriptions from subjective or downright indefinite ones.

8. Following is a list of items for description. How would each description best be organized? When you have decided on your purpose and audience needs, choose the descriptive sequence that best describes the overall item. Next, identify the best sequence for describing each part of the item.

An overall description of a ski binding as a mechanism at rest, for the recreational skier, might follow a spatial sequence that emphasized its appearance and function. In contrast, an overall description of the ski binding as a working mechanism, for the technician who will repair and adjust bindings, might follow a functional sequence. In both cases, individual parts would be described according to their logic of spatial organization. Choose a plan of attack that works best for your readers.

a handsaw	an accident
a pogo stick	a flower
a sailboat	a courtroom
a textbook	the members of your family
a retractable ball-point pen	a cigarette lighter
a steeplechase track	a Bunsen burner
a pocketknife	a microscope
a pair of binoculars	a football team

9. a. Choose an item from the following list, from exercise 8, or from your major. Using the general outline in this chapter as a model, outline a descriptive report. Write for a specific use by a specified audience. (Attach your written audience analysis to your report.) Write one draft of the report, revise following the revision checklist, and exchange reports with a classmate for further suggestions about revision.

a soda-acid fire extinguisher	a 100-amp electrical panel
a breathalyzer	a computer
a sphygmomanometer	a Skinner box
a ditto machine	a simple radio
Rubik's Cube	a flat-plate solar collector
a transit	a distilling apparatus
a saber saw	a bodily organ
a drafting table	a Wilson cloud chamber
a hazardous-waste site	a programming flowchart
a blowtorch	an accomplishment

Remember, you are simply describing the item, its parts, and function; do not provide directions for its assembly or operation.

b. As an optional assignment, describe a place you know well. You are trying to convey a visual image, not a mood; therefore, your description should be objective, discussing only the observable details.

10. The bumper-jack description in this chapter is aimed toward a general reading audience. Evaluate it by using the revision checklist. In one or two paragraphs, discuss your evaluation, and suggest revisions.

11. *In class:* Compose an outline of the bumper-jack description, subdividing for subparts. Identify the various descriptive sequences.

11

Explaining
a Process

DEFINITION

A process is a series of actions or changes leading to a product or result. A process explanation, then, is an account of how these actions or changes occur.

PURPOSE OF PROCESS EXPLANATION

A process explanation is designed to answer the reader's question "How?" — in one of three ways:

1. How do I do (or make) something?
2. How did you do something?
3. How does something happen?

When you explain a process, you discuss causes and effects. On the job, for example, you might instruct a colleague or customer on how to do something:

how to analyze a soil sample for chlordane contamination, how to gain access to a database, using a microcomputer; how to assemble a precut log cabin.

Besides instructions, you may have to give two other kinds of process explanation. You might tell an employer or client how you did something: repaired a piece of equipment; inspected a building site; conducted an experiment. Or, you might explain to a new employee how things happen: how the electronic mail system works; or how the budget for various departments is determined. Each type is discussed in the following section.

TYPES OF PROCESS EXPLANATION

Depending on your purpose and your readers' needs, you can design your explanation to answer one of the "how" questions, respectively, with *how to do it; how I did it,* or *how it happens.* As an illustration of how purpose and audience dictate your approach, consider Figure 11–1. Written communication occurs when these elements interact, but each type of explanation emphasizes a different element. Instructions emphasize the reader's performance; a process narrative, the writer's performance; and a process analysis, the subject's performance.

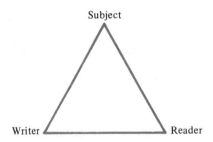

FIGURE 11–1 A Rhetorical Triangle

Instructions

Instructions tell how to do (or make) something. Almost anyone with a responsible job writes instructions, and almost everyone has to read some sort of instructions. The new employee needs instructions for operating the office machines; the employee going on vacation writes instructions for the person filling in. Owners of new cars read the operator's manual for servicing and operating instructions. The person who buys a personal computer reads instructions for connecting peripheral apparatus, such as a modem or a printer, for adding a disk drive or running a graphics package, and so on. Instructions *emphasize the reader's role,* explaining each step in enough detail so readers can complete the procedure safely and efficiently.

Narrative

Related to an explanation of *how to do something* is *how you did something,* and your results, as in a lab report. Police and fire personnel write accounts of investigations and inspections. Laboratory technicians write accounts of tests and results. Construction supervisors write daily accounts of work on a project. This type of explanation *emphasizes your role* so readers will understand the steps you followed and the results you achieved.

Analysis

An explanation of how things work or happen divides the process into its parts and principles. Colleagues and clients need to know how stock and bond prices are governed, how fermentation occurs, how your bank reviews a mortgage application, how your town decides on its zoning laws, how transmitters propagate radio waves, how optic fiber conducts electricity, and so on. Process analysis emphasizes *the process itself* instead of the reader or writer. The process is dissected so readers can follow it as "observers."

ELEMENTS OF USEFUL INSTRUCTIONS

Your purpose in writing instructions is to answer all possible questions about a procedure. After all, your reader may be operating complex machinery or surrounded by hot wires, with nothing but words on the page for directions. Foggy instructions can be disastrous. Make your instructions readable by including the following elements.

Clear, Limiting, and Inclusive Title

Make your title promise exactly what your instructions deliver — no more and no less. This title, "Instructions for Cleaning the Carriage, Key Faces, and Exterior of an Electronic Typewriter," is effective because it tells readers what to expect: instructions for performing a specific procedure on selected parts. A title like "The Electronic Typewriter" would give readers no forecast: the document might be a history of the typewriter, typing instructions, or a description of each part. On the other hand, a title like "Instructions for Cleaning an Electronic Typewriter" would be misleading if the instructions were limited to selected parts.

Informed Content

To write good instructions you must know the specific procedure down to the smallest detail. Unless you have performed the task successfully, don't try to write instructions for it.

Logically Ordered Steps

Good instructions not only divide the process into steps; they also guide readers through the steps in *order*, so mistakes won't occur. For example,

> You can't splice two wires to make an electrical connection before you have removed the insulation. To remove the insulation, you will need. . . .

The logic of a process requires that you explain the steps in chronological order, as they are performed. Thus, instructions for building a house begin with the foundation, and move to the floor, walls, and roof.

Appropriate Level of Technicality

Unless you know that readers have both a relevant technical background *and* skills, write for general readers, and do two things. First, give them enough background to understand *why* they should follow the instructions. Second, give sufficiently detailed explanations so readers can understand the instructions.

Background Information

Begin by explaining the reason for the procedure:

> Some people think they don't have the time or skill to change their car's motor oil and filter. They have the changes done at a garage, service station, or car dealership. At today's prices, however, you may spend $15 or more for a service you can perform yourself in 30 minutes. These instructions will help you change your oil and filter, so you can save up to $8.

Also, state your assumptions about your reader's level of technical understanding:

> To build a wind harp, you need a familiarity with hand tools and woodworking techniques. Also, if you are to use power tools, and use them safely, a working knowledge of their operation is essential.

Before writing the step-by-step instructions, define any special terms:

> Jogging is a kind of running — a brisk trot consuming not more than 10 minutes per mile.

When your reader understands *what* and *why*, you are ready to explain *how*.

Details

Most of us know the frustration of buying a disassembled item, opening the box, reading the instructions for assembly, and being lost because the instructions provide inadequate explanation. Even some do-it-yourself books and cookbooks are notorious for lack of detail and overly technical instructions. For instance, how does one "treat a rack of lamb gently with truffle essence"? Vague instructions are based on the writer's failure to consider the intended readers' needs. Consider this set of fog-bound instructions for treating an electrical-shock victim:

1. Check vital signs.
2. Establish an airway.
3. Administer external cardiac massage if needed.
4. Ventilate, if cyanosed.
5. Treat for shock.

These instructions might be clear to the medical expert, but they mean little to the average reader. Not only are the details inadequate, but terms such as "vital signs," "cyanosed," and "ventilate" are too technical for the layperson. Such instructions for employees in a high-voltage industrial area would be useless. The instructions need to be rewritten with clear illustrations and detailed explanations, as in a Red Cross First Aid manual.

It's easy to assume that others know more than they do about a procedure; this is especially true when you can perform the task almost automatically. Think about the days when a relative or friend was teaching you to drive a car; or perhaps you have tried to teach someone else. When you write instructions, remember that the reader knows less than you do. A colleague will know at least a little less; a layperson will know a good deal less — maybe nothing — about this procedure. Analyze your reader's needs carefully.

Provide enough details for readers to understand and perform the task successfully. But omit general information the average person can be expected to know in advance (for example, the difference between a hand tool and a power tool). Excessive details clutter the steps. The following instructions contain just enough details for the general reader:

BREAKING INTO A JOG

After completing your warming-up exercises, set a brisk pace walking. Exaggerate the distance between steps, making bountiful strides and swinging your arms freely and loosely. After roughly one hundred yards of this pace, you should feel lively and ready to jog.

Immediately break into a slow trot: let your torso lean forward and let one foot fall in front of the other (one foot leaving the ground while the other is on the pavement) at the slowest pace possible, just above a walk. *Do not bolt out like a sprinter!* The worst thing is to start fast and injure yourself. Increase your pace gradually, to the point where you feel comfortable enough to cover at least one mile.

Relax your body while jogging. Keep your shoulders straight and your head up, and enjoy the scenery — after all, it is one of the joys of jogging. Keep your arms low and slightly bent at your sides. Move your legs freely from the hips in an action that is easy, not forced. Make your feet perform a heel-to-toe action: land on the heel; rock forward; take off from the toe.

These instructions are clear and uncluttered because the writer has correctly assessed the general reader's needs. Terms like "bountiful stride," "torso," and "sprinter" should be clear to the general reader, without definitions or illustrations. In contrast, "a slow trot" is given a working definition because some people could have differing interpretations of this term. When in doubt about the need for certain details, you are safer to overexplain than to underexplain.

Visual Aids

Whenever you can, incorporate a visual within your discussion of a step, as in this procedure for raising the roof section of an umbrella tent:

> Lift the ridge pole until the pin of the support post can be inserted into the hole 2 inches behind the cap of the ridge pole (Figure 11–2). Push the ridge pole up until the support post is perpendicular to the ground.

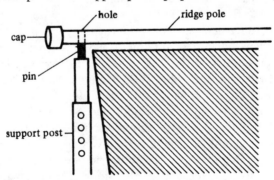

FIGURE 11–2   A Lateral View of the Support-Post Insertion into the Ridge Pole

Warnings, Cautions, and Notes

The only items that properly interrupt the steps are warnings and cautions about a particular step, or notes explaining it further. Refer to these in your introduction, advising readers of their locations in the text. Place the warnings and cautions themselves, *clearly marked,* immediately before the respective steps.

> *Caution:* No one over thirty should be jogging without a doctor's approval. A person of any age who is overweight or who has not recently exercised should also have a physical checkup. Jogging, for most people, is healthful, but for some it can be deadly.

Overuse of warnings and cautions, however, might cause readers to ignore a crucial warning.

Appropriate Terms, Phrasing, and Paragraph Structure

Of all communication, instructions have the most demanding requirements for clarity. Here, the words on the page lead to *immediate action.* Most readers are impatient and unlikely to read the entire set of instructions before plunging into the first step. Poorly phrased and misleading instructions could have frustrating — or even tragic — results. Imagine astronauts receiving unclear instructions during launching or reentry. Closer to home, the possible results of pilot confusion as a result of garbled landing or takeoff instructions are all too clear.

Like descriptions, instructions name parts, use location and position words, and state exact measurements, weights, and dimensions. Additionally, four elements of phrasing and paragraph development require unusually close attention for instructions: (1) consistent voice and mood — normally, active voice and imperative mood; (2) time- and sequence-marking words like "first," "next," "after drying," etc.; (3) parallel phrasing; and (4) short sentences and paragraphs.

Active Voice and Imperative Mood

The active voice ("He opened the door") speaks more directly than the passive ("The door was opened by him"). Likewise, the imperative mood ("Open the door") lends more authority to your instructions than the indicative mood ("You open the door").

Weak	The rudder arm is moved toward the sail to cause the boat to "come about."
Stronger	*Move* the rudder arm toward the sail to cause the boat to "come about."
Weak	The airflow valve should be turned fully to the right.
Stronger	*Turn* the airflow valve fully to the right.

The imperative makes instructions more definite because the action verb — the crucial word that tells what the next action will be — comes first. Instead of burying your verb in mid-sentence, *begin* with an action verb ("raise," "connect," "wash," "insert," "open") to give readers an immediate signal.

Transitions to Mark Time and Sequence

Transitional words are like bridges between related ideas (Appendix A). Some transitional words ("in addition," "next," "meanwhile," "finally," "on Tuesday morning," "in ten minutes," "the next day," "before," "the following afternoon") are designed to mark time and sequence. They help readers understand the task as a step-by-step process, as shown in these instructions for preparing the ground before pitching a tent:

PREPARING THE GROUND

Begin by clearing and smoothing the area that will be under the tent. This step will prevent damage to the tent floor and eliminate the discomfort of sleeping on uneven ground. *First,* remove all large stones, branches, or other debris within a level 10 × 13-foot area. Use your camping shovel to remove half-buried rocks that cannot easily be moved by hand. *Next,* fill in any large holes with soil or leaves. *Finally,* make several light surface passes with the shovel or a large, leafy branch to smooth the area.

Parallel Phrasing

Like any other items in a series, steps in a set of instructions should be in identical grammatical form, that is, in parallel construction (Appendix A). Parallelism is important in all writing but particularly in instructions, because the repetition of the same grammatical forms can further emphasize the step-by-step organization of the instructions:

Incorrect

The major steps in sprouting an avocado pit are as follows:

1. Peeling the pit.
2. Insert toothpicks to hold the pit in position with the flat end down.

3. The water level should be maintained to cover ⅔ of the pit.
4. Keep the pit in a warm, dark place until the roots are developed.

Notice that step (1) is expressed as an "ing" phrase; steps 2 and 4 are expressed as complete sentences in the imperative mood; and step 3 is expressed as a complete sentence in the indicative mood. These should be rewritten so that each step is expressed as an "ing" phrase. (The instructions for performing each step will, of course, be written as imperative sentences.)

Correct

The major steps in sprouting an avocado pit are as follows:

1. Peeling the pit
2. Inserting toothpicks to hold the pit in position with the flat end down
3. Maintaining the water level so it covers ⅔ of the pit
4. Keeping the pit in a warm, dark place until the roots are developed

Parallel construction makes the steps more readable and lends continuity to the instructions.

Short Sentences and Paragraphs

As a rule, each paragraph should explain one major step, and individual sentences within the paragraph should explain individual minor steps. This way, each activity is clearly separated. The previous instructions titled "Preparing the Ground" are a good example of effective paragraph and sentence division.

Brief is not always best, however, especially if readers get stuck with filling in gaps. Do not shorten sentences by omitting articles ("a," "an," "the"). Instead of writing "Cover opening with half-inch plywood sheet," write "Cover the opening with a half-inch plywood sheet."

Introduction-Body-Conclusion Structure

Introduce readers to the procedure, explain the steps, and review your explanation. This structure is discussed in the next section.

COMPOSING INSTRUCTIONS

You can adapt the introduction-body-conclusion structure to any instructions. Here is a general outline:

      I. INTRODUCTION
        A. Definition (or Background) and Purpose of the Procedure
        B. Intended Audience

    C. Knowledge and Skills Needed
    D. Brief Overall Description of the Procedure
    E. Principle of Operation
    F. Materials, Equipment (in order of use), and Special Conditions
    G. Working Definitions (*Note:* Because definitions are vital to effective performance, always place them in your introduction.)
    H. Warnings, Cautions, and Notes (mentioned here, and spelled out in the Body)
    I. List of Major Steps

II. INSTRUCTIONS FOR PERFORMANCE
    A. First Major Step
        1. Definition and purpose
        2. Materials, equipment, and special conditions needed for this step
        3. Substeps (if applicable)
           a.
           b.
           c.
    B. Second Major Step
        etc.

III. CONCLUSION
    A. Summary of Major Steps
    B. Interrelation of Steps

This outline is only tentative because you might modify, delete, or combine some elements, depending on your subject, purpose, and audience.

Introduction

In your introduction, give the background (when, where, why). To avoid later interruptions, define the procedure, identify your audience, and state your assumptions about the knowledge and skills needed to follow your instructions. Also, describe the process briefly and the item's principle of operation (especially if these are repair instructions). Next, identify required materials, tools, and special conditions. Then define specialized terms, and alert readers to warnings, cautions, and notes. Finally, list the major steps.

    Below is an introduction from instructions for *making something.* Written for interested laypersons as well as accomplished gardeners, the instructions will be printed in an organic gardening magazine.

HOW TO SET UP A HYDROPONIC FLOODING SYSTEM

Definition, Purpose, and Audience

Hydroponics is the process of growing plants without soil. This process is particularly useful in environments where soil and space are limited. The instructions below give the knowledgeable gardener an alternative to soil gardening and also provide directions for anyone interested in gardening.

Description of the Procedure

You can assemble an automatic hydroponic flooding system and make it function easily (Figure 1).

Operating Principle

Hydroponic gardening operates on the principle that plants will grow quickly and abundantly without soil, as long as their roots are oxygenated, kept moist at all times, and supplied with the elements essential for healthy plant growth.

Materials and Equipment

1. foundation (3 x 2 ft cart for rollaway, with two shelves)
2. plant container (waterproof, nontoxic)
3. inert aggregate (growth medium other than soil)
4. 2 ft of rubber tubing (1-in diameter)
5. sealed reservoir (½ to 1 gallon capacity)
6. nutrient solution (mixture of water and 13 elements)
7. aquarium vibrator air pump (with polyethylene tubing)
8. automatic household timer

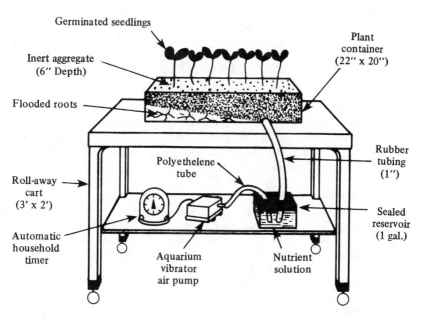

FIGURE 1   An Automatic Hydroponic Flooding System

Working Definitions

Inert aggregate: a growth medium that will support plant roots, keep them moist between feedings, and allow aeration (e.g., gravel, chipped bricks, cinders, sand, marble chips, perlite, vermiculite). *Note:* Perlite and vermiculite are soft media often used for below-ground plants such as carrots, turnips, and beets.

Nutrient solution: a solution of water and 13 elements essential for plant growth (iron, potassium, calcium, magnesium, nitrogen, phosphorus, sulfur, manganese, boron, zinc, copper, molybdenum, and chlorine). You can mix with water and use commercial mixtures such as Hyponex or RA-PID-GRO.

Major Steps

The major steps for setting up an automatic hydroponic system are (1) placing the cart, (2) arranging the top shelf, (3) arranging the lower shelf, (4) mixing the nutrient solution, (5) sowing the plant seeds, and (6) designating the hour of feeding.

Body

In your body section (labeled *Instructions for Performance*), give the actual instructions, explaining each step and substep in order. Insert warnings, cautions, and notes as needed. Begin your discussion of each step by stating its definition and/or purpose. Readers who understand the reasons for a step can do a better job performing it. A numbered list (like the one below) is an excellent way to segment the steps.

Here is the body section giving instructions for the hydroponic system:

INSTRUCTIONS FOR PERFORMANCE

1. Placing the Cart

Use the rollaway cart as the foundation for the assembly. Place the cart in an area where sunlight or artificial light (high-intensity, fluorescent, or grow light) will reflect on the upper shelf. With the foundation in place, you can begin to assemble the other parts. (Use Figure 1 as a reference for materials, equipment, and assembly.)

2. Arranging the Top Shelf

On the top shelf, place a plant container that is waterproof and nontoxic to plants (preferably concrete, earthenware, ceramic, or plastic).

Bore a 1-inch drainage/feeding hole in the plant container for accommodating the rubber tubing. Insert one end of the rubber tube a half inch into the container and seal the fitting with hot wax.

Fill the container with an inert aggregate. *Note:* Depending on the type of plant to be grown, use a hard or soft growth medium. (See *Working Definitions*.)

3. Arranging the Lower Shelf

On the lower shelf, place a sealed reservoir to hold the nutrient solution. Next to the reservoir, place an aquarium vibrator air pump and an automatic household timer.

To assemble the lower unit, drill two holes through the cover of the sealed reservoir: one to accommodate the polyethylene tube from the aquarium pump, and the other, the 1-inch rubber tubing from the plant container.

Next, take the polyethylene tube attached to the aquarium pump and insert it through the hole in the cover of the sealed reservoir so it reaches bottom. Do likewise with the free end of the 1-inch rubber tubing from the plant container. Seal both tubes with wax at their point of entry into the reservoir cover.

4. Mixing the Nutrient Solution

Fill the sealed reservoir with a nutrient solution. (See *Working Definitions*.) A number of good commercial products (Eco-Grow, Hyponex, RAPID-GRO) work well. After filling the reservoir with nutrient solution, replace the cover and seal its edges with wax.

5. Sowing the Plant Seeds

Sow preferred seeds in the inert aggregate according to instructions on the seed package. Planting seeds two inches apart in consecutive rows is a most effective method for hydroponic systems.

6. Designating the Hour of Feeding

Connect the plug from the aquarium pump to the automatic timer, and set the timer to the desired hour for feeding. Your hydroponic system is now operational.

Conclusion

In your conclusion, allow readers to review their performance by summarizing the major steps and discussing how the steps interrelate to bring about the desired result.

Summary

Setting up a hydroponic system is simple. After you obtain and assemble the materials and equipment (foundation, plant container, inert aggregate, rubber tubing, sealed reservoir, nutrient solution, aquarium pump, and automatic timer), growing plants becomes effortless.

Interrelation of Steps

At the designated time, the aquarium pump will activate air through the polyethylene tube into the nutrient solution. In turn, the solution will ascend the rubber tubing into the plant container, flooding the inert aggregate and plant roots. When the plant roots become saturated, the nutrient solution will drain back down the rubber tubing into the sealed reservoir, ready to be recycled at the next designated time.

Throughout your instructions, use headings generously and appropriately to serve as sign posts and to segment your information into digestible portions. Add well-labeled visuals whenever they clarify your explanations. Use transitional sentences to provide logical bridges between sections.

APPLYING THE STEPS

The instructions for *doing something* (felling a tree) in Figure 11–3 are patterned after our general outline, shown earlier. They will be printed in one of a series of brochures for forestry students about to begin summer jobs with the Idaho Forestry Service.

AUDIENCE-AND-USE ANALYSIS

These instructions are aimed at partially informed readers who know how to use chainsaws, axes, and wedges but who are approaching this dangerous procedure for the first time. Therefore, no visuals of cutting equipment (chainsaws, etc.) are included because the audience already knows what these items look like. Basic information (such as what happens when a tree binds a chainsaw) is omitted because the audience already has this knowledge also. Likewise, these readers need no definition of general forestry terms such as *culling* and *thinning*, but they *do* need definitions of terms that relate specifically to tree felling (*barberchair, undercut, holding wood*, etc.).

To ensure clarity, the final three steps are illustrated with visuals. The conclusion, for these readers, is kept short and to the point — a simple summary of major steps with an emphasis on safety.

COMPOSING A PROCESS NARRATIVE

With a few exceptions, a process narrative has the same elements as a set of instructions (pages 199–205). Because you are describing what you did, however, use the indicative (instead of the imperative) mood. Also, use (but don't overuse) the passive voice occasionally ("The experiment was begun" rather than "I began . . ."), to emphasize the process itself. Otherwise, overuse of "I" seems egocentric.

The introduction-body-conclusion structure — what you did, and why, when,

INSTRUCTIONS FOR FELLING A TREE

INTRODUCTION

Definition and Purpose

Felling is the process of cutting down trees. Forestry Service personnel may be called upon to fell trees for several purposes: to cull or thin a forested plot; to eliminate the hazard of standing, dead trees near powerlines; to clear an area for road or building construction; to furnish wood for heating; etc.

Intended Audience and Required Skills

These instructions are written for personnel who need to remove moderate-sized trees and who already know how to use a chainsaw, axe, and wedge safely.

General Description

Felling always follows the same procedure. Of the total time devoted to felling, more is given to planning the operation and preparing the surrounding area. Then two cuts are made with a chainsaw, severing the tree from its stump. Finally, depending on the direction of the cuts and the weather and terrain, the tree falls into a predetermined

FIGURE 11–3 A Set of Instructions

2

clearing on the ground, where it can be cut into smaller
sections and removed.

Cautions

    Although these instructions cover the basic process,
felling can be a dangerous operation. Trees, felling
equipment, and terrain can vary greatly. Therefore, have
a skilled tree feller demonstrate the process before you
try it.

    The main consideration is safety. Chainsaws, axes, and
trees are all potentially dangerous. A number of professional fellers are killed each year because of errors in
judgment or misuse of tools.

Materials and Equipment

    A 3- to 5-horsepower chainsaw with a 20-inch blade, a
single-blade splitting axe, and one or more 12-inch steel
wedges are the basic tools. Keep fuel and minor repair
tools available for maintaining the chainsaw.

    If conditions prohibit the use of the chainsaw, the
axe may serve as a substitute, but it slows the operation.
Otherwise, use the axe to drive wedges or to clear brush for
an escape path

FIGURE 11-3   (*Continued*)

3

List of Major Steps

The major steps in felling a tree are (1) choosing the
lay, (2) making provisions for an escape path, (3) making the
undercut, and (4) making the backcut.

NOTE: For the safest and most successful operation,
give the most effort to planning and the most concentration
to cutting.

INSTRUCTIONS FOR PERFORMANCE

1. Choosing the Lay

To "choose the lay" is to decide where you want the
tree to fall. If the tree stands on level ground in an open
field, it makes little difference which way you direct the
fall, but since these ideal conditions are rare, you will be
limited by special conditions.

Obstacles on the ground, surrounding trees, ground
topography, and the tree's condition should all be considered. Decide on your escape path and the location of
your cuts by considering surrounding houses, electrical
wires, and trees. Since most trees lean downhill, try to
direct the fall downhill. If the tree is leaning away

FIGURE 11–3 (*Continued*)

4

from the desired felling direction, use wedges to deflect
the fall.

CAUTION: If the tree is dead and leaning substantially,
do not try to fell it without professional help. Many dead,
leaning trees have a tendency to "barberchair" (split up the
trunk, causing a massive slab to fall spontaneously).

Use the following steps for choosing the lay:

a. Make sure the tree is still living.

b. Determine the direction and the amount of lean.

c. Find an appropriate opening in the direction of lean,
   or as close to the direction of lean as possible.

2. Providing for an Escape Path

After determining where to fell the tree, clear a path
through which you can exit if necessary. Because trees are
unpredictable, the best way to prevent injury is to build an
escape path by following the steps below:

a. Locate the path in the opposite direction from which
   you plan to direct the fall (Figure 1).

b. With the axe, clear a path roughly two feet wide and
   twenty feet long.

FIGURE 11-3 (*Continued*)

5

c. Place the axe where you will not stumble on it, but
   close enough to the tree for later use.

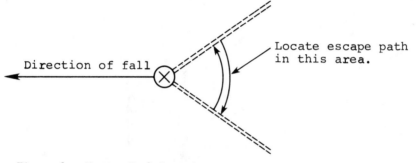

Figure 1.   Escape-Path Location

3. Making the Undercut

Making the undercut is the process of cutting a slab of
wood from the tree trunk on the side where you want the tree
to fall.   The undercut both directs and controls the fall.
Use the following steps for making the undercut:

a. Start the chainsaw.

b. Holding the saw with the blade parallel to the
   ground, make the first cut 2 to 3 feet above the

FIGURE 11–3   (*Continued*)

6

ground, cutting horizontally into the tree to one-
third of the diameter (Figure 2).

c. Make a downward sloping cut, starting 4 to 6 inches
above the first so that the cuts intersect at 1/3
of the diameter (Figure 2).

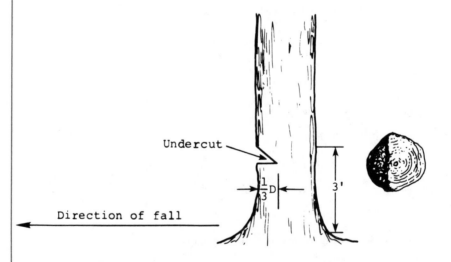

Figure 2. Making the Undercut

FIGURE 11-3 (*Continued*)

7

4. Making the Backcut

After the undercut is completed, the backcut is made to sever the stem from the stump. Follow these steps for the backcut:

a. Start the cut 1 to 3 inches above the undercut, on the opposite side of the trunk (Figure 3).

b. Do not cut completely through to the undercut, but leave a narrow strip of "holding wood" as a hinge.

CAUTION: Observe tree movement closely while cutting. If the tree starts to bind the chainsaw, hammer a wedge into the backcut with the blunt side of the axe head. Then continue cutting.

c. As the tree begins to fall, withdraw the chainsaw and turn it off. Step back a foot or two and watch the tree movement closely.

WARNING: If the tree comes your way, use your escape path.

FIGURE 11–3 (*Continued*)

8

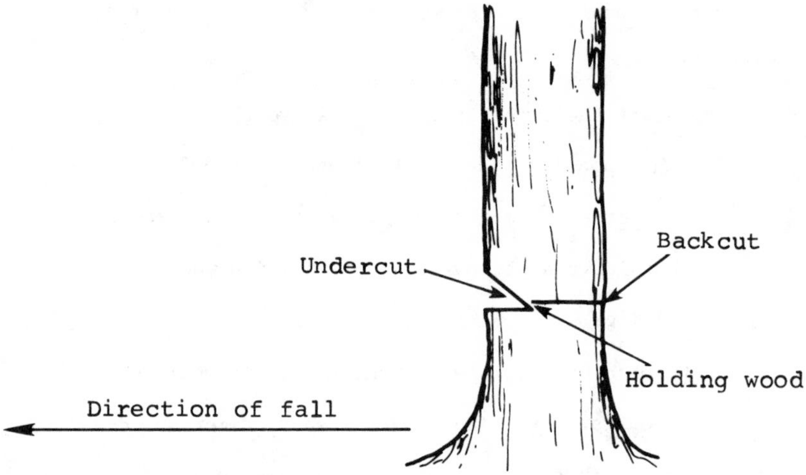

Figure 3. Making the Backcut

SUMMARY

Felling can be a complex and dangerous operation.
Choosing the lay, providing for an escape path, making the
undercut, and making the backcut are steps for the most
basic procedure, but trees, terrain, and other circumstances
can vary greatly. For the safest operation, seek professional assistance whenever you foresee complications.

FIGURE 11-3 (*Continued*)

and where you did it; how you did it; and what results you achieved — can be adapted to any process narrative. Here is the suggested outline:

I. INTRODUCTION
    A. Purpose of the Procedure
    B. Intended Audience
    C. Brief Description of the Procedure
    D. Principle of Operation
    E. Materials, Equipment (in order of use), and Special Conditions
    F. List of Major Steps

II. STEPS IN THE PROCEDURE
    A. First Major Step
        1. Purpose (include a definition when writing for the general reader)
        2. Materials, equipment, and special conditions
        3. Substeps (if applicable)
            a.
            b.
            c.
    B. Second Major Step
        etc.

III. CONCLUSION
    A. Summary of Major Steps
    B. Description of Results

Depending on your subject and audience, you might modify, delete, or combine some elements in the outline. Revise it to suit your purpose.

The following process narrative is patterned after our sample outline. Because it was written by a student for her instructor and classmates, the report reflects clear assumptions about the readers' levels of technical understanding (expert and informed). This writer chose to add a subsection titled "Significance of the Experiment" to her conclusion; in it she describes her perception of the lab exercise as a learning experience.

AUDIENCE-AND-USE ANALYSIS

The informed and expert readers of this report will need a minimum of background explanation. A brief discussion of theory is included, however, to demonstrate the *writer's* understanding of the procedure. Items commonly used by chemists (crucible, Bunsen burner, analytical scale, etc.) are not illustrated or defined for this audience. A version written for laypersons would include theoretical explanations, definitions of specialized terms, equipment, and materials, as well as visuals to illustrate certain equipment and techniques. The less readers know, the more explanation they need.

THE PROCESS OF DETERMINING THE PERCENT COMPOSITION OF A COMPOUND

INTRODUCTION

Purpose of the Procedure

On January 28, 1981, I performed a laboratory experiment to determine the relative amounts of barium chloride salts ($BaCl_2$) and water ($2H_2O$) in barium chloride hydrate. This report describes that procedure and its results.

Intended Audience

The report is written for readers who have at least a basic knowledge of chemistry principles and procedure.

Principle of Operation

All compounds are homogeneous substances whose chemical compositions are determined by the definite proportions of elements they contain. These elements can be separated through appropriate techniques. In the case of barium chloride hydrate, the water can be separated from the salt through heating and evaporation.

Brief Description of the Procedure

A specific amount of barium chloride hydrate was weighed, heated, and reweighed. The difference in the two weighings equaled the amount of water in the compound. With these data, the percentage of water (hydrate) and the percentage of barium chloride salt in the compound were calculated. All data are recorded on the data sheet later in this report.

Materials, Equipment, and Special Conditions

The following materials and equipment were used:

1. roughly three grams of barium chloride hydrate
2. one 1½-inch diameter crucible and cover
3. one Bunsen burner
4. an analytical scale

I performed all steps under supervision in a chemistry laboratory where I took appropriate safety precautions.

List of Major Steps

This procedure comprised three major steps: (1) determining the weight of the compound, (2) heating and reweighing the compound, and (3) calculating the percentages of salt and water in the compound.

STEPS IN THE PROCEDURE

Determining the Weight of the Compound

The purpose of the first step was to determine the precise weight of the compound.

First, I heated the crucible and its cover over the Bunsen burner until red hot, to burn off any particles that may have added to their mass weight. After the crucible and cover cooled to room temperature, they were weighed on the analytical scale: *weight of crucible and cover = 18.4128 grams.*

Next, a sample of barium chloride hydrate weighing approximately three grams was added to the preweighed crucible. The covered crucible and contents were then weighed on the analytical scale: *weight of crucible, cover, and sample = 21.4373 grams.*

The weight of the barium chloride hydrate sample was then calculated by subtracting the weight of the crucible and cover from the weight of the crucible with contents covered (as shown on the data sheet): *weight of the barium chloride hydrate sample = 3.0245 grams.*

Heating and Reweighing the Compound

After heating the compound to drive out all the water, I reweighed the crucible, calculating the difference in weight.

First, the covered crucible containing the compound was gradually heated over the Bunsen burner for ten minutes. After cooling to room temperature, the covered crucible and contents were weighed on the analytical scale: *weight of crucible, cover, and sample after primary heating = 20.9912 grams.*

Next, the covered crucible and contents were heated for five minutes, allowed to cool to room temperature, and weighed on the analytical scale: *weight of crucible, cover, and sample after secondary heating = 20.9912 grams.*

Calculating Percentages of Water and Salt

I determined the weight of water in the compound before calculating the percentage of water. The weight of water was determined by subtracting the weight of the sample after the second heating from the weight of the sample before heating (as shown on the data sheet): *weight of water = 00.4461 grams.*

The percentage of water in the sample was calculated by dividing the weight of the water by the weight of the compound before heating and multiplying the quotient by 100 (as shown on the data sheet): *percentage of water in the compound = 14.76 percent.*

Finally, the percentage of barium chloride salt in the compound was determined by subtracting the percentage of water from 100 percent (as shown on the data sheet): *percentage of barium chloride salt = 85.24 percent.*

CONCLUSION

Summary of Major Steps

In this experiment, the weight of a sample of barium chloride hydrate was determined; the sample was then heated and reweighed twice to ensure that all water had been evaporated; from these data, the percentages of water and barium chloride in the compound were calculated.

Significance of the Experiment

The experiment provided practical insights into the physical chemistry of a compound and its formation, along with practice in the use of selected laboratory equipment.

Description of Calculations and Results

All data, calculations, and results of this experiment are listed in the data sheet that follows.

Data Sheet for Calculating Percent Composition of
Barium Chloride Hydrate

Data:

Weight of crucible, cover and sample	21.4373 grams
Weight of crucible and cover	18.4128 grams
Weight of sample before heating	3.0245 grams
Weight after primary heating	20.9912 grams
Weight after secondary heating	20.9912 grams

Final Calculations:

Weight of water (equals loss of weight from heating)

weight of crucible, cover, and sample	21.4373 grams
weight after secondary heating	− 20.9912
	00.4461 grams of water

Percentage of water

$$\frac{\text{weight of water}}{\text{weight of compound before heating}} \times 100 = \text{percentage of water}$$

$$\frac{0.04661 \text{ grams}}{3.0245 \text{ grams}} \times 100 = 14.76\% \text{ water}$$

Percentage of $BaCl_2$

100.00	percent
− 14.76	(percentage of water)
85.24%	barium chloride

COMPOSING A PROCESS ANALYSIS

A process analysis has virtually the same elements as instructions (pages 199–205); however, the analysis is written in the third person, to keep the emphasis on the process, not on the writer or reader. In the introduction, tell what the process is, and why, when, and where it happens. In the body, tell how it happens, analyzing each step in sequence. In the conclusion, summarize the steps, and describe one full cycle of the process.

The following general outline can be adapted to any process analysis.

I. INTRODUCTION
   A. Definition, Background, and Purpose of the Process
   B. Intended Audience
   C. Knowledge Needed to Understand the Process as Explained
   D. Brief Description of the Process
   E. Principle of Operation
   F. Special Conditions Needed for the Process to Occur
   G. Definitions of Special Terms
   H. List of Major Steps

II. STEPS IN THE PROCESS
   A. First Major Step
       1. Definition and purpose
       2. Special conditions (needed for the specific step)
       3. Substeps (if applicable)
          a.
          b.
   B. Second Major Step
      etc.

III. CONCLUSION
   A. Summary of Major Steps
   B. One Complete Process Cycle

Modify, delete, or combine the elements in this outline to suit the structure of the process you are explaining. The following report is patterned after the sample outline.

AUDIENCE-AND-USE ANALYSIS

Our writer, Bill Kelly, belongs to an environmental group whose members are studying the problem of acid rain in their Massachusetts community. (Massachusetts is among the states most affected by acid rain.) To gain community support, the environmentalists must make citizens aware of the problem. So Bill's group is publishing and mailing a series of brochures. The first brochure explains how acid rain is formed.

Bill's audience will consist of general readers. Some will already be interested in the problem; others will have no awareness (or interest) at all. So Bill decides to keep his explanation at the lowest level of technicality (no chemical formulas, equations, etc.). But the explanation needs to be vivid enough, as well, to appeal to less aware or less interested readers. So Bill uses visuals to create interest and to illustrate simply. To give an explanation thorough enough for broad understanding, Bill partitions the process into three major steps: how acid rain develops, spreads, and destroys.

HOW ACID RAIN DEVELOPS, SPREADS, AND DESTROYS

INTRODUCTION

Background

Acid rain is a phenomenon that occurs when fossil fuels are burned, releasing nitrogen and sulfur oxides into the atmosphere. Simply put, acid rain changes the pH level of waterways because these nitrogen and sulfur oxides combine with the normal moisture in the air. The resulting precipitation is far more acidic than normal rainfall. Acid rain is a silent threat because its effects, although slow, are cumulative. This report explains the cause, the distribution cycle, and the effects of acid rain.

Overview of the Problem

Most research shows that power plants burning oil or coal are the primary cause of acid rain. Basically, the fuel used to create energy is not completely expended, and some of the residue enters the atmosphere. Although this residue contains several potentially toxic elements, sulfur oxide and, to a lesser extent, nitrogen oxide are the major concerns, since they are transformed when they combine with moisture. This chemical reaction forms sulfur dioxide and nitric acid, which then rain down to earth.

List of Major Steps

The major steps discussed in this explanation are (1) how acid rain develops, (2) how acid rain spreads, and (3) how acid rain destroys.

STEPS IN THE PROCESS

How Acid Rain Develops

Once fossil fuels have been burned, their usefulness is over. Unfortunately, at this point, the acid rain problem begins.

Fossil fuels contain a number of elements that are released during combustion. Two of these, sulfur oxide and nitrogen oxide, combine with normal moisture to produce sulfuric acid and nitric acid. (Figure 1 illustrates how acid rain develops.) The released gases undergo a chemical change as they combine with atmospheric ozone and water vapor. The resulting rain or snowfall is consistently more acid than normal precipitation.

Acid level is measured by pH readings. The pH scale runs from 0 through 14 — a pH of 7 is considered neutral. Distilled water, for example, has a pH of 7. Numbers above 7 indicate increasing degrees of alkalinity (household ammonia has a pH of 11), and numbers below 7 indicate increasing acidity. However, movement in either direction on the pH scale means multiplying by 10. For instance, lemon juice, which has a pH value of 2, is 10 times more acidic than apples, which have a pH of 3, and is 1,000 times more acidic than carrots, which have a pH of 5.

Because of carbon dioxide (an acid substance) normally present in air,

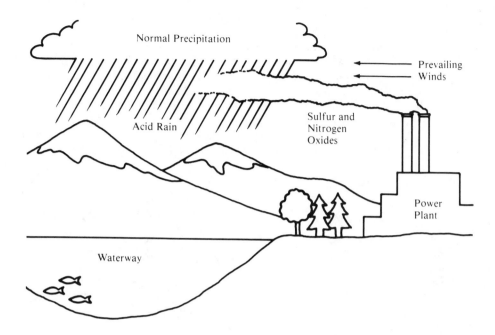

FIGURE 1   How Acid Rain Develops

unaffected rainfall has a pH of 5.6. At this time, the pH of precipitation in the northeastern United States and Canada is between 4.5 and 4. In Massachusetts, for example, rain and snowfall have an average pH reading of 4.1. A pH reading below 5 is considered to be abnormally acidic, and therefore a threat to aquatic populations.

How Acid Rain Spreads

Although it might seem that areas containing power plants would be most severely affected by acid rain, in fact, acid rain can travel thousands of miles from its source. Stack gases escape and drift with the wind currents. The sulfur and nitrogen oxides are thus able to travel great distances before they return to earth as acid rain.

For an average of two to five days, the gases follow the prevailing winds far from the point of origin. For example, estimates show that about 50 percent of the acid rain that affects Canada originates in the United States; at the same time, 15 to 25 percent of the U.S. acid rain problem has its origin in Canada.

The tendency of stack gases to drift is what makes acid rain such a widespread problem. For instance, over 200 lakes in the Adirondacks, hundreds of miles from any industrial center, are unable to support life because the water has become so acidic.

How Acid Rain Destroys

Acid rain causes extensive damage wherever it falls. It erodes various types of rock used as building material. Limestone, marble, and mortar, for example, are gradually eaten away by the constant bathing of acid. Damage to buildings, houses, monuments, statues, and cars is widespread and costly. In some cases, priceless monuments and carvings have been destroyed. Even certain kinds of trees are dying.

More important, though, is acid rain damage to waterways in the affected areas. Figure 2 illustrates how a typical waterway is infiltrated.

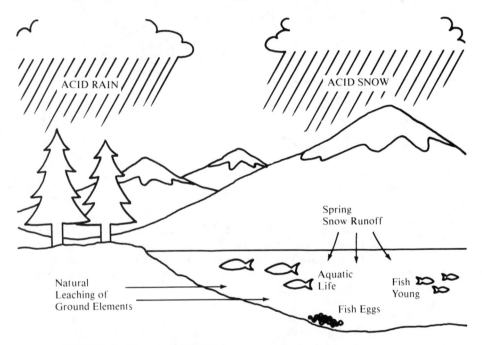

FIGURE 2   How Acid Rain Destroys

Because of its high acidity, acid rain dramatically lowers the pH of lakes and streams. Although its effect is not immediate, acid rain can eventually make a waterway so acidic it dies. In areas where there are natural acid-buffering elements like limestone, the dilute acid has less effect. The northeastern United States and Canada, however, lack this natural protection, and so are continually vulnerable.

The pH level in an affected waterway drops to a point where some species cease to reproduce. In fact, a pH level of 5.1 to 5.4 means that fisheries are threatened; once a waterway reaches a pH level of 4.5, no fish reproduction occurs. Since each creature is part of the overall food chain, the loss of one part disrupts the whole cycle.

In the northeastern United States and Canada, the problem of excess acidity is compounded by the runoff from acid snow. During the cold winter months, acid snow sits with little melting, so that by spring thaw, the acid released is greatly concentrated. In addition, aluminum and other heavy metals normally present in soil are released by acid rain and runoff. These toxic substances leach into waterways in heavy concentrations, affecting fish in all stages of development.

SUMMARY

Acid rain develops from nitrogen and sulfur oxides emitted by industrial and power plants burning fossil fuels. In the atmosphere, these oxides combine with ozone and water to form acid rain: precipitation with a lower-than-average pH. This acid precipitation returns to earth many miles from its source, causing severe damage to waterways that lack natural buffering agents. The northeastern United States and Canada are the most severely affected areas in North America.

CHAPTER SUMMARY

There are three types of process explanations: instructions (how to do it), narrative (how you did it), and analysis (how it happens).

For instructions, begin with a forecasting title. Explain each step, in order, at a level of technicality appropriate for the audience. Use visuals, and indicate warnings, cautions, and notes. Write in the active voice and imperative mood, using transitions to mark time and sequence, and parallel phrasing (as appropriate). Explain each major step in a single paragraph, and each minor step in a single sentence (as appropriate).

1. In your introduction, define the process, explain its purpose, and specify your audience. Next, describe the process briefly and mention all needed materials, equipment, and special conditions, along with the location of warnings, cautions, and notes. Finally, preview the major steps in the body.

2. In your body, give the instructions describing each step and substep in order. State the definition and purpose of each step, and insert warnings, cautions, and notes as required.

3. In your conclusion, summarize and explain how the steps interrelate.

In a process narrative, follow the guidelines for instructions. But use the indicative mood and the passive voice occasionally. Emphasize the results.

1. In your introduction, tell what you did and when, where, and why.
2. In your body, explain how you did it — step by step.
3. In your conclusion, summarize what you did, and describe results.

In a process analysis, follow the guidelines for the process narrative. Write in the indicative mood, and use the passive voice selectively.

1. In your introduction, tell what it is — and why, when, and where it happens.
2. In your body, explain how it happens — step by step.
3. In your conclusion, summarize the steps and describe one complete cycle.

Modify or delete subsections in this structure to suit your purpose.

REVISION CHECKLIST FOR PROCESS EXPLANATIONS

Use this list to refine your process explanation.

Content

1. Is the explanation keyed to your stated purpose (instructions, narrative, analysis)?
2. Does the report title promise exactly what you deliver?
3. Have you given enough background for readers to understand and be interested in the process?
4. Have you defined each step and discussed its purpose before explaining it (except in a narrative for informed or expert readers)?
5. Have you included visuals whenever they can clarify your explanation?
6. In a process narrative, does your conclusion emphasize results?
7. For instructions, have you placed cautions and warnings *before* the steps where they apply?

Arrangement

1. Does your report have a full introduction, body, and conclusion?
2. Does your report follow your final outline?
3. Does your explanation follow the exact sequence of steps in the process or procedure?
4. Have you used a single paragraph for each major step and a single sentence for each minor step (as appropriate)?
5. Have you used adequate transitions as time and sequence markers?
6. Are your headings appropriate and adequate?

Style

1. Does the level of technicality connect with your intended audience?
2. Have you expressed items of equal importance in parallel form?
3. For instructions, have you used the imperative mood and active voice?
4. For a process narrative or analysis, have you used the indicative mood throughout and the passive voice selectively?
5. Is each sentence clear, concise, and fluent?
6. Is the report written in correct English (Appendix A)?

Now list those elements of your process explanation that need improvement.

EXERCISES

1. Use the revision checklist to evaluate the sample reports in this chapter. Compare your evaluation with those of your classmates in a classroom-workshop session.

2. In a short report, discuss the types of process explanations you would write on the job. As specifically as you can, identify typical subjects, your expected audience, and its level of technical understanding.

3. Identify a situation from your own experience where you received faulty, incomplete, or unclear instructions, either written or spoken. Describe the situation, the specific deficiency, and its consequences.

4. In your textbooks or library, locate a one-page process explanation written for the general reader (or a beginning student). Using the revision checklist, evaluate its effectiveness. Include suggestions for improvements. Submit your report and a copy of the explanation to your instructor.

5. *In class:* From your major field, select a simple but specialized process that you understand well (how gum disease develops, how a savings bank makes a profit, how the heart pumps blood, how an earthquake occurs, how steel is made, how various petroleum fuels are made, how electricity is generated, how a corporation is formed, how a verdict is appealed to a higher court, how a bankruptcy claim is filed, how a Xerox machine produces copies, how nicotine affects the body, how paper is made, etc.). Write an essay explaining the process. Exchange your essay with a classmate in another major. Study your classmate's explanation for fifteen minutes and then write an explanation of that process in your own words, referring back to your classmate's paper as needed. Now, evaluate your classmate's version of your original explanation for accuracy of content. Does it show that your explanation was understood? If not, why not? Discuss your conclusions in a short essay and submit all samples to your instructor.

6. *In class:* Draw a map of the route from your classroom to your dorm, apartment, or home — whichever is closest. Be sure to include identifying landmarks. When your map is completed, write a set of instructions for a classmate who will try to duplicate your map from the information given in your written instructions. Be sure that your classmate does not see your map! Exchange your instructions and try to duplicate your classmate's map. Compare your results with the original map. Discuss your conclusions.

7. Select a recent event in which you participated (a lab experiment, a meeting, etc.). Write a process narrative, using one paragraph for each major step of your activities. Include enough details for a general reader to understand your role in this event.

8. Make the following instructions more readable by rewriting them in the appropriate voice, mood, phrasing, and sentence or paragraph division. Insert transitions wherever necessary.

WHAT TO DO BEFORE JACKING UP YOUR CAR

Whenever the misfortune of a flat tire occurs, some basic procedures should be followed before the car is jacked up. If possible, your car should be positioned on as firm and level a surface as is available. The engine has to be turned off, the parking brake set, and the automatic transmission shift lever placed in "park," or the manual transmission lever in "reverse." The wheel diagonally opposite the one to be removed should have a piece of wood placed beneath it to prevent the wheel from rolling. The spare wheel, jack bar, jack stand, hook, and lug wrench should be removed from the luggage compartment.

9. Locate a technical manual in your field or instructions written for a general reader. Using the revision checklist, evaluate the sample. In a paragraph, discuss the strong and weak points of the instructions.

10. Choose a topic from the following list, your major, or an area of interest. Using the general outline in this chapter as a model, outline a set of instructions for a procedure that requires at least three major steps. Write for a general reader and include a title page; a clear title; major topic headings; transitional sentences between major topics; cautions, notes, and warnings; and strict chronological organization. Write one draft, modify your outline as needed, and write a second draft in accordance with the revision checklist. Exchange instructions with another class member, preferably in your field, for revision suggestions.

achieving a golden tan
growing tomatoes
filleting a fish
hot-waxing skis
making beer or wine at home
safely losing ten pounds in four weeks
taking a job interview
jump-starting a car battery
hanging wallpaper

preparing logs for log cabins
using an on-line computer terminal
   to search for a book
removing the rear wheel of a
   12-speed bicycle
adding an electrical outlet
creating a computer file
avoiding hypothermia

12

Designing
an Effective Format
and Supplements

DEFINITIONS

Report Length and Complexity: Informal or Formal

Depending on your specific job and reporting responsibilities, you may be writing brief (informal) reports, longer and more complex (formal) reports, or both.

Informal reports vary in length and arrangement. They might be memos of less than a page, or progress reports and short proposals several pages long. Informal reports usually are written for readers within your organization. In contrast, formal reports usually are at least ten pages long and contain information too complex to be covered in an informal report. Accordingly, formal reports contain parts not found in informal reports. Formal reports often are written for readers outside (as well as inside) your organization and may form a permanent record. Both types follow certain format conventions that make their information accessible and their appearance pleasing.

There is no rule for deciding whether data should be presented in an informal or a formal report. For instance, a short proposal to improve safety in your laboratory might be a memo, whereas a proposal for merging with another company probably would be a formal report to be read by colleagues, executives, investors, etc. Likewise, instructions for a colleague who will fill in as department supervisor during your absence might be informally written in a few pages, whereas instructions for new equipment in your inhalation therapy unit would likely be formally written and filed as a manual. The situation determines the formality of the report. Some reports may not fit neatly into either the informal or formal category, but may fall somewhere in between. Make your report only as detailed and formal as it has to be to serve the readers' information needs.

Whether the report is informal or formal, you will need a professional format. In a formal report, include some or all supplements discussed in this chapter.

Format

Format is the mechanical arrangement of words on the page: indentation, margins, spacing, typeface, headings, page numbering, and division of report sections. Format determines the physical appearance of your report.

Supplements

Supplements are parts added to the report proper, to make it more accessible. The title page, letter of transmittal, table of contents, and abstract give summary information about the content of the report. The glossary and appendixes either provide supporting data or help readers follow certain technical sections. Readers can refer to these supplements or skip them, according to their needs. Endnotes identify sources of data. All supplements are written *after* you complete your report proper.

PURPOSE OF A GOOD FORMAT

Writing should be impressive in appearance and readability as well as content. *What your report looks like* and *how it is arranged* may be just as important as what it *says*. A good format helps *you* look good and invites the reader's attention.

No matter how vital your information, a ragtag presentation surely will alienate readers. How seriously would you take a textbook that looked like Figure 12–1? How would you judge its author, publisher, and credibility?

Giving the IM Injection

First of all, the preparation for giving the injection must be ~~taken care of~~ carried out. This includes: selecting the correct medication, preparing the needle, and drawing the medication. In selecting the medication, it must be triple cheecked to ensure that the right medication and dosage is being given. This is done by checking the order against the medication card, against the label on the drug container. Have~~ing~~ the needle ready to go ~~is important because it~~ in order to prevent fumbling with the needle and medication bottle when drawing up the medication. ~~It is important to~~ Make sure that the needle is tight to the ~~syri~~ syringe and that it is the right size. Freeing the plunger so that it will draw back and push forward easily is a good idea as it prevents fighting with it when it may be in an awkward position, like in the patient's leg. Drawing up the medication has several points that are important in avoiding contamination of the needle, or medication, and in ensuring that the right dosage is being given.

For ease of expl anation I am assuming that the medication is in liquid form in a container with a rubber seal that is supposed to be the correct dose.

FIGURE 12–1 An Ineffective Format

GIVING THE INTRAMUSCULAR INJECTION

Selecting the Correct Medication and Dosage

CAUTION: Triple-check the physician's order against the medication card and the label on the medication container, to ensure that you administer the correct medication in precise dosage.

After selecting the correct medication and dosage, prepare your needle and syringe.

Preparing Your Needle and Syringe

1. Choose a twenty-six (26)-gauge needle and affix it tightly to the neck of your syringe.

2. Free the syringe plunger so it draws back and pushes forward easily, to avoid later difficulties when the needle is in the muscle.

With your needle and syringe prepared, you are ready to draw up the medication.

Drawing up the Medication

CAUTION: Use aseptic technique to avoid contamination of the needle and/or medication; recheck the correct dosage on the container label.

*   *   *   *   *

FIGURE 12-2   An Effective Format

Imagine trying to follow such instructions! Figure 12–2 shows the same information after a format overhaul. As readers, we take good format for granted; that is, we hardly notice format *unless* it is offensive.

Your format is the wrapping on your information package. Just as there are various techniques and styles for wrapping a package, there are various formats. In fact, many companies have their own requirements. Here you will study one style, which you may later modify according to your needs.

CREATING A GOOD FORMAT

Format requirements vary from organization to organization. In the absence of specific reader requests, however, you will find the following advice broadly applicable.

High-Quality Paper

Type your final draft on 8½-by-11-inch plain, white paper. Use heavy (20 lb. or higher) bond paper with a high fiber content (25 percent minimum). Erasures with typewriter correction tape are the neatest. Use onionskin for carbon copies only.

Neat Typing

Keep erasures to a minimum and retype all smudged pages. Use a fresh ribbon and keep typewriter keys clean.

Uniform Margins, White Space, and Indentation

White space (created by margins, spacing, and indentation) is vital to the appearance and readability of any document. Leave the following margins on each page: top margin, 1¼ inches; bottom, 1 inch; left 1½ inches; right 1 inch (the larger left margin leaves space for binding the report). Make your report easy to scan by double spacing within and between paragraphs.

Indent the first line of all paragraphs five spaces from the left margin. Be sure no illustrations extend beyond the inside limits of your margins. As a rule, avoid hyphenating a word at the end of a line; if you must hyphenate, check your dictionary for the correct syllabic breakdown of the word.

Consistently Numbered Pages

Count your title page as page i, without numbering it, and number all pages up to and including your table of contents and abstract with small roman numerals. Use arabic numerals (1, 2, 3, etc.) for subsequent pages, numbering the first page of your report proper as page 1. Place all page numbers in the upper right corner, two spaces below the top edge of the paper and five spaces to the left of the right edge.

Introduction-Body-Conclusion Structure

Organize your report like any well-structured communication: orientation, discussion, and review sections.

Section Length

The length of each section depends on your subject. Instructions, for instance (pages 199–218), usually begin with a detailed introduction listing materials, equipment, cautions, and so on. The body enumerates each step and substep. Only a brief conclusion follows; the key information was in the procedure itself.

On the other hand, a problem-solving report (pages 476–481) often has a brief introduction outlining the problem. The body may be quite long, explaining the possible and probable causes of the problem. Because the conclusion contains a summary of findings, an overall interpretation of the evidence, and definite recommendations, it will likely be detailed. Only when your investigation uncovers one specific answer or one definite cause will the body section be relatively short.

Examples of varying section length, according to subject and purpose, are found in the sample reports throughout this text.

Effective Headings

Like textbooks, many reports are not read in linear order, from cover to cover, as a novel would be read. Just as you refer back or jump ahead to specific sections in a textbook, so do readers of a report. Without headings, the text is confusing — and often useless, because readers can't find what they need without wasting time.

Headings help you as well as your readers. Like points on a road map they keep you on course and provide transitions. Headings signal readers that a part of your discussion has ended and another is beginning. They make a report less intimidating by partitioning large and diverse blocks of data into smaller, more readable parts. And headings are attention-getters, because they help readers focus on parts of your report they find most important. Guidelines for headings are discussed below.

Make Your Headings Informative

Phrase all headings so they are short but informative, and serve to advance thought. Preview your topic. Compare, for instance, these versions of a heading in a set of instructions for handling diskettes:

Uninformative   The Diskette

Informative   Precautions for Handling Diskettes

The second version orients readers, showing them what to expect.

Express your headings as phrases ("Frequency of Cleaning") rather than as sentence fragments ("Need Frequent Cleaning").

Make Your Headings Comprehensive

Headings must be comprehensive in two ways: make individual headings inclusive, and provide enough headings to cover all categories. As a frame or fence encloses a specific visual or geographic space, your heading "encloses" or contains an information category. When framing a picture, you would logically frame the entire visual area, without leaving out, say, the upper right corner. Follow the same logic of inclusiveness with headings. If you are discussing the effects of acid rain on lake trout, be sure your heading is not simply "Acid Rain."

Also, provide enough headings to contain each category. If, say, chemical, bacterial, and nuclear wastes are three *separate* discussion items, provide a heading for each. Do not simply lump them together under the single heading, "Hazardous Wastes." Take major and minor headings from your outline.

Make Your Headings Parallel

All major (or minor) topics in a report generally are considered of equal importance; to emphasize that equality, express topics at the same level in the same grammatical form.

Nonparallel Headings

1. Cleaning the Disk-Drive Heads
2. Keeping the Disks away from Magnets
3. Write on Diskette Labels with a Felt-Tip Pen

   4. Keeping Diskettes away from Heat
   5. Your Diskettes Should Be Kept out of Direct Sunlight
   6. Formatting a Diskette
   7. It is Crucial that Diskettes Be Kept in Their Protective Jackets

Headings 1, 2, 4, and 6 are participial phrases, whereas 3, 5, and 7 are sentences. Also, each nonparallel heading is in a different mood: heading 3 gives a command in the imperative mood; heading 5 makes a declaration in the indicative mood; heading 7 makes a recommendation in the subjunctive mood. This mood shift obscures the relationship between individual steps, and confuses readers. These steps have been rephrased to be parallel:

   Parallel Headings
   3. Writing on Diskette Labels with a Felt-Tip Pen
   5. Keeping Diskettes Out of Direct Sunlight
   7. Keeping Diskettes in Their Protective Jackets

Parallelism is discussed further in Appendix A.

Lay Out Headings by Rank

Within its three general areas, a report will contain major topics; major topics often contain subtopics; in turn, subtopics may contain sub-subtopics, depending on the amount of detail the report requires. (Use the logical divisions in your outline as a model for heading arrangement.) The logic of your divisions will be clear if your headings reflect the rank of each item. The headings in Figure 12–3 vary in typeface, indentation, and position, according to rank. Notice that the sentence immediately following your heading stands independent of the heading. Don't begin your sentence with a pronoun such as "this" or "it" to refer back to the heading.

   To divide logically, be sure each higher-level heading yields at least *two* lower-level headings. Use indentation and varying typescript in your table of contents to reflect parallel relationships among major or minor headings.

PURPOSE OF REPORT SUPPLEMENTS

The substance of a formal report is in its introduction-body-conclusion sections. Report supplements help readers grasp the substance by providing an overview or by adding related details. Supplements can be time savers: a busy person may read only your letter of transmittal, table of contents, and abstract for key information. If you have ever written a research paper, you probably used the same approach — skimming prefaces, tables of contents, and introductions to many books for information.

MAJOR AREA HEADING

Write major headings in full capitals, centered horizontally on the page, two spaces above your following text (and
three spaces below your preceding text, for subsequent area
headings). Do not underline or italicize.

Major Topic Heading

Begin each word (except for articles and prepositions)
in your major topic heading with a capital letter. Make the
heading abut the margin, two spaces below the preceding text
and two spaces above the following text. Underline or italicize this heading.

    Subtopic Heading

Begin each word in your subtopic heading with a capital
letter. Indent it five spaces from your left margin and
place it two spaces below the preceding text and two spaces
above the following text. Underline or italicize this
heading.

    Sub-Subtopic Heading. Begin each word in your sub-
subtopic heading with a capital letter. Indent it five
spaces from the left margin and place it two spaces below
your preceding text and on the same line as the first
sentence of your following text (separated by a period).
Underline or italicize this heading.

FIGURE 12-3   A Sample Heading System

Formal reports are likely to be read by many people for a variety of purposes. Technically qualified persons may be studying the actual body of the report and the appendixes for supporting data such as maps, graphs, or charts. Executives and managers may read only the letter of transmittal and the abstract. If they read any of the report proper, they are likely to read only the recommendations. Most supplements, then, accommodate readers with various purposes.

COMPOSING REPORT SUPPLEMENTS

Report supplements can be classified in two groups:

1. *supplements that precede your report* (front matter): cover, title page, letter of transmittal, table of contents (and figures), and abstract

2. *supplements that follow your report* (end matter): glossary, footnotes, endnote pages, appendix(es)

Cover

Use a sturdy, plain, light cardboard cover with page fasteners. With the cover on, the open pages should remain flat. Use covers only for longer, formal reports, not for short pieces.

Center the report title and your name four to five inches from the upper edge:

AN ANALYSIS OF THE EFFECTIVENESS OF THE COMPUTER FLUENCY

COURSES AT CALVIN COLLEGE

by

Francis Freeman

Title Page

The title page signals readers by giving the report title, author's name, name of person or organization to whom the report is addressed, and date of submission.

Title

Your title promises what the report will deliver by stating the report's purpose and content. The following title is effective because it is clear, accurate, comprehensive, and specific:

An Effective Title   AN ANALYSIS OF THE EFFECTIVENESS OF
THE COMPUTER FLUENCY COURSES
AT CALVIN COLLEGE

Notice how slight word changes can distort this title's signal.

An Unclear Title   COMPUTER FLUENCY COURSES
AT CALVIN COLLEGE

This version does not state clearly the report's purpose. Its signal is confusing.
Is the report *describing* the courses, *proposing* such courses, *giving instructions*
for setting them up, or *discussing their history?*

An Inaccurate Title   THE EFFECTIVENESS OF THE
COMPUTER FLUENCY COURSES
AT CALVIN COLLEGE

This version does not state accurately the purpose of the report. In fact, it
might present a distorted signal by suggesting the program's effectiveness is
already proven, instead of being in question. Insert key words in your title
("analysis," "instructions," "proposal," "feasibility," "description," "progress,"
"proposal," etc.) which accurately state your purpose.

A Noncomprehensive   AN ANALYSIS OF THE EFFECTIVENESS OF
Title   THE INTRODUCTORY PASCAL COURSE
AT CALVIN COLLEGE

The above title fails to name all that the report will cover: namely, an analysis
of *all* computer fluency courses, including Logo, BASIC, and Pascal.

A Nonspecific Title   AN ANALYSIS OF THE EFFECTIVENESS OF
COMPUTER COURSES AT CALVIN COLLEGE

This version has the reverse deficiency of the previous one: it promises an
analysis of all computer courses without focusing on its proper subject — computer fluency courses. The signal is too broad and imprecise. To be sure that
your title promises what your report delivers, write its final version *after* completing your report.

Placement of Title Page Items

Do not number your title page, but count it as page i of your prefatory pages.
Center the title horizontally on the page, three to four inches below the upper
edge, using all capital letters. If the title is longer than six to eight words,
center it on two or more single-spaced lines, as in Figure 12–4. Place the items
in the following spacing, order, and typescript:

1. seven spaces below the title and horizontally centered, the word "for" in
small letters

AN ANALYSIS OF THE EFFECTIVENESS
OF
COMPUTER FLUENCY COURSES
AT CALVIN COLLEGE

for

Professor John Johnson
Technical Writing Instructor
Calvin College
Kalamazoo, Minnesota

by

Sarah Jane Robertson
Student in English 266

December 20, 1984

FIGURE 12–4   A Title Page for a Formal Report

2. two spaces below, your reader's name, position, organization, and address, horizontally centered, the first letter of each word capitalized.

3. seven spaces below, the word "by" in small letters

4. two spaces below, your name, position, and organization on each of three separate single-spaced lines with the first letter of all words capitalized

5. ten spaces below, the date of report submission written out in full

You may work out your own system, as long as your page is balanced.

Letter of Transmittal

Your letter of transmittal usually comes immediately after the title page and is bound as part of your report. Include a letter of transmittal with any formal report or proposal addressed to a specific reader. Your letter adds a note of courtesy besides giving you a place for adding personal remarks or opinions. Figure 12–5 shows a sample letter of transmittal.

Whereas your informative abstract (discussed later in this chapter) summarizes all major findings, conclusions, and recommendations in the report, your letter of transmittal calls attention to items of special interest to a specific reader. Depending on the situation, your letter might:

– Acknowledge those who helped with the report.

– Refer readers to sections of special interest: unexpected findings, key charts or diagrams, major conclusions, special recommendations, and the like.

– Discuss the scope and limitations of your study, along with any special problems gathering data.

– Discuss the need and approaches for follow-up investigations.

– Describe any personal (or "off-the-record") observations.

– Suggest some special uses of the information.

– Urge the reader to immediate action.

The letter of transmittal can be tailored to a particular reader. So if a report is being sent to a number of people who are variously qualified and bear various relationships to the writer, the letters of transmittal may vary. The letter itself has an introduction-body-conclusion structure.

Introduction

Open with a cordial statement referring to the reader's original request. Briefly review the reasons for your report.

Maintain a confident and positive tone throughout the letter. Indicate pride and satisfaction in your work. Avoid implied apologies, such as "I hope this information is adequate," or "I hope this report meets your expectations."

654 Harbor Road
Little Compton, RI   09812
May 8, 1983

Company Officers
ComTech
2 Paradise Lane
No. Dartmouth, MA 02747

Dear Beth and Jeanne:

Here is the report you requested for ComTech in January.
It compares the following six personal computer systems,
all bought and serviced locally: TRS-80 Color Computer,
Commodore 64, TRS-80 Model III, Apple IIe, IBM PC, and the
Victor 9000.

My comparison covers eight features important for the type of
work you do: screen format, numeric keypad, memory capacity,
documentation, monitor quality, microprocessor capacity,
service and support, and available software.

My conclusion is that the IBM PC would best suit your needs.
Although buying the PC would deplete your budget before you
can buy accounting software, the PC offers the best value
and the most features mentioned above. And as ComTech grows,
you can be sure your PC has the potential to grow too. Should
you buy the PC, I cannot foresee a need to invest in another
computer for many years.

If you have any questions about my findings, please contact
me.

                              Sincerely,

                              Alan Greene

                              Alan Greene

FIGURE 12–5   A Letter of Transmittal for a Formal Report

Body

In the letter body, include items from our prior list of possibilities (acknowledgments, special problems, limitations, unexpected findings, etc.). Although your abstract summarizes major findings, conclusions, and recommendations, your letter gives an overview of the *entire project*, from the moment you receive the request to the moment you submit the report.

Conclusion

State your willingness to answer any questions or to discuss findings. End on a positive note, with something like "I believe that the data in this report are accurate, that they have been analyzed rigorously and impartially, and that the recommendations are sound." Figure 12–5 shows a sample letter of transmittal (for the report on pages 508–516).

Table of Contents

Your table of contents serves as a road map for readers and an inventory checklist for you. Derived from your outline, this supplement is easy to compose: simply assign page numbers to headings in your outline.

Follow these guidelines for a table of contents:

1. List front matter (transmittal letter, abstract), numbering the pages with small roman numerals. (The title page, though not listed, is counted as page i.) List glossary, appendix, and endnotes; number these pages with arabic numerals, continuing the page sequence of your report proper, where page 1 is the first page of the report text.

2. Include no headings in the table of contents not listed as headings or subheadings in the report; your report text may, however, contain certain subheadings not listed in the table of contents.

3. Phrase headings in the table of contents exactly as they appear in the report text.

4. To reflect their rank, list headings at various levels in varying typescript and indentation.

5. Use horizontal dots (. . . . .) to connect heading to page number.

The table of contents in Figure 12–6 is adapted from the outline for "An Analysis of the Advisability of Converting Our Office Building from Oil to Gas Heat." The formal outline was shown in Chapter 9.

Table of Figures

Following your table of contents is a table of figures, if needed. If your report contains more than four or five figures, place this table on a separate page.

iii

TABLE OF CONTENTS

FIGURE 12–6   A Table of Contents for a Formal Report

iv

TABLE OF CONTENTS (continued)

FIGURE 12–6 (*Continued*)

Figure

Plateau."

For

abstract.

Writing

—

you

revision.

Follow

abstract:

1.

mini-report.

2.

data.

3.

report.

4.

order.

5.

abstract.

The

9.

Glossary

A

Assembling

Figure 12–7 shows a table of figures from a report titled "The Negative Effects of Strip Mining on Kentucky's Cumberland Plateau."

Informative Abstract

For some readers, your abstract (Chapter 6) is the most important part of your report. The abstract is always written *after* your report proper. Sometimes a busy person will read only your abstract.

Writing the abstract gives you the chance to measure your own control over the material. Because you are compressing your message to roughly 10 percent of its original length — or much less for an abstract of a long report — you must establish explicit connections. If you cannot effectively summarize your report, you probably need more homework and revision.

Follow these guidelines for your abstract:

1. Make your abstract able to stand alone in meaning — a mini-report.
2. Make your abstract intelligible to the general reader. Readers of the abstract will vary widely in expertise, perhaps much more than those who read the report itself. So translate all technical data.
3. Add no new information. Simply summarize the report.
4. Stick to the chronology of your report. If it discusses chemical, bacterial, and radioactive contamination, in this order, follow the same order.
5. Emphasize only major points. Omit prefaces, minor details, computations, and lengthy arguments. Your table of contents or report headings provide a good outline for your abstract.

The informative abstract in Figure 12–8 accompanies the heating conversion report outlined in Chapter 9.

Glossary

A glossary is an alphabetical listing of specialized terms and their definitions, immediately following your report. Many specialized reports contain glossaries, especially when written for both technical and nontechnical readers. A glossary makes key definitions available to nontechnical readers without interrupting technical readers. If fewer than five terms need to be defined, place them instead in the introduction of your report, listing them as working definitions (as discussed in Chapter 7) — or use footnote definitions. If you use a separate glossary, inform readers of its location: "(See the glossary at the end of this report)."

TABLE OF FIGURES

FIGURE 12–7   A Table of Figures for a Formal Report

Follow these guidelines for a glossary:

1. Define all terms unfamiliar to a general reader (i.e., the intelligent layperson).

2. Define all terms that have a special meaning in your report (e.g., "In this report, a small business is defined as . . .").

3. Define all terms by giving their class and distinguishing features (Chapter 7), unless some terms need expanded definitions.

4. List your glossary and its first page number in your table of contents.

5. List all terms in alphabetical order. Underline each term and use a colon to separate it from its single-spaced definition.

6. Define only terms that need explanation. In doubtful cases, however, overdefining is safer than underdefining.

7. On first use, place an asterisk in the text by each term defined in the glossary.

Figure 12–9 shows part of a glossary for a comparative analysis of two techniques of natural childbirth, written by a nurse practitioner for expectant mothers and student nurses. The term *natural childbirth* receives a more expanded definition because it is the subject of the report.

INFORMATIVE ABSTRACT: AN ANALYSIS OF THE ADVISABILITY
OF CONVERTING OUR OFFICE BUILDING FROM OIL TO GAS HEAT

The rising cost and declining availability of heating oil
have led our company to consider the advisability of converting our office from oil to gas heat. Our present heating
needs are supplied by circulating hot air generated by an
electrically fired Twinflame oil burner, which is fed by a
275-gallon oil tank. Both burner and tank are twelve years
old and in good condition. In 1982 our total heating costs
were $1508.28.

Conversion to gas heat would first require removal of the oil
burner and tank from the basement. Jamison Salvage Company
will remove these items at no cost, provided they keep both
burner and tank. Next, a gas line would need to be installed from the street to the building. Although the gas
company will provide free installation, our landscaping
costs, after installation, would be $225. Added to this
figure is the labor and materials charge of $570 for in-
stalling the gas burner. These conversion costs amount to
$79.50 yearly over ten years. With a projected gas-supply
and maintenance cost of $1780 yearly, the overall yearly cost
of gas heating would be roughly $351.22 higher than that of
oil heating. Therefore, we should retain our present system
and consider installing an auxiliary 500-gallon underground
oil tank. This tank would ensure adequate supply during peak
heating months.

FIGURE 12–8   An Informative Abstract

Endnote Pages

The endnote pages list each of your outside references in the same numerical
order as they are cited in your report proper. See Chapter 15 for a discussion
of documentation.

Appendix

An appendix comes at the very end of your report. The appendix expands on
items discussed in the report without cluttering the report text. Typical items
in an appendix include:

GLOSSARY

Analgesic: a medication given to relieve pain during the
        first stage of labor.

Anesthetic: a substance administered to cause loss of
        consciousness or insensitivity to pain in a region of
        the body.

Cervix: the neck-shaped anatomical structure which forms
        the mouth of the uterus.

Childbirth Education: the process of instruction providing
        parents with information about childbirth and preparing
        them for active participation in the delivery.

Dilation: the act of cervical expansion occurring during
        the first stage of labor.

Episiotomy: an incision of the outer vaginal tissue, made by
        the obstetrician just before the delivery, to enlarge
        the vaginal opening.

First Stage of Labor: the stage in which the cervix dilates
        and the baby remains in the uterus.

Induction: the process of stimulating labor by puncturing
        the membranes around the baby or by giving an oxytoxic
        drug (uterine contractant), or both.

Natural Childbirth: (also called Prepared Childbirth, Par-
        ticipating Childbirth, Educated Childbirth, and Coopera-
        tive Childbirth) a process of giving birth in which
        parents actively participate and which is based on an
        understanding of the body, muscular relaxation,
        breathing techniques, and emotional support. It is not
        a primitive process where modern knowledge plays no
        part, or a rigid system forbidding obstetrical inter-
        ference regardless of circumstances. Medication may or
        may not be used, according to individual need. The
        natural childbirth experience is presently defined
        by the woman's preparation and knowledge of how to
        cooperate actively in her delivery.

FIGURE 12–9   A Glossary Page for a Formal Report

– details of an experiment
– statistical or other measurements
– maps
– complex formulas
– long quotes (one or more pages)
– photographs
– texts of laws, regulations, etc.
– related correspondence (letters of inquiry, etc.)
– interview questions and responses
– sample questionnaires and tabulated responses
– sample tests and tabulated results
– some visuals occupying more than one full page

The appendix is a "catch-all" for items that are important but difficult to integrate in your text.

Do not misuse appendixes by including needless information or by excluding vital material from the report proper. Readers should not have to turn to appendixes to understand the report itself. Follow these guidelines:

1. Include only material that is relevant.
2. Use a separate appendix for each major item.
3. Title each appendix clearly: "Appendix A: A Sample Questionnaire."
4. Do not use too many appendixes. Four or five appendixes in a 10-page report would indicate a poorly organized report.
5. Limit an appendix to a few pages, unless greater length is absolutely necessary.
6. Mention your appendix early in your introduction and refer readers to it at appropriate points in the report: "(See Appendix A)."

Use an appendix for any item that is essential, but would harm the unity and coherence of your report. Remember that readers should be able to understand your report without having to turn to the appendix. Distill the essential facts from your appendix and place them in your report text.

> Improper Reference   The whale population declined drastically between 1976 and 1977 (see Appendix B for details).

> Proper Reference   The whale population declined *by 16%* from 1976 to 1977 (see Appendix B for a statistical breakdown).

Figure 12–10 shows the first page of an appendix to a report on modern whaling techniques.

APPENDIX A

A CLASSIFICATION OF WHALE SPECIES ACCORDING TO SIZE, DIET, RANGE,
GESTATION PERIOD, AND LEVEL OF ENDANGERMENT BY MODERN WHALING

Species	Size(ft.)	Diet	Range	Gestation Period	Level of Endangerment
Sperm	M 47 F 37 baby 14	Giant squid, cuttlefish.	Cosmopolitan. F & baby remain in warm water. M have harems. Bachelors in polar seas in summer. Travels singly or in schools.	11-16 months. One calf every three years.	Threatened.
Grey	M 43 F 46 baby 16	Unknown. Main item thought to be bottom-dwelling amphipods.	Eastern No. Pacific. Summer feeding grounds off Alaska coast. Winter breeding grounds off California. Travels singly or in pairs.	13 months. One calf every two years.	At threshold of extinc-tion.
Minke	M 27 F 27 baby 9	Krill in southern seas. Cod, herring, whiting, mackerel in northern seas.	Cosmopolitan. Highly migratory. M in deep waters. F & young in coastal waters. Travels in schools of up to 20.	12 months. One calf every two years.	At lowest ebb in history. Bordering threatened.

FIGURE 12-10   A Partial Appendix to a Formal Report

CHAPTER SUMMARY

Format and supplements determine your report's appearance, the arrangement of its parts, and its accessibility. For an effective format, follow these guidelines:

1. Use high-quality paper.
2. Type your page neatly.
3. Use uniform margins, spacing, and indentation.
4. Number your pages consistently.
5. Use an introduction-body-conclusion structure.
6. Use headings effectively and extensively.

A formal report is prefaced by supplements: cover, title page, letter of transmittal, table of contents, and informative abstract. Include the following supplements — glossary, endnotes, and appendixes — only as needed. Compose your supplements *after* you've written the report proper.

EXERCISES

1. (a) In a paragraph, explain the role of format. (b) In a second paragraph, explain the role of supplements.

2. Find an example of effective formatting in one of your textbooks. (Concentrate on titles, headings, appearance, and ease of use.) Make a copy of a representative section and append a brief explanation of why it is effective. Next, find an example of a poor text format. Make a copy and append an explanation of why it is ineffective. (Note: Your copies should be brief but representative of the overall text format.)

3. The following titles are intended for investigative, research, or analytical reports. Revise all inadequate titles.

   a. "The Effectiveness of the Prison Furlough Program in Our State"
   b. "The Small Business Institute and the University"
   c. "The Effects of Nuclear Power Plants"
   d. "Woodburning Stoves"
   e. "Interviewing"
   f. "An Analysis of Vegetables" (for a report assessing the physiological effects of a vegetarian diet)
   g. "Wood as a Fuel Source"
   h. "Oral Contraceptives"

4. The following headings are from a set of instructions for listening. Write them in parallel form.

- You must focus on the message
- Paying attention to nonverbal communication
- Suppress your biases
- You should listen critically
- You should provide verbal and nonverbal feedback
- Listen for main ideas
- Avoiding distractions
- Make use of lagtime
- Are you allowing the speaker time to make his point?
- You should remain open-minded about the speaker and the message

5. Using the format principles in this chapter, revise an earlier assignment. Submit the revision and the original to your instructor.

6. Prepare a title page, a letter of transmittal (for a definite reader who can *use* your information in a definite way), a table of contents, and an informative abstract for a report you have written earlier (for example, a research report). If needed, revise the heading system.

13

Visuals

VISUALS DEFINED

A visual is any pictorial representation used to clarify a discussion. The common visuals used in report writing are (1) tables and (2) figures. Types of figures include graphs, charts, diagrams, photographs, and samples of material.

PURPOSE OF VISUALS

Visuals attract attention and increase reader understanding by emphasizing certain information. Translate prose into visuals whenever you can, *as long as the visuals make your point more clearly than the prose does.* Use visuals to clarify your discussion, not simply to decorate it. And keep them simple.

Visuals work in several ways to improve your report:

1. They increase reader interest by providing a view more vivid and clear than a prose equivalent. They are easier to follow and grasp. A visual satisfies the reader's demand to be shown.

256

2. They set off and emphasize significant data. A bar graph showing that the price of a microcomputer with a 64K memory has dropped 50 percent in four years is more dramatic than a prose statement. Some readers, in fact, might only skim the prose in a report, concentrating instead on the visuals.

3. They condense information. A simple table, for instance, can often replace a long and difficult prose passage.

4. Certain visuals (tables, charts, and graphs) can pull together diverse data on the basis of their similarities or contrasts. Thus they are easy to interpret.

Suppose you are researching a possible relationship between chemicals and cancer deaths. From various sources, you collect the following data:

1. In 1900, the leading cause of death was pneumonia and influenza, accounting for 11.8 percent of all deaths. Tuberculosis was second (11.3 percent), followed by gastritis (8.3 percent), heart disease (8.0 percent), blood vessel lesions affecting the central nervous system (6.2 percent), chronic nephritis (4.7 percent), accidents (4.2 percent), cancer (3.7 percent), infantile diseases (3.6 percent), and diphtheria (3.2 percent). These ten leading causes accounted for 64 percent of all deaths.

2. In 1960, the leading cause of death was heart disease, accounting for 38.7 percent of all deaths. Cancer was second (15.6 percent), followed by blood vessel lesions affecting the central nervous system (11.3 percent), accidents (5.5 percent), infantile diseases (3.9 percent), pneumonia and influenza (3.5 percent), arteriosclerosis (2.1 percent), Diabetes mellitus (1.8 percent), birth defects (1.3 percent), and liver cirrhosis (1.2 percent). These ten causes accounted for 85 percent of all deaths.

3. In 1970, the . . . .

Clearly, this prose version is repetitious and difficult to interpret. When arranged in Table 13–1, these data are much more readable and easy to compare. Visuals can be great time-savers for readers.

A full-scale study of visuals would require a course in drafting and technical illustration. So this chapter discusses only the most common and most easily composed.

TABLES

Tables are displays of data that can be numerical (as in Table 13–1) or nonnumerical (as in Table 8–4 on page 141). The data are arranged in vertical columns under category headings so they can be easily compared and contrasted.

TABLE 13–1   Leading Causes of Death, 1900, 1960, and 1970

Rank	Cause of death	Deaths per 100,000 population	Percent of all deaths
	1900		
	(All causes)	(1,719)	(100)
1	Pneumonia and influenza	202.2	11.8
2	Tuberculosis (all forms)	194.4	11.3
3	Gastritis, etc.	142.7	8.3
4	Diseases of the heart	137.4	8.0
5	Vascular lesions affecting the central nervous system	106.9	6.2
6	Chronic nephritis	81.0	4.7
7	All accidents[a]	72.3	4.2
8	Malignant neoplasms (cancer)	64.0	3.7
9	Certain diseases of early infancy	62.5	3.6
10	Diphtheria	40.3	2.3
	Total		64
	1960		
	(All causes)	(955)	(100)
1	Diseases of the heart	366.4	38.7
2	Malignant neoplasms (cancer)	147.4	15.6
3	Vascular lesions affecting the central nervous system	107.3	11.3
4	All accidents[b]	51.9	5.5
5	Certain diseases of early infancy	37.0	3.9
6	Pneumonia and influenza	36.0	3.5
7	General arteriosclerosis	20.3	2.1
8	Diabetes mellitus	17.1	1.8
9	Congenital malformations	12.0	1.3
10	Cirrhosis of the liver	11.2	1.2
	Total		85
	1970		
	(All causes)	(945.3)	(100)
1	Diseases of heart	362.0	38.3
2	Malignant neoplasms (cancer)	162.8	17.2
3	Cerebrovascular diseases	101.9	10.8
4	Accidents	56.4	6.0
5	Influenza and pneumonia	30.9	3.3
6	Certain causes of mortality in early infancy[c]	21.3	2.2

7	Diabetes mellitus	18.9	2.0
8	Arteriosclerosis	15.6	1.6
9	Cirrhosis of the liver	15.5	1.6
10	Bronchitis, emphysema, and asthma	15.2	1.6
	Total		85

[a] Violence would add 1.4 percent; horse, vehicle, and railroad accidents provide 0.8 percent.

[b] Violence would add 1.5 percent; motor vehicle accidents provide 2.3 percent; railroad accidents provide less than 0.1 percent.

[c] Birth injuries, asphyxia, infections of newborn, ill-defined diseases, immaturity, etc.

Source: President's Science Advisory Committee Panel on Chemicals, *Chemicals and Health* (Washington, D.C.: Government Printing Office, 1973), p. 152; U.S. Department of Health, Education, and Welfare, Public Health Service, *Facts of Life and Death*, DHEW Pub. No. (HRA) 74–1222 (Washington, D.C.: Government Printing Office, 1974), p. 31.

Levels of Complexity

A table can be as simple as Table 13–2, which includes only one basis of comparison. Notice the explanatory notes that limit the categories. A more complex table, like 13–3, contains several bases of comparison.

Although not as visually dramatic as a graph or chart, a table is best for numbers and units of measurement that must be displayed precisely.

Construction

To make a table, follow these guidelines:

1. Number each table in order of its appearance (for easy reference), and give it a clear title that forecasts exactly what the table contains.

TABLE 13–2   1984 American Subcompacts Classified in Descending Order on the Basis of Gas Mileage

Make of Car [a]	Miles per Gallon [b]
Midgo II	33.4
Vampira ST	32.9
Locomoto	32.5
Zoomer	32.3

[a] All models tested were two-door sedans with three-speed manual transmissions.
[b] Mileage figures are based on EPA averages for combined city and highway driving.

TABLE 13–3   Five Popular Sandwiches Classified on the Basis of Caloric, Protein, and Fat Content, and Estimated Cost

Sandwich	Calories	Protein (in grams)	Fat (in grams)	Estimated Cost
Bologna (3½ oz., 1 T. mustard)	436	17	30	.64
Hamburger (4 oz. cooked, 1 T. catsup)ᵃ	331	25	16	.69
Tuna salad (3½ oz., 1 T. mayonnaise)	422	33	21	.73
Egg salad (1 large egg, 1 T. mayonnaise)	313	11	19	.31
Peanut butter (1 oz.)	296	12	16	.28

Source: Based on Bureau of Labor Statistics estimates found in the *Retail Food Price Index, October 1982* (Washington, D.C.: U.S. Department of Labor).
[a] The hamburger is served on a bun. All other sandwiches are on white bread.

2. Begin each vertical column with a heading that identifies the types of items listed (e.g., "Make of Car") and specific units of measurement and comparison (e.g., "Miles per Gallon," "Grams per Ounce," "Percentage"). Use only the approved abbreviations and symbols in Appendix A. Give all items in the same column the same units of measurement (inches, sq. ft., etc.) and keep decimals vertically aligned.

3. Use footnotes to clarify certain entries. Whereas notation in a discussion is in arabic numerals (1, 2, 3), in a table it is in small letters (a, b, c).

4. Set your table off from the discussion by framing it with adequate white space all around. Be sure the table does not extend into the page margins.

5. Try to keep the table on a single report page. If it does take up more than one full page, write "continued" at the bottom and begin the second page with the full title and "continued." Also, place the same headings at the tops of each column that are on the first page of the table. If you need to total your columns, begin second-page columns with subtotals from the first page.

6. If your table is so wide that you need to turn it to the vertical plane of your page, place the top against the inside binding (Figure 13–9).

7. Relate your table to the surrounding discussion. Refer specifically to the table by number and title in the report text. Introduce it and discuss any special features about the data. Do not make readers interpret raw data.

8. If the table clarifies a part of your discussion, place it in that area of your text. If, however, it simply provides supporting information of interest only to some readers, place it in an appendix so those readers can refer to it if necessary. Avoid cluttering up your discussion.

9. Identify your data sources below the table, beginning at the left margin. If the table itself is borrowed, so indicate. And list your sources even if you make your own table from borrowed data.

FIGURES

Any visual that is not a table is classified as a figure and should be so titled (e.g., "Figure 1: An Aerial View of the Panhandle Building Site"). The most common figures are graphs, charts, diagrams, photographs, and samples.

Graphs

You construct graphs by plotting a set of points on a coordinate system. A graph provides a picture of the relationship between two variables, and it shows a comparison, a change over time, or a trend. When you decide to use a graph, choose the best type for your purpose: bar graph or line graph.

Bar Graphs

A bar graph, as in Figure 13–1, illustrates comparisons. In this case, the visual impact of the bar graph makes it a clear choice over a prose or tabular version. Percentage figures are recorded above each bar to increase clarity. Here the independent variable[1] range extends only from 0 to 40 percent. This range creates enough space between vertical increments to dramatize the comparison without taking up too much of the page. Units of measurement on the vertical line are clearly identified.

The scale in a bar graph (such as 10 percent per inch) is crucial. Try different scales until your graph represents all quantities in accurate proportion. If, for instance, the vertical scale in Figure 13–1 were extended to 100 percent, the bars would seem dwarfed and space would be wasted. Other distortions

[1] In all graphs, the horizontal line (abscissa) lists those items whose value is fixed (independent variables); the vertical line (ordinate) lists those values that change (dependent variables). The dependent variable changes according to the specific activity of the independent variable (e.g., an increase in percentage over time, as shown in Figure 13–1).

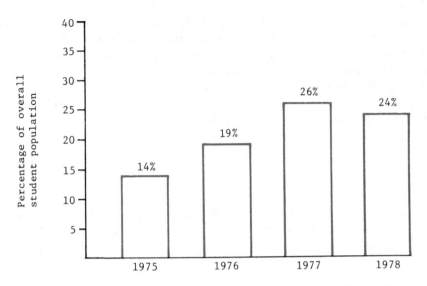

FIGURE 13–1   A Bar Graph, Showing the Percentage of Students Making the
Dean's List at X College, 1975–78

would occur if the vertical increments, for example, were increased to 5 percent per inch or decreased to 30 percent per inch.

A bar graph can also contain multiple bars (up to three) at each major point
on the horizontal line, as in Figure 13–2. In a multiple-bar graph, include a
legend to explain the various bars.

Another common graph is the segmented-bar graph, which breaks down
each bar into its components. Notice that the vertical scale in Figure 13–3 is
large enough to allow comparison.

When comparing a horizontal quantity (such as distance traveled), you
can use a horizontal bar graph.

Make all graphs on graph paper, so lines and increments are evenly spaced.
Begin a bar graph directly on the horizontal line. To express negative values,
extend the vertical line below the horizontal, following the same incremental
division as above it, only in negative values. Make all bars the same width
so as not to mislead readers about relative values.

Line Graphs

Whereas a bar graph provides units of measurement for visual comparison, a
line graph, like Figure 13–4, shows a change or trend over time. Unlike a bar
graph, which must begin on the horizontal line, a line graph can begin at any

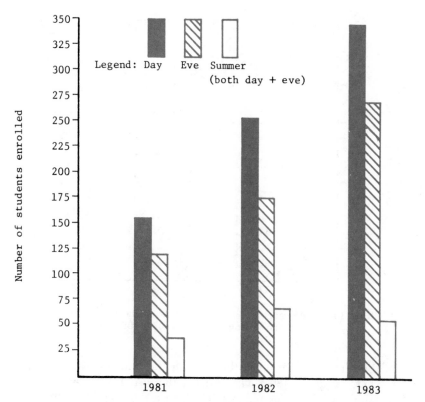

FIGURE 13–2   A Multiple-Bar Graph, Showing the Number of Students Enrolled in Technical Writing, 1981–83

intersection on the coordinate grid.[2] Select a readable scale, and stipulate the units of measurement (e.g., building permits issued).

A line graph is particularly useful for comparing trends or changes among two or three related dependent variables, as in Figure 13–5. These pictorial data give an instant overview of daily shopping patterns in various locations. In this kind of multiple-line graph, your choice of scale is crucial. Imagine, for example, that the vertical scale in Figure 13–5 were condensed to $1500 per increment, as in Figure 13–6. With this reduced scale, the graph becomes almost impossible to interpret. Conversely, an overly expanded vertical scale would yield another distortion. Figure 13–7 shows the same data on a graph whose scale has been increased to $250 per increment and whose vertical line begins at $1000 instead of 0, to try to save space. This error distorts the quan-

[2] You may have noticed that we can make line graphs from bar graphs, simply by connecting the tops of bars (say, in Figure 13–2).

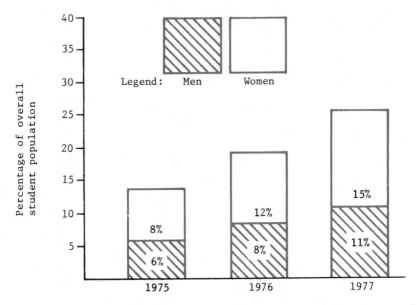

FIGURE 13–3   A Segmented-Bar Graph, Showing the Breakdown, by Sex, of
Students Making the Dean's List, 1975–77

titative relationship between lines: the high for the Midwest is $2500, and for
the East, $4500 (roughly 180% higher); yet the visual relationship between
these lines suggests the sales volume for the East is roughly 230% higher.
Visual relationships should parallel the numerical relationships.

Construction

1. Number the graph in order of its appearance and give it a clear title.
2. Label the items on your horizontal and vertical lines. State units of measurement along your vertical coordinate (hundreds of dollars, pounds per
square inch, etc.). In a multiple-bar or line graph, include a legend.
3. Experiment with scales until you find the most faithful representation.
4. Keep the graph simple and easy to read. Never plot more than three different lines or types of bars.
5. Because you're using graph paper, plan carefully for integrating a graph
into your discussion. If the graph is only a few inches high, you might trim
the excess graph paper and paste or glue the graph in the appropriate section
of your discussion. Otherwise, place the full graph page immediately after the
related discussion page or in an appendix.
6. Introduce, discuss, and interpret your graph, referring specifically to its
number and title. Do not leave the reader with a page full of raw data.

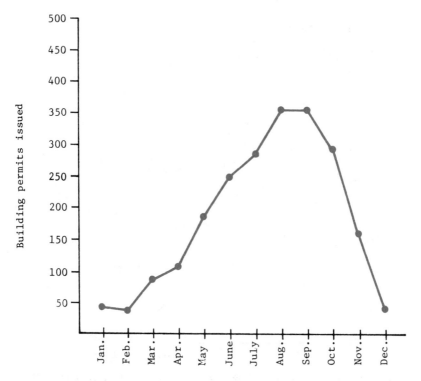

FIGURE 13–4   A Line Graph, Showing Building Permits Issued in Dade County
in 1977

7. If the graph must be presented on the vertical plane of your page, place
the top against the inside binding.

8. Credit your data sources two spaces below your figure number and title.

Charts

The terms *chart* and *graph* often are used interchangeably. We define a chart
as a figure that illustrates relationships (quantitative or cause-and-effect) but
is not plotted on a coordinate system. So no graph paper is used. Common
charts are pie charts, organizational charts, and flowcharts.

Pie Charts

A pie chart partitions a whole into its parts and provides an image of the
parts-whole relationship. The parts of a pie chart must add up to 100 percent,
as shown in Figure 13–8.

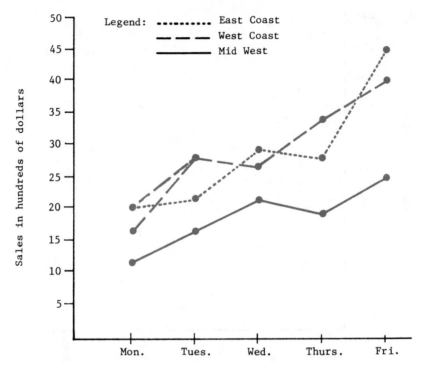

FIGURE 13–5   A Multiple-Line Graph, Showing Total Sales in Three Major
Outlets for the Week of June 2, 1978

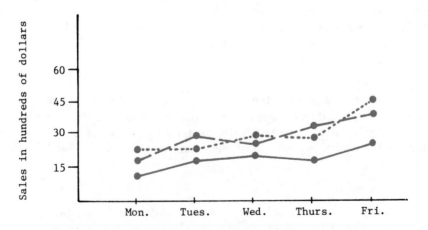

FIGURE 13–6   A Poorly Scaled Graph (Increments Condensed)

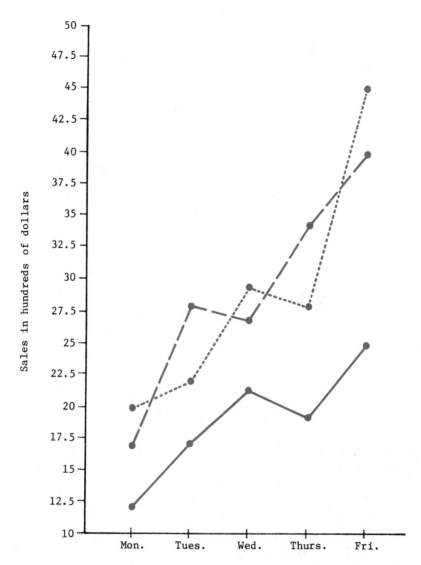

FIGURE 13–7   A Poorly Scaled Graph (Increments Expanded)

Follow these guidelines in making your pie chart:

1. Number it in order of its appearance with other figures, and give it a clear and precise title. Place figure number and title two spaces below your chart.

2. Use a compass to draw a perfect circle and to locate its center. Use a protractor for precise segmentation. Each 3.6 degrees equals 1 percent.

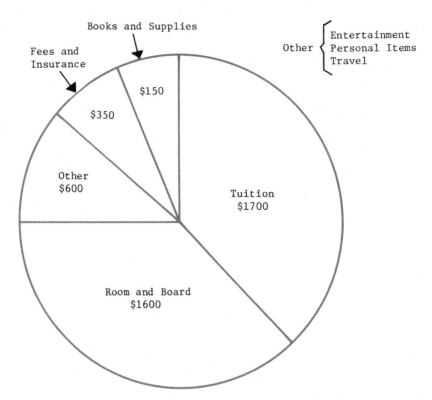

FIGURE 13–8   A Pie Chart, Showing Yearly Cost Breakdown for Attending
Calvin College (Total Cost = $4400/Year)

3. Begin segmenting your chart by locating your first radial line at twelve o'clock. Move clockwise, in descending order, from largest to smallest segments.

4. Use at least three, but no more than seven, segments. Combine several small segments (1 percent to 5 percent each) under the heading "Other." Explain these combined items, as in Figure 13–8.

5. Write section headings, quantities, and units of measurement horizontally.

6. Place your pie chart where it belongs in your discussion. Introduce it, explain it, and credit data sources.

Though not as precise as a tabular list, a pie chart draws attention more dramatically than a list of numbers would.

Organizational Charts

An organizational chart partitions the administrative functions of an organization. It ranks each member in order of authority and responsibility as that member relates to other members and departments. Figure 13–9 shows the partial organizational chart of a typical management structure of a college.

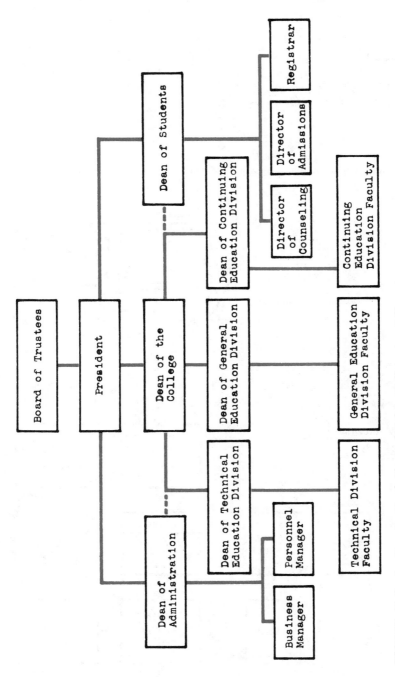

FIGURE 13–9 An Organizational Chart of Calvin College (partial)

Flowcharts

A flowchart traces a process from beginning to end. In outlining the specific steps of a manufacturing or refining process, the flowchart moves from raw material to finished product. In showing how something happens, it moves through the specific steps, as in Figure 13–10.

Diagrams

Diagrams are sketches or drawings of the parts of an item or the steps in a process.

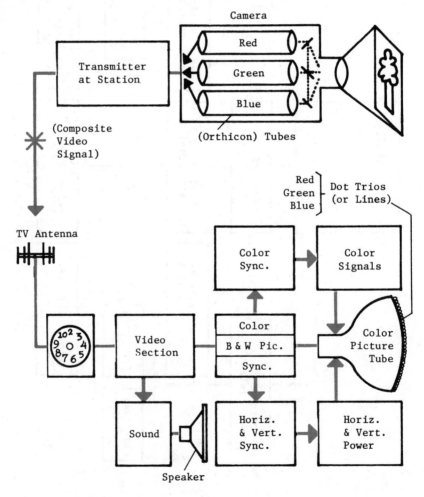

FIGURE 13–10   A Flowchart, Showing How Color Television Works

Diagrams of Mechanical Parts

A mechanism description should be accompanied by diagrams that show its parts and illustrate its operating principle, such as Figure 13–11. Always name your perspective: front view, lateral view, anterior, superior, cross-sectional, and so on.

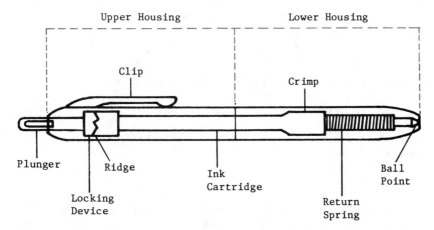

FIGURE 13–11   A Diagram of a Retractable Ball-Point Pen with Point Retracted (Cutaway)

Exploded Diagrams

Exploded diagrams, like Figure 13–12, show how the parts of an item are assembled; they often appear in repair or maintenance manuals.

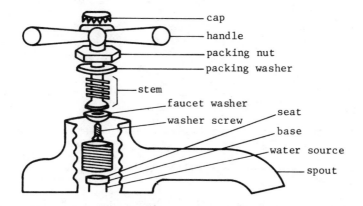

FIGURE 13–12   An Exploded Diagram, Showing a Faucet

FIGURE 13–13   Frontal View of a Camera Held Correctly

Diagrams of Procedures

Diagrams are useful for clarifying instructions by illustrating certain steps, as in Figure 13–13.

Block Diagrams

Block diagrams are simplified sketches that represent the relationship between the parts of an item or process. Since block diagrams are designed to illustrate *concepts* (such as current flow in a circuit), the parts are represented as symbols or geometric shapes. The block diagram in Figure 13–14 illustrates how

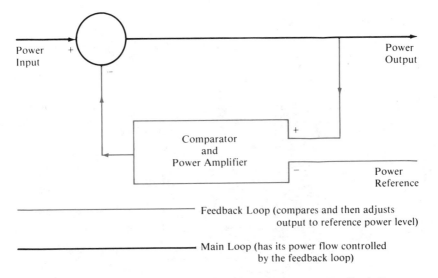

FIGURE 13–14   A Block Diagram, Illustrating the Basic Feedback Concept

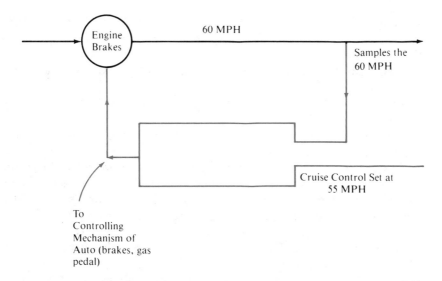

FIGURE 13-15   A Block Diagram, Showing a Cruise-Control Mechanism

any process can be controlled automatically through a feedback mechanism. Figure 13-15 shows the feedback concept applied as the cruise-control mechanism on a motor vehicle.

Photographs

Photographs provide an accurate overall view, but sometimes they can be too "busy." By showing all details as more or less equal, a photograph sometimes fails to emphasize important areas.

When you use photographs, keep them distinct, well focused, and uncluttered. For a complex mechanism, you probably should rely on diagrams instead, unless you intend simply to show an overall view. Lend a sense of scale by including a person or a familiar object (such as a hand) in your photo.

COMPUTER GRAPHICS

With one or two commands, many computer systems can now create highly sophisticated, six-color graphics programs. The system analyzes and displays the data in the appropriate visual form within seconds. With the right software, all the visuals in this chapter could be created by computer. And many more complex visuals are possible with computer graphics, as in these few examples:

– With an electronic stylus (a pen with an electronic signal), you can draw pictures on a graphics tablet to be displayed on the monitor, stored, or sent to other computers.

– You can map different concentrations in different colors (say, in a mineral map), to differentiate data more clearly.

– You can model structures that can't be seen, such as a DNA helix or other protein molecules.

– You can create three-dimensional effects, showing an object from different angles by using shading, shadows, etc.

– You can recreate the visual effect of a mathematical model, as in writing a set of equations to explain what happens when high winds strike a tall building. (As the wind deforms the structure, the equation changes. Then you can take those new equations and represent them visually.)

– You can design a device, build a model, simulate the physical environment, and let the computer forecast what will happen with different variables (say, Space Shuttle simulation).

– You can integrate a computer-assisted design process (CAD) with a computer-assisted manufacturing process (CAM), so the design will actually direct the machinery that makes the parts themselves (CAD/CAM).

– You can create animations, to see how bodies move (say, in a car crash or in athletics).

– You can practice dealing with toxic chemicals, operating sophisticated machines, or making other rapid, high-level decisions in various medical or technical environments, without the cost or danger of actual situations.

Once you've created a model, you can do any number of "what-if" projections. Because the computer can generate and evaluate many possibilities rapidly, it enables you to test various hypotheses without doing the calculations, thereby generating *more* hypotheses.

Computer graphics have countless applications. A firm charts its projected yearly profit in minutes. A manufacturer gets a visual display of inventory. A marketing department displays consumer patterns and regional trends. An engineer in research and development projects "what-if" scenarios to check adjustments in projected figures. Sales managers compare employees' performances or check sales trends. The company president tracks progress on major projects. Computer graphics organize raw data so it can be interpreted at a glance.

These applications imply obvious benefits. For one, computer graphics save time. Rather than having to analyze and interpret masses of data, managers can generate graphics from their data bases and interpret the *visual* results in minutes. These graphic results can, in turn, be used for presentations at staff, sales, and stockholders' meetings, conferences, or when presenting important findings or reports. In these situations, a picture often is worth a thousand words.

Faster interpretation creates a third benefit: faster decision making, since trends or deviations are much easier to spot in graphic form. Faster decisions, in turn, give companies using graphics a competitive edge. These advantages will cause computer graphics to grow into a multi-billion-dollar industry by 1990.

CHAPTER SUMMARY

A visual is any pictorial device that clarifies your discussion. In report writing, the most common visuals are tables and figures.

In making a table, follow these guidelines:

1. Number it chronologically and give it a clear title.
2. Begin each column with a clear heading, and list all items in that column according to the same unit of measurement.
3. Use footnotes to explain complex data.
4. Leave plenty of white space between the table and your text.
5. Try to keep the table on a single page.
6. Introduce, discuss, and interpret the table.
7. Place the table in the text or an appendix, as appropriate.
8. Place the top of a wide table on the vertical plane, against the binding.
9. Identify any data sources.

Figures include such items as graphs, charts, diagrams, photographs, and samples. Any visual is a figure if it isn't a table.

In a graph, numbers are plotted as points on a coordinate system. A bar graph shows comparisons, while a line graph shows change over time.

In making a graph, follow the guidelines for tables, with these additions:

1. Include a legend in a multiple-bar or line graph.
2. Experiment with scales until you find the most accurate.
3. Never plot more than three variables on one graph.
4. Use graph paper; trim the excess; paste the figure on the text page.

Charts are not plotted on a coordinate system. Pie charts, organizational charts, and flowcharts are most common. Each partitions a whole into its parts to represent the relationship of part to part and of part to whole.

Diagrams are sketches or drawings, and they are usually better than photographs for emphasizing parts. Photographs, however, present overall views.

Computer graphics are transforming the role of visuals in decision making.

REVISION CHECKLIST FOR VISUALS

1. Does the visual serve a real purpose; that is, does it clarify, not simply decorate, your report?

2. Have you chosen the best form of visual for your purposes?

3. Is the visual titled and numbered appropriately?

4. Can it stand alone in meaning, if necessary?

5. Does each vertical column in the table begin with a clear heading?

6. Are units of measurement in the same column identical (inches, etc.)?

7. Are all decimal points in each tabular column vertically aligned?

8. Are explanatory notes added as needed (in lowercase letter notation)?

9. Are the margins clear?

10. Is the visual set off from the text by adequate white space?

11. Does the top of an excessively wide visual abut the inside binding?

12. Is the visual introduced, discussed, and interpreted as needed?

13. Is it in the best location for its purpose in your report (within the text if it clarifies the discussion; in an appendix if it merely supports it)?

14. Are all sources of data identified?

15. In a graph, are the independent variables plotted along the horizontal line and the dependent along the vertical?

16. Does a multiple-bar or line graph have a legend?

17. Does the graph have a clear and accurate scale?

18. Is the graph restricted to three or fewer lines or types of bars?

19. Does the segmentation in the pie chart begin at twelve o'clock?

20. Does the pie chart have at least three, but no more than seven, segments?

21. Do the segments add up to 100 percent?

22. Are any small segments in the pie chart (1 percent to 5 percent each) integrated under the heading "Other"?

Now list those elements of your visuals that need improvement.

EXERCISES

1. The following statistics are based on data gathered from three competing colleges in a large western city. They give the number of applicants to each college over six years.

– In 1978, X College received 2341 applications for admission, Y College received 3116, and Z College received 1807.

– In 1979, X College received 2410 applications for admission, Y College received 3224, and Z College received 1784.

– In 1980, X College received 2689 applications for admission, Y College received 2976, and Z College received 1929.

– In 1981, X College received 2714 applications for admission, Y College received 2840, and Z College received 1992.

– In 1982, X College received 2872 applications for admission, Y College received 2615, and Z College received 2112.

– In 1983, X College received 2868 applications, Y College received 2421, and Z College received 2267.

Illustrate these data in a line graph, a bar graph, and a formal table. Which form seems most effective? Include a brief prose interpretation with the most effective illustration.

2. Devise a flowchart for a process in your field or area of interest. Include a title and a brief prose discussion.

3. Devise an organizational chart showing the lines of responsibility and authority in an organization where you hold a part-time or summer job.

4. Devise a pie chart to illustrate the partition of one of your typical weekdays. Include a full title and a brief prose discussion of your data.

5. Call or visit your town accountant for a breakdown of town income and expenditures (where each part of the revenue dollar comes from; how each part of the revenue dollar is spent). Compose visuals to illustrate these partitions. Include prose explanations.

6. Obtain enrollment figures for the past five years at your college on the basis of sex, age, race, or any other pertinent category. Construct a segmented bar graph to illustrate one of these relationships over the five years.

7. Keep track of your pulse and respiration at thirty-minute intervals over a four-hour period of changing activities. Record your findings in a line graph, noting both times and specific activities below your horizontal coordinate. Write a prose interpretation of your graph and give it a title.

8. In textbooks or professional journal articles, locate each of the following visuals: a table, a multiple-bar graph, a multiple-line graph, a diagram, and a photograph. Evaluate each according to the revision checklist, and discuss the most effective visual in class.

9. We have discussed the importance of choosing an appropriate scale for your graph, and choosing the most effective form for presenting your data. Study the following presentation carefully:

Strong evidence now indicates that not only nicotine and tar in cigarette smoke can be lethal, but also carbon monoxide. Experts have learned that a high percentage of cigarette smoke is composed of carbon monoxide. The bar graph in Figure 13–16 (next page) lists the ten leading U.S. cigarette brands according to the carbon monoxide given off per pack of inhaled cigarettes.

Is the scale effective? If not, why not? Can these data best be presented in a bar graph? Present the same data in some other form that seems most effective.

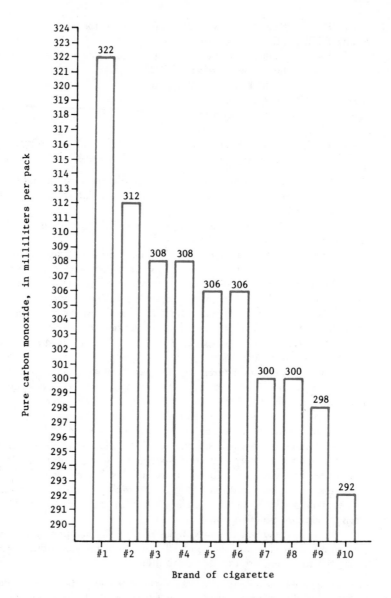

FIGURE 13–16   Ten Leading U.S. Cigarette Brands in Order of CO Content

10. Choose the most appropriate visual aid for illustrating each of the following. Justify each choice in a short paragraph.

   a. a comparison of three top brands of Fiberglas ski, according to cost, weight, durability, and edge control
   b. a breakdown of your monthly budget

    c. the changing cost of an average cup of coffee, as opposed to that of an average cup of tea, over the past three years

    d. the percentage of college graduates finding desirable jobs within three months after graduation, over the last ten years

    e. the percentage of college graduates finding desirable jobs within three months after graduation, over the last ten years — according to sex

    f. an illustration of automobile damage for an insurance claim

    g. a breakdown of the process of radio-wave transmission

    h. a comparison of five cereals on the basis of cost and nutrition

    i. a comparison of the average age of students enrolled at your college in summer, day, and evening programs, over the last five years

    j. a comparison of sales for three items made by your company

14

Researching Information

RESEARCH DEFINED

Research is the effort to discover any fact or set of facts: the cost of building the first x-ray machine or the price range of half-acre building lots in Boville in January 1984. Any kind of significant research is for some purpose — to answer a question, to make an evaluation, to establish a principle. We set out to discover whether diesel engines are efficient and dependable for a reason: we're thinking about buying a car equipped with one, or we're trying to decide whether to begin producing them, or the like. Research is the way to find your own answers, to submit your opinions to the test of fact.

Depending on its information sources, research may be classified as *primary* or *secondary*. Primary research is a firsthand study of the subject; its sources are memory, observation, questionnaires, interviews, letters of inquiry, and records of business transactions or scientific and technological activities. Secondary research is based on information that other researchers — by their primary research — have compiled in books, articles, reports, brochures, and other publications. Most research calls for both primary and secondary approaches.

PURPOSE OF RESEARCH

The purpose of all research is to arrive at an informed opinion, to establish a conclusion that has the greatest chance of being valid.

We might have uninformed opinions about political candidates, kinds of cars, controversial subjects like abortion and capital punishment, or anything else. Opinions are beliefs that are not proven but seem to us to be true or valid. Without a basis in fact, opinions are uninformed, disputable, and subject to change in the light of new experience. Sometimes we forget that many of our opinions rest on no objective data. Instead, they are based on a chaotic collection of the beliefs reiterated around us, notions we've inherited from advertising, things we've read but never checked, and so on. Commercials especially are designed to manipulate the consumer's uninformed opinion.

Any claim is valid only insofar as it is supported by facts, what we know from observation or study. An opinion based on fact is more legitimate than an uninformed opinion. In many cases, we must consider a variety of facts. Consider, for instance, a commercial claim that Brand X toothpaste makes teeth whiter. Although this claim may be true, a related fact may be that Brand X toothpaste contains tiny particles of ground glass, thereby harming more than helping teeth. The second fact may change your opinion. Similarly, a United States senator may claim he is committed to reducing pollution, but he may have voted against important environmental legislation. So, if you want to know whether to believe his claim, you need to establish the facts. You want your opinion to rest on verifiable information, not merely on his claim.

Facts also affect the quality of personal decisions. Before studying for a career, investigate job openings, salary range, and requirements; better yet, interview someone who has such a job. Conduct the same kind of research before buying a new car or emigrating to Australia. Although the facts you discover may contradict your original opinion, they will enable you to make an informed decision.

On the job, virtually no important decision is made without some research. And the results of any research are almost inevitably recorded in a report.

IDENTIFYING INFORMATION SOURCES

Personal Experience

Begin with what you already know about your subject. Search your memory, and brainstorm (Chapter 2).

The Library

Make the library your first stop. You may even learn that your problem has already been studied and solved by others, which could save you time.

Libraries have one or more of these means of organizing their information: card catalogs, reference works, indexes to periodicals, government publications indexes, and electronic information services.

The Card Catalog

In many libraries, every book, film, filmstrip, phonograph album, and tape is indexed in the card catalog under three separate designations: author, title, and subject. So you have at least three possible ways of locating the item.

Your library may place author, title, and subject cards in a single alphabetical file or may provide individual catalogs labeled "Author," "Title," and "Subject." In your search, first decide whether you are looking for a specific title, a book by a certain author, or material about a subject. Then check the arrangement of the card catalog to determine where to look.

Locating the Card. Say you are looking for a book on food technology by Norman W. Desrosier. If your library has a divided catalog, locate the D cards in the author section. If not, locate the D cards in the combined catalog. Flip through the cards in the drawer until you locate those listed under "Desrosier, Norman W." Figure 14–1 shows a typical author catalog card.

As entry F on the author card indicates, you can find the same book listed alphabetically under its title, as shown in Figure 14–2. All other information on the title card is identical to that on the author card.

If you know neither the authors nor the titles of books on a subject, turn to the subject listing. Figure 14–3 shows the card for the same book found under the subject heading "Food — Preservation." Here again, all other items on the card remain the same.

Locating the Book on the Shelf. The call number is your key for locating the book. In the card catalog area, in stairways and elevators, and on doors to individual floors, you will see call-number maps.

Reference Books

Reference books include technical encyclopedias, almanacs, handbooks, dictionaries, histories, and biographies. These can be a good place to begin research, because they provide background information and bibliographies that can lead to more specific information. The one drawback here is that some

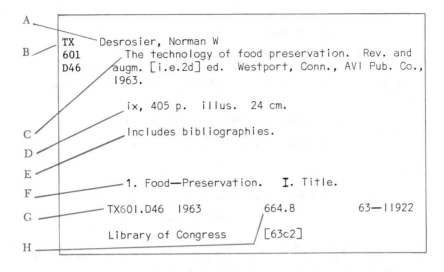

A: The author's name, listed last name first. (On some cards the author's name is followed by his date of birth — and death, if he was deceased at the time of the book's last printing.)

B: The call number: each book has a different call number under the coding system by which books are classified. Books are arranged on the shelves in the order of their respective call numbers. This number is your key for locating the book.

C: The title of one book written by this author, followed by the number of the edition, city of publication, publisher, and publication date.

D: Technical information: This book has 9 chapters, 405 pages, illustrations, and is 24 centimeters high.

E: Special information about the book: This book includes bibliographies listing related sources of information.

F: Other headings under which this book is listed in the card catalog: In the subject section, the book is listed under "Food — Preservation"; it is also listed alphabetically by title.

G: Library of Congress call number.

H: Dewey Decimal System call number.

FIGURE 14–1   A Catalog Card Classified by Author

reference books not revised within the last few years may be out of date. Always check the last copyright date.

Reference works are in a special section marked "Reference" — usually on the main floor of the building. All works will be indexed in the "Subject" card

```
            The technology of food preservation.

TX     Desrosier, Norman W
601        The technology of food preservation.  Rev. and
D46     augm. [i.e.2d] ed.  Westport, Conn., AVI Pub. Co.,
1963    1963.

           ix, 405 p.  illus.  24 cm.

           Includes bibliographies.


           1. Food—Preservation.   I. Title.

       TX601.D46  1963          664.8           63—11922

       Library of Congress      [63c2]
```

FIGURE 14–2   A Catalog Card Classified by Title

```
            Food - Preservation

TX     Desrosier, Norman W
601        The technology of food preservation.  Rev. and
D46     augm. [i.e.2d] ed.  Westport, Conn., AVI Pub. Co.,
1963    1963.

           ix, 405 p.  illus.  24 cm.

           Includes bibliographies.


           1. Food—Preservation.   I. Title.

       TX601.D46  1963          664.8           63—11922

       Library of Congress      [63c2]
```

FIGURE 14–3   A Catalog Card Classified by Subject

catalog, with a "Ref." designation immediately above the call number. Here is
a partial list of reference works:

- *Civil Engineering Handbook*
- *Dictionary of Scientific Biography*
- *The Encyclopedia of How It Works*
- *Encyclopedia of Food Technology*

- *Encyclopedia of the Social Sciences*
- *Fire Protection Handbook*
- *Handbook of Chemistry and Physics*
- *The Harper Encyclopedia of Science*
- *McGraw-Hill Encyclopedia of Environmental Science*
- *McGraw-Hill Encyclopedia of Science and Technology*
- *The New Dictionary and Handbook of Aerospace*
- *Oxford English Dictionary*
- *Paramedical Dictionary*
- *The Way the New Technology Works*
- *The Way Things Work*

There are reference works for every discipline. In researching the effects of food additives, for instance, you would want to start with titles such as the *McGraw-Hill Encyclopedia of Food, Agriculture, and Nutrition* or the *RC Handbook of Food Additives*. Look for your subject in the card catalog; then check for any subheadings such as "Handbooks, Manuals, etc." or "Dictionaries." These will be books that get you started.

Periodical Indexes

Just as the card catalog is an index to the library's books and audiovisual materials, periodical indexes are alphabetical guides listing names, titles, and subjects of material in journals. Through periodical indexes, you can locate the latest information. The most comprehensive index is the *Reader's Guide to Periodical Literature*, which indexes articles from over 150 popular magazines and journals. Because a new volume of the *Readers' Guide* is published every few weeks, you often can locate articles less than one month old. The *Readers' Guide* indexes items alphabetically by subject, author, and book or film title. Figure 14–4 shows a section from one of its pages.

Locating the Entry. Assume you are researching how awareness of the effects of food additives and preservatives has grown in the last decade. As you turn to a 1976 issue of the *Readers' Guide*, you find many entries under the general heading "Food." Scanning the entries classified under "Food Additives," you spot an item that looks useful: "Food Additives and Hyperactive Kids" — item *A*. From this entry, you gain the following information: because the author's name is not given, the article probably was written by a writer on the magazine staff; the name of the periodical is *Science Digest* (abbreviations are explained in the opening pages of each *Readers' Guide*); the volume number is 80; the article is found on page 13 of the November 1976 issue. Item *B* refers to other headings under which you might find relevant articles.

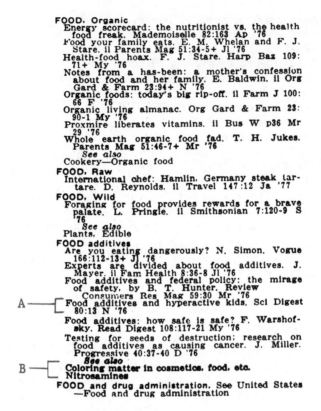

FOOD, Organic
Energy scorecard: the nutritionist vs. the health food freak. Mademoiselle 82:163 Ap '76
Food your family eats. E. M. Whelan and F. J. Stare. il Parents Mag 51:34-5+ Jl '76
Health-food hoax. F. J. Stare. Harp Baz 109:71+ My '76
Notes from a has-been: a mother's confession about food and her family. E. Baldwin. il Org Gard & Farm 23:94+ N '76
Organic foods: today's big rip-off. il Farm J 100:66 F '76
Organic living almanac. Org Gard & Farm 23:90-1 My '76
Proxmire liberates vitamins. il Bus W p36 Mr 29 '76
Whole earth organic food fad. T. H. Jukes. Parents Mag 51:46-7+ Mr '76
    See also
Cookery—Organic food
FOOD, Raw
International chef: Hamlin, Germany steak tartare. D. Reynolds. il Travel 147:12 Ja '77
FOOD, Wild
Foraging for food provides rewards for a brave palate. L. Pringle. il Smithsonian 7:120-9 S '76
    See also
Plants, Edible
FOOD additives
Are you eating dangerously? N. Simon. Vogue 166:112-13+ Jl '76
Experts are divided about food additives. J. Mayer. il Fam Health 8:36-8 Jl '76
Food additives and federal policy: the mirage of safety, by B. T. Hunter. Review Consumers Res Mag 59:30 Mr '76
A ——[Food additives and hyperactive kids. Sci Digest 80:13 N '76
Food additives: how safe is safe? F. Warshofsky. Read Digest 108:117-21 My '76
Testing for seeds of destruction; research on food additives as causing cancer. J. Miller. Progressive 40:37-40 D '76
    See also
B ——[Coloring matter in cosmetics, food, etc.
Nitrosamines
FOOD and drug administration. See United States —Food and drug administration

FIGURE 14–4 A Page Section from *Readers' Guide to Periodical Literature.* (Copyright © 1976, 1977 by the H.W. Wilson Company. Material reproduced by permission of the publisher.)

Locating the Indexed Article.   Each work listed in the card catalog is held by that library. So, unless the item has been borrowed by someone else, you should locate it under its call-number designation. Periodicals are more of a problem, however; not all works in periodical indexes will be held by any one library. The indexes list articles from hundreds of journals, newspapers, and magazines, but your library subscribes only to a cross-section. If you find an index reference to an article not held by your library, check with other libraries or ask your librarian to see if the article is available through an interlibrary loan.

Scan the periodicals holdings list to determine which periodicals are held by your library. Copies are available in the area where indexes are shelved.

Earlier, you located an entry in the *Readers' Guide* titled "Food Additives and Hyperactive Kids," printed in *Science Digest.* Now, as you scan the periodicals holdings list, you learn your library subscribes to this journal (see item *A*), and that all back issues to 1960 are on microfilm.[1] Your article is probably not recent enough to be found in the actual journal (many libraries keep only one year of back issues as bound copies).

If the article is very recent, scan the periodical shelves, usually arranged alphabetically by title, to find your issue. In some libraries older issues are bound together in hard-cover bindings instead of being microfilmed. Each bound volume has a call number and is listed in the card catalog under "Title."

In addition to indexes for general periodicals, specialized indexes are available in nearly every discipline. A researcher in the business field, for instance, might consult the following indexes of specialized periodicals:

- *Business Periodicals Index*
- *Editor and Publisher Market Guide*
- *Editorial Research Reports*
- *F and S Index*
- *Predicasts*
- *Public Affairs Index*

Here are some indexes in other fields:

- *Agricultural Index*
- *Applied Science and Technology Index*
- *Biological Abstracts*
- *Chemical Abstracts*
- *Education Index*
- *The Energy Index*
- *Engineering Index*
- *Environment Index*
- *General Science Index*
- *Index Medicus*
- *International Nursing Index*
- *Pollution Abstracts*
- *Psychological Abstracts*

Two most valuable sources for quickly building exhaustive bibliographies (lists of publications on your subject) are the *Science Citation Index* and its counterpart, the *Social Sciences Citation Index*. Your library may have other indexes. Ask the librarian to help you identify those for your subject.

Having looked in the reference books that provided a good overview of the psychological effects of food additives, you might now turn to *Psychological Abstracts* or *Index Medicus* for journal literature.

---

[1] Many libraries subscribe to a microfilm service that photographs back issues of periodicals on microfilm. They store several periodicals on one small roll of film, and you read the original on a microfilm reader.

The Reference Librarian

Your best bet for help with library research is the reference librarian. This person's job — as the title implies — is to *refer* people to information sources. So don't be afraid to ask for assistance. Even professional researchers sometimes need help. The reference librarian can save hours of wasted time by showing you how to use indexes or microfilm and microfiche readers and how to locate reference books or bibliographies. And he or she can order items from other libraries, but this may take time.

Access Tools for United States Government Publications

The federal government publishes maps, periodicals, books, pamphlets, manuals, monographs, annual reports, research reports, and a bewildering array of other information. Kinds of information available to the public include presidential proclamations, congressional bills and reports, judiciary rulings, certain reports from the Central Intelligence Agency, and publications from all other government agencies (Departments of Agriculture, Commerce, Transportation, etc.). Below is a brief sampling of the countless titles available in this gold mine of information:

> *Effects of New York's Fiscal Crisis on Small Business*
> *Economic Report of the President*
> *Major Oil and Gas Fields of the Free World*
> *Decisions of the Federal Trade Commission*
> *Journal of Research of the National Bureau of Standards*
> *Siting Small Wind Turbines*

Your best bet for tapping this valuable but complex resource is to ask the librarian in charge of government documents for help. If your library does not own the publication you seek, it can be obtained through an interlibrary loan.

The basic access tools for documents issued or published at government expense are discussed below.

– *The Monthly Catalog of the United States Government,* the major access to government publications, is indexed by author, subject, and title. These indexes provide you with the catalog entry number that leads you, in turn, to a complete citation for a given work (Figure 14–5).

– *Government Reports Announcements & Index* is a listing published every two weeks by the National Technical Information Service (NTIS),[2] a federal clearinghouse for scientific and technical information — all stored in a computer database. The collection contains summaries of some 900,000 federally

---

[2] A branch of the U.S. Department of Commerce.

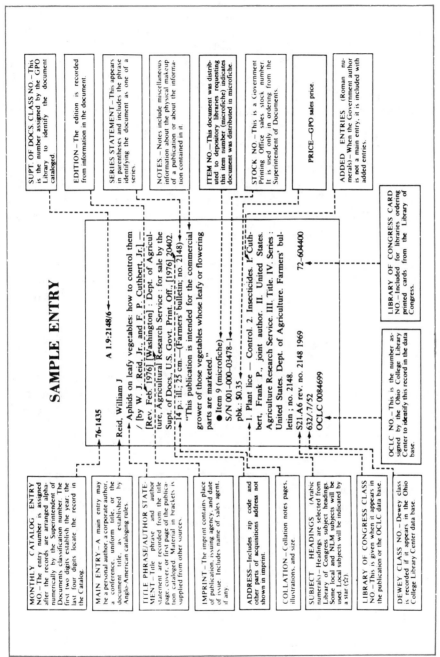

FIGURE 14–5   A Sample Entry in *The Monthly Catalog*

sponsored research reports published since 1964. About 70,000 new summaries are added annually in 22 subject categories ranging from aeronautics to medicine and biology. Copies of the full reports are available from NTIS.

– *The American Statistics Index,* a yearly guide to the statistical publications of the U.S. Government, is divided into two sections: *Index* and *Abstracts* (an index with summaries). The *Index* volume lists material by subject and provides geographic (U.S., state, etc.), economic (income, occupation, etc.), and demographic (sex, marital status, race, etc.) breakdowns. The *Index* volume refers you to a number and entry in the *Abstract* volume.

In addition to these three basic access tools, the government issues *Selected Government Publications,* a monthly list containing roughly 150 titles (with descriptive abstracts). These titles range from highly general (*Questions About the Oceans*) to highly technical (*An Emission-Line Survey of the Milky Way*). While this list may be of interest, it is by no means a refined research tool.

The government also publishes bibliographies on hundreds of subjects, ranging from "Accidents and Accident Prevention" to "Home Gardening of Fruits and Vegetables." Ask your librarian for information about these subject bibliographies.

Electronic Information Services

The information revolution has created the need for modern cataloguing and researching methods. In the United States alone, from thirty-five to forty thousand books are published *each year.* Add to this number foreign publications and thousands of special-interest magazines, journals, newspapers, films, recordings, and reports. Clearly, the problem of finding information quickly — or finding where it's available — becomes acute.

To meet rapid information needs, more and more libraries offer computerized information services. Through electronic card catalogs, you quickly can search your own library's holdings for specific information. And through database retrieval services, you have access to information from countless indexes, journals, books, monographs, dissertations, and reports. In a bibliographic database system, entries are arranged according to different fields. Titles and abstracts of the various works help you find useful documents. Two of these databased services are discussed below.

The On-Line Computer Library Center (OCLC)

You can speed up your research tremendously if a nearby library belongs to the On-Line Computer Library Center. The OCLC database, in Columbus, Ohio, has over eight million records containing the same information found in a conventional card catalog. Using the library's computer terminal, you simply key in author, title, or subject. Within seconds, you get a listing of the publication you seek. If your library doesn't have the publication, your librarian can activate the Interlibrary Loan System (ILS). The system forwards requests to libraries holding the material. Once a lender indicates (via its terminal) that it will supply the material, the system stops forwarding the request, and notifies your librarian that the request has been filled. Your order will arrive at the library by mail within a week or two.

Dialog

Some libraries also belong to Dialog, a comprehensive technical database in Palo Alto, California. Through Dialog, you can access information from many different fields by typing in key terms that enable the computer to scan bibliography lists for titles containing those terms. Say you needed information on the potential *health hazards* of *video display terminals* (i.e., monitors). You would instruct the computer to search Medline, a medical database (one of over 150 Dialog databases in science, technology, medicine, business, etc.), for titles containing the words italicized above (or synonymous words, such as *risk, danger, cathode-ray tubes,* etc.). The system would provide you with full bibliographies and abstracts of the most recent medical articles on your topic.

In addition to bibliographic information on published works, Dialog can provide financial and product information about companies, names and addresses of company officers, and various statistical data. Here are just a few of Dialog's databases:

> *Career Placement Registry*
> *Electronic Yellow Pages* (for Retailers, Services, Manufacturers, etc.)
> *Environmental Bibliography*
> *Foreign Traders Index*
> *International Software Database*
> *Oceanic Abstracts*
> *U.S. Exports*
> *Water Resources Abstracts*

Because of its expense ($150 an hour for many of its databases), Dialog is not yet widely used by college libraries. But many major companies do subscribe to Dialog, in addition to having their own management information departments with full-time database researchers.

BRS

Bibliographic Retrieval Services (BRS) is another popular database providing bibliographies and abstracts from a broad variety of fields in life sciences, physical sciences, business, or social sciences. Here are just a few of the more than fifty BRS databases:

American Chemical Society Journals
Dissertation Abstracts
Harvard Business Review
International Pharmaceutical Abstracts
Military and Federal Specifications and Standards
Pollution Abstracts
Robotics Information Database

Many college libraries now offer BRS service.

A Sample Automated Search.   Assume you are continuing research on the effects of food additives on children. You have done a brief search through the
"manual" indexes, and for a comprehensive view, you decide on an automated
search, using your library's BRS service. You ask a librarian for help, and the
two of you sit at the terminal and begin.

After logging into the BRS system, you instruct the computer to search all
the databases under the group headings *Life Sciences* and *Social Sciences,*
using the key words *food, additive, children,* and *behavior.* The computer responds with a listing of each database under Life Sciences and Social Sciences,
and the number of articles (containing the key words in each). For instance,
the Drug Information/Alcohol Use-Abuse database contains no articles; Medline contains nine; International Pharmaceutical Abstracts contains one article; and so on. You notice that the Psycinfo Database (PSYC) lists sixteen
articles (since 1967). Since you are most interested in the psychological effects
of additives (versus, say, the physiological effects), you instruct the computer
to list the titles of all sixteen articles. Here is the list (partial):

```
        1
TI THE FEINGOLD DIET: A CURRENT REAPPRAISAL
        2
TI FOOD ADDITIVES: THE CONTROVERSY CONTINUES
        [The list continues to number 15.]
        15
TI THE FUNCTIONAL RELATIONSHIP BETWEEN ARTIFICIAL
   FOOD COLORS AND HYPERACTIVITY
        16
TI HYPERKINESIS AND LEARNING DISABILITIES LINKED TO
   THE INGESTION OF ARTIFICIAL FOOD COLORS AND FLAVORS
*END OF DOCUMENTS IN LIST
```

On this list are several titles of interest, but number 15 seems the most relevant. So you now instruct the computer to print the full bibliographic information (including the abstract) on this article. The computer responds:

Accession Number	**AN 03410 64-2.**
Author	**AU ROSE-TERRY**
Institution	**IN NORTHERN ILLINOIS U.**
Title	**TI THE FUNCTIONAL RELATIONSHIP BETWEEN ARTIFICIAL FOOD COLORS AND HYPERACTIVITY**
Source	**SO JOURNAL OF APPLIED BEHAVIOR ANALYSIS 1978 WIN VOL 11(4) 439–446**
Abstract	**AB THE PRESENCE OF A FUNCTIONAL RELATIONSHIP BETWEEN THE INGESTION OF ARTIFICIAL FOOD COLORS AND AN INCREASE IN THE FREQUENCY AND/OR DURATION OF SELECTED BEHAVIORS THAT ARE REPRESENTATIVE OF THE HYPERACTIVE BEHAVIOR SYNDROME WAS EXPERIMENTALLY INVESTIGATED. TWO 8-YR-OLD FEMALES, WHO HAD BEEN ON B. F. FEINGOLD'S (1975–1976) K-P DIET FOR A MINIMUM OF 11 MONTHS, WERE THE SUBJECTS STUDIED. [The abstract continues.]**

*END OF DOCUMENT

After reading the abstract, you decide it would be worthwhile to read the entire article, so you make a note to check your library's holdings or to order a reprint or photocopy through the library's interlibrary loan system. (Certain reprints can be ordered right through the computer terminal.) You turn again to the list of titles to see if any others seem promising.

Once you've searched the PSYC database to your satisfaction, you might decide to look at Medline or some other database on the list that was shown to contain one or more articles with your key words in the title.

Advantages and Limitations of Automated Searches. On-line systems have several distinct advantages over manual search techniques (e.g., flipping pages by hand). The most obvious is that automated searches are fast: ten or fifteen years of an index can be reviewed in minutes. Automated searches are also more detailed: beyond listing titles, they often provide abstracts for a fuller view of the material. And these searches are more timely as well: the index usually comes on line about six weeks before published copies. This feature is especially handy in the sciences and high-technology fields, where developments occur so rapidly. Finally, automated searches are efficient: bibliographic databases allow users to extract just the precise information they need (as shown in the sample search above), without any wasted effort tracking material that turns out to be of no use.

But automated retrieval systems have limitations as well. Most computerized bibliographies include no entries before the mid-1960s. So, for information published earlier, a manual search would be required. Also, in a manual search, you have the whole database (i.e., the bound index or abstracts) right in front of you. As you flip pages and browse, you often *randomly* discover something useful. Such randomness, of course, is lost with an automated search.

For any database search, keep in mind that a manual (random) search is almost always an essential adjunct.

Personal Information Retrieval Services

On a personal computer, you can do some research at home. For instance, for a small fee, you can join *Compuserve Information Service* or *The Source Telecommunications Service,* and have access to energy news, corporate news releases, new product news, business reports, medical news, and a variety of other reference sources.

A service such as *Trade Plus* gives subscribers the latest business-oriented reports on stock portfolios and financial analysis. And the *Dow Jones News Retrieval Service* gives instant access to the stock market and world finance.

Link Services provides a link between business, technical, or scientific communities and information stored in over 600 databases nationwide. Link gives excellent reports on industrial planning and forecast, research on competing companies, analysis of markets, and other data essential to decision makers.

Computerized information retrieval will play a major role in future research. Ask your librarian for information about commercial databases in your field.[3]

Informative Interviews

Although libraries are a good secondary research source, you may also have to consult firsthand, or primary, sources. One excellent primary source can be the personal interview.

---

[3] For a major source of information on databases in various disciplines, see Martha E. Williams et al., eds., *Computer-Readable DataBases: A Directory and Data Sourcebook* (Urbana, Ill.: American Society for Information Science, 1982). In late 1984, *The Database Book,* a directory of the world's databases, will be published by Online, Inc.

Conducting a good interview is not easy. Some of my own students who are reporters for the campus newspaper have complained that their informative interviews are often disappointing and embarrassing. They tell of not knowing what questions to ask and having to suffer through long, terrible silences. Afterward, they become confused about writing up their findings. Much of this frustration results from one or more of the following errors:

1. the wrong choice of interviewee (respondent)
2. inadequate preparation for the interview
3. failure to control the interview
4. failure to record vital information immediately after the interview

Each of these errors derives from the interviewer's inability to identify a clear purpose and to formulate a plan. You must know *exactly* what you are looking for. An informative interview is not a bull session. When planned and directed it can be pleasurable, rewarding, and a solid example of your talent for professional dialogue. The following suggestions should help you conduct a good interview: identify your purpose; choose the most knowledgeable interviewee; prepare well; control the interview; record your responses.

Identification of Purpose

Keep in mind the statement of purpose discussed in earlier chapters. Suppose you are thinking of opening a motorcycle shop. Determine what information you hope to obtain:

> The purpose of my interview with Ms. Jones, city clerk, is to learn how licensing requirements, fees, and zoning restrictions would affect the operation of my proposed motorcycle shop. Also, I need to know how to meet the legal requirements for this business.

When you are sure enough about your purpose that you can write it clearly in one or two sentences, you are ready for your next step.

Choice of Respondent

Choose your respondent carefully. Don't waste time interviewing someone who knows little about the subject. As part of your motorcycle-shop feasibility study, for instance, you might select these respondents: the head of your chamber of commerce, a representative for the brand of machine you expect to sell, a member of your local retail licensing authority board, the loan officer at a local bank. Any of these people might give helpful leads for interviews with other key people.

One way to ensure that you will interview a knowledgeable person is to state the purpose of your interview when you call for an appointment:

> I am studying the possibility of opening a motorcycle shop in Winesburg, and I would appreciate the opportunity to meet with you, at your convenience, because I feel that you could answer some important questions about the future of such a business in our town.

If you communicate your purpose clearly, your intended respondent can immediately tell you whether he or she knows enough to make an interview worthwhile. If you're planning several interviews, begin with the least crucial so you can sharpen your skills for the more important ones. The next step is to do your homework.

Preparation

Preparation is the step that requires imagination. After you have identified your purpose and chosen a respondent, you have to plan questions that will get the information you need. Begin by collecting all available information. To ask useful questions, you have to know your subject! Complete your background investigation *before* the interview, and when your homework is done, write out your questions, *on paper.* Too many beginners set out for an interview armed only with a few blank notecards, a pencil, and a confident feeling that "the questions are all in my head." This mistake leads to embarrassing silences. Don't expect your respondent to keep the ball rolling.

Write out your questions on three-by-five notecards, one question per card, and arrange them in order. You then can summarize a response on the appropriate card.

Make your questions specific and direct. When interviewing a motorcycle wholesale representative, for example, you might phrase a question in one of two ways:

> 1. Does your company have a provision for buying back unsold stock from the retailer?
> 2. What provisions, if any, does your company have for buying back unsold stock from the retailer?

Version 1 can yield a simple yes or no answer, whereas version 2 requires a detailed explanation of the yes answer. Likewise, a question from a problem-solving interview can be worded in one of two ways:

> 1. Do you think this problem can be solved?
> 2. What steps would you recommend to solve this problem?

Version 2 again calls for hard information. The answer you receive will be only as good as the question you ask.

Also, be sure your questions cover all important parts of your subject. Brainstorming can help here (see Chapter 2). Your questions for an interview about financing for your shop, for instance, might cover these subtopics: interest rate, required down payment, other required security or collateral, credit background, other provisions (such as cosigner), application procedure, time required to process the application, prepayment penalties, length of loan period, total amount to be repaid (including yearly interest), amount of monthly payments, emergency loan provisions, and so on.

Avoid questions whose answers can be found in books, newspapers, and town bylaws. Phrase your questions so as not to influence the answers:

1. Don't you think this is a good plan?
2. What do you think are the principal (faults, merits) of this plan and how should they affect my success?

Version 1 suggests that you want your respondent to agree with you, whereas version 2 encourages a candid response.

Organize your questions by following the outlining procedure in Chapter 9. Move logically from one consideration to another. When your questions are specific, comprehensive, and written out in logical order, you can face your next step with confidence.

Maintaining Control

You — not your respondent — are responsible for conducting a professional and productive interview. These suggestions should help:

1. Dress neatly and arrive on time.
2. Begin by thanking your respondent, in advance.
3. Restate the purpose of your interview.
4. Tell your respondent why you feel he or she can be helpful.
5. Discuss your plans for using the information.
6. Ask (preferably before you arrive) if your respondent objects to being quoted or taped. (Although taping is the most accurate means of recording responses, it makes some people uncomfortable.)[4]
7. Avoid small talk.
8. Ask your questions clearly and directly, in the order you prepared them.
9. Be assertive but courteous. Ask pointed questions, but remember that the respondent is doing you a favor.

---

[4] In an interview about your prospective business, it would be inappropriate and unnecessary to tape the responses of the bank representative, sales representative, or licensing official. You might, however, wish to tape an interview with a member of the Small Business Advisory Services who is giving detailed advice about how to proceed with your plan.

10. Let your respondent do most of the talking. Keep your opinions to yourself.

11. Guide the interview. If your respondent begins to wander, politely bring the conversation back on track (unless the additional information is useful).

12. Be a good listener. Don't stare out the window, doodle, or ogle office staff while your respondent is speaking.

13. Be prepared to explore new areas of questioning, if needed. Sometimes a respondent's answers will uncover new directions for the interview.

14. Keep your note taking to an absolute minimum. Record all numbers, statistics, dates, names, and other precise data, but don't transcribe responses word-for-word. This would annoy the respondent. Simply record key words and phrases that later can refresh your memory.

15. If interviewing several people about the same issue, standardize your questions. Ask each respondent the same questions in the same order and in the same phrasing. Avoid random comments that could affect one person's responses.

16. Don't hesitate to ask for clarification or further explanation.

17. When your questions have all been answered, ask for any additional comments.

18. Offer to send a copy of the document in which this information will be used.

19. Finally, thank your respondent, and leave promptly.

Recording Responses

As soon as you leave the interview, find a quiet place to write out a summary of the responses. Don't count on your memory. Get material on paper while the responses are still fresh in your mind.

If you seek opinions on a controversial subject for a newspaper or magazine, ask respondents to review the final draft before you quote them in print.

Here is an interview text for the report titled "Survival Problems of Television Service Businesses" in Chapter 19. Notice how well-planned questions generate high-information responses that lead to more specific questions.

APPENDIX A: FIRST INTERVIEW

(This interview was with Al Jones, the owner of Al's TV Shop, which grosses under $100,000 per year.)

Q. I see you have an excellent location. Is location of a service shop important?

A. Sure. If I ever get back to selling, it will be very important. Doing just service minimizes the importance since the customer will search you out if he really wants you. I suppose the same holds true if I start selling.

Q. Well, you seem to be selling car radios and stereos. Isn't that "selling"?

A. Yes. But by selling I mean TVs. Floor plans, etc.

Q. Why don't you sell now? What's a floor plan?

A. The floor plan is why I won't sell now. I used to sell with a floor plan. A floor plan means a TV distributor will sell me, say, $10,000 worth of TVs for only 10 percent down payment. I pay the balance as I sell the sets. The problem is that the temptation to use the distributor's money is very great. These large amounts of money in the checking account give you a false sense of security.

Q. In other words, you've had a bad experience?

A. Not only me. Most TV businesses fail because they start using someone else's money. Where can a small business come up with six, eight, and even ten thousand dollars after all the stock has been sold?

Q. So you just sell car radios and stereos and parts as needed for repairs?

A. That's right. I pay for them as I use them and stay out of trouble. I realize it's better to have thirty-day charge accounts and use the distributor's money to capitalize my business, but this can also cause trouble.

Q. OK. What about advertising?

A. I use the yellow pages. That's all I can afford. Sometimes the local radio stations offer attractive deals which cost me half the usual rate. I take advantage of these. Radio is good for the stereo radio sales.

Q. Wouldn't it be a good idea for a group of you technicians to get together and do some group advertising?

A. TV technicians don't want to get together. They just don't have the time.

Q. Tell me something about your manpower situation.

A. OK. Here's something. My TV man came up to me a few days ago and told me he was giving me thirty days' notice. He wants to go into business for himself. It's always like this. You hire and train a technician for years, and when he gets the itch, off he goes. Better give him a copy of your report; maybe he can avoid the mistakes I made over the years.

Q. Can you think of any mistakes you made along manpower lines?

A. I don't think I'll ever hire a beginner anymore. They just aren't productive enough. From now on they must have at least formal training and the technician's license.

Q. What do you mean by production?

A. I have to get so much production for every dollar I pay a technician. Unfortunately it has taken me many years to come to a good interpretation of what a reasonable return is. The minimum return I must have is $3.00 for every wage dollar.

Q. Your sign on the wall says "Minimum Labor Charge, $21.00." Does that mean you pay your technicians at least $7.00 per hour?

A. That's close enough. Sometimes technicians take longer than an hour to fix a unit. But it averages out to that.

Q. So your customers realize that they must be charged this amount so you can cover your wages?

A. [Laughter] Customers don't realize anything. They complain no matter what we charge and no matter how long we take. I sincerely believe that customers start complaining even before their sets break.

Q. Have you ever tried discussing these notions with your customers?

A. Sometimes. I think they are beginning to learn that good shops and good technicians are getting scarcer. It pays for them to be cordial as it does for us.

Q. Can you think of anything else you can say about customers?

A. Yes. As you know, most homes have more than one TV set. This has created an unusual problem for us. With money being tight it seems as if the customer leaves one set at home and one in the shop. I have a time getting people to pick up their merchandise. This puts a money crunch on me since I have invested labor and money in the repair.

Q. Have you thought of any solution to this problem?

A. I have tried taking substantial deposits and threatened to sell the set for the price of the repairs (which my attorney says I can't do). As you can see, the place is filled with unclaimed merchandise. This is getting a little out of hand, to say the least.

Q. How about the money situation? How much did you gross last year?

A. A little over $60,000. This wasn't too good at all. I have two technicians full time with occasional part-time help. When I'm not busy doing desk work, I help with pickups, deliveries, and antenna work. At $21.00 per hour for forty hours per man that comes out just right to cover their labor, but there are many other expenses to be covered. Also my labor wasn't covered. The figure should have been closer to $90,000.

Q. How much of this was gross margin?

A. Gross margin? Let's see. That's what's left over after the cost of the parts used in the repairs or sales. Oh, about $40,000.

Q. And how much of this turned out to be net income?

A. Very little! I only paid a couple hundred dollars in taxes this year.

Q. One last question. Is it worth it? Do you think you are getting a return on your investment?

A. It beats working for someone else. I believe I have made all the mistakes there are to be made in this business. I don't intend to make them again. Don't forget to show me the completed report.

Questionnaires

An interview has some advantages over a questionnaire: face-to-face, you can tell whether your respondent understands your questions, and your questions are more likely to be answered. Also, you may get unexpected information However, an interview sometimes seems like an interrogation. Respondents may feel threatened by certain questions. So a questionnaire is often a better alternative.

A questionnaire has several advantages. It saves time and is an inexpensive way of surveying a large cross-section. Respondents can answer privately and anonymously — thus more candidly. And they have plenty of time to think about their answers.

These advantages, however, can be offset by the problem of getting people to take time to fill out and return a questionnaire. Whether you mail or otherwise distribute a questionnaire, expect less than a 30 percent response, perhaps far less — and give plenty of time for responses to return. You can increase your chances for response by following these suggestions:

1. Introduce your questionnaire, so respondents will appreciate the significance and purpose of their answers. Also, thank the respondent.

> Your answers to the following questions about your views on proposed state handgun legislation will be appreciated. Your state representative will tabulate all responses so she may continue to speak accurately for *your* views in legislative session. Thank you.

Some writers include a cover letter with a questionnaire (as shown on page 302).

2. Keep it short. Try to limit both questions and response space to two sides of a page. Also, phrase questions concisely.

3. Place easiest or more interesting questions first, to attract attention.

4. Make each question precise and unambiguous. *This is crucial*. If readers can interpret a question in more than one way, their responses are useless — or worse, misleading. Consider this ambiguous question: "Do you favor foreign aid?" *Foreign aid* can mean military, economic, or humanitarian aid, or all three. Some people might support one form or another, but not all three. Rephrase the question to allow for these differences:

> Do you favor:
> a. Our current foreign aid program of both military and humanitarian assistance to other countries?
> b. A greater emphasis on military assistance in our foreign aid program?
> c. A greater emphasis on economic and humanitarian assistance in foreign aid programs?
> d. No foreign aid program at all?

5. Avoid influencing your reader's responses with leading questions:

> Do you think foreign aid is a waste of taxpayers' money?
> Do you think foreign aid is a noble gesture by a wealthy nation toward the less fortunate?

Notice that each of these questions is phrased to elicit a definite response.

6. When possible, phrase questions so you can easily classify and tabulate responses: yes-or-no; multiple-choice; true-false; fill-in-the-blank.

7. Include an "Additional Comments" section at the end.

8. Enclose a stamped, self-addressed envelope along with the questionnaire.

9. If possible, do the questioning by telephone to save time and ensure a response.

The following questionnaire, sent to presidents of 612 businesses, is prefaced by a cover letter explaining the questionnaire's purpose. The clearer the rationale for a questionnaire, the better the response will be.

Note that the questions themselves are phrased to elicit specific responses which then can be easily classified and tabulated.

SAMPLE COVER LETTER

As part of my course work in Professional Communication, I am preparing an analytical report on Southeastern Massachusetts University's professional communication program. The purpose of this survey is to determine how SMU's communication program can better serve students and the community.

I plan to assess program services and area needs by:

1. determining what local businesses and industries see as their communication needs;
2. studying the feasibility of offering on-campus and in-house seminars in professional communication;
3. studying the feasibility of offering consulting services in writing and editing;
4. determining whether a broader variety of professional communication courses should be offered at SMU.

Would you please take a few minutes to respond to this survey? *Your response is important* since, without it, I will not have the necessary data to write an effective report. The results of the survey, along with conclusions and recommendations, will be published in the fall issue of *The Business and Industry Newsletter.*

One last favor: Kindly mail the survey to:

> Lynne Taylor
> Box 9624 House 10
> SMU Dorms
> No. Dartmouth, MA 02747

I regret I cannot include an addressed, stamped envelope for your convenience, but my operating budget for this project has been depleted on printing and mailing the survey.

Cordially,

Lynne Taylor

Lynne Taylor

SAMPLE QUESTIONNAIRE

PROFESSIONAL COMMUNICATION QUESTIONNAIRE

1. Describe the type of business you run. For example, manufacturing, high tech, banking, etc.

2. How many people do you employ? Please check.

   5–25 _____   100–150 _____
   25–50 _____   150–300 _____
   50–100 _____   300–450 _____

3. How do you usually communicate information to employees?

   memo _____
   in person _____
   through supervisors _____

4. Which of the following types of writing are done in your company? (Please rank in order of frequency: e.g., never, rarely, sometimes, often, most often.)

   interoffice memos _____   catalogues _____
   letters _____   procedures _____
   reports _____   handbooks _____
   operating manuals _____   reference manuals _____
   house organs _____
   advertisements _____
                    other (please specify) _____

5. Who does most of the writing in your organization? (Please give title.)
   _____

6. How would you characterize the effectiveness of writing in your organization?

   Good _____   Fair _____   Poor _____

7. Does your company have guidelines for written communication?

   Yes _____   No _____   If so, could you briefly describe them.

8. What percentage of your operating budget is spent on written communication?

   1–4% _____   5–10% _____   10–25% _____   25–40% _____
   more than 40% _____

9. Do you hire outside writers when you are marketing new products? (for example, sales brochures, promotional material, and the like)

   Yes _____   No _____

10. Do you have an in-house program to teach employees to communicate more effectively? Yes_____ No _____

11. If you do have a program, list the types of writing skills you emphasize.

12. Which of the following topics would you consider most important in a writing program? (with 1 as most important through 9, least important)

    – Organizing information _____
    – Precise phrasing and word choice _____
    – Summarizing information _____
    – Formatting various reports _____
    – Editing for clarity, conciseness, and fluency _____
    – Adapting various messages for various audiences _____
    – Grammar and punctuation _____
    – Sales and promotional writing _____
    – Other (Please specify) _____

13. Rate the following skills in order of importance (with 1 as most important, through 6).

    – Reading _____
    – Writing _____
    – Listening _____
    – Speaking to groups _____
    – Speaking face-to-face _____
    – Dictating _____

14. Do you provide tuition reimbursement for employees who wish to improve their communication skills? Yes _____ No _____

15. If not, would you consider tuition reimbursement for your employees? Yes _____ No _____

16. Have you ever used the services of a communication consultant? Yes _____ No _____

17. If not, would you be interested in such a service? Yes _____ No _____

18. Should SMU offer Saturday seminars in Business Communications? Yes _____ No _____

19. Would you be interested in participating in a contract learning program whereby SMU students taking a business communication course would work with you? Yes _____ No _____

20. At present, SMU offers only one semester of Professional Communication. If more Professional Communication courses were offered, which would you prefer? (Rate in order of preference, with 1 as the most preferred)

    – Report writing _____
    – Proposal writing _____
    – Effective public speaking _____
    – Procedure writing _____

    – Computer documentation _____
    – Sales & promotional writing _____
    – Others (Please specify) _____

Additional Comments:
    How could SMU better serve your specific communication needs? Please
    be specific.

Caution: Interpret statistical results carefully. With a political questionnaire, for example, more people with strong views will respond than people with moderate views. So your data may not truly reflect a cross-section. Supplement questionnaire responses with other sources whenever possible.

Include the full text of the interview or questionnaire (with questions, answers, and tabulations) in an appendix (Chapter 12). You can then refer to these data in your report discussion as necessary.

Letters of Inquiry

Letters of inquiry are handy for obtaining specific information. These are discussed in Chapter 16.

Organizational Records and Publications

Company records are a good primary data source. Most organizations also publish pamphlets, brochures, annual reports, or prospectuses for consumers, employees, prospective investors, voters, and so on. Be alert for bias, however, in company literature. Your local nuclear power company, for example, may publish pamphlets explaining the nuclear systems and listing their advantages. If you were evaluating the safety measures at the plant, you would want information representing the complete picture. Along with the power company's literature, you would want publications of federal, state, and local environmental groups, to discover *their* assessment of these safety measures. A pamphlet can be a good beginning for your research.

Personal Observation

If possible, amplify and verify your research findings with a firsthand look. Conclude your motorcycle-shop feasibility study, for instance, by visiting possible shop sites. Measure consumer traffic by observing each site at identical hours on typical business days.

Observation should be your final step, because you now know what to look for. Without understanding the issues, you could be wasting time. A visit to the nuclear power plant, for example, will be useful if you understand how meltdown can occur or how safety systems work. Have a plan. Know what to look for, and jot down observations immediately. You might even take photos or make drawings, but don't rely on your memory.

Informed observations can pinpoint serious problems. Here is an excerpt from a report investigating poor management-employee relations at an electronics firm. This researcher's observations are crucial in defining the problem:

> The survey of employees did not reveal they were aware of any major barriers to communication. The 75 percent of employees who have positive work attitudes said they felt free to talk to their managers, but the managers questioned said only 50 percent of employees felt free to talk to managers.
>
> The discrepancy appears to arise from the problems discussed in the section on morale: 50 percent of the employees don't know how management feels about them, and managers feel they don't hear everything that employees have to say. From questioning both sides, I conclude that both employees and management are assuming messages have gotten across, instead of checking to make sure they have. Managers are assuming that employees know they want to hear complaints and suggestions, so they aren't asking for them. Employees, on the other hand, mention that they would like to know how they are doing, but they never ask managers for their evaluations.
>
> The problem involves perception and distortion from misinterpretation. Because managers don't ask for complaints, employees are afraid to make them, and because employees never ask for an evaluation, they never get one. Both sides have improper perceptions of what the other wants, and because of ineffective communications they don't know that their perceptions are wrong.

Effective research requires informed observation.

TAKING NOTES

Purpose

Your finished report will only be as good as the notes you take. Notecards are best because they are easy to organize. Put one idea on each card and give each card a code number in the upper right corner. You can pretty well figure out the organization of your report before you actually write it by arranging the cards in logical order on a large table.

Instructions for keying cards to the outline are discussed in the next chapter, "Writing and Documenting the Research Report."

Procedure

1. Begin by making bibliography cards, listing the books, articles, and other works you plan to consult. Using a separate notecard for each source, write the complete bibliographic information (Figure 14–6). This entry is identical to the citation you will include in the final draft of your report.

2. Skim the entire book, chapter, article, or pamphlet.

3. Go back and decide what to record. Use one card for each item.

4. Decide how to record the item: as a quotation or as a paraphrase. In quoting, copy the statement word for word (Figure 14–7). Place quotation marks around all directly quoted material, even a phrase or a word used in a special way. Otherwise, you could forget to give proper credit to the author, and thereby face a charge of plagiarism (borrowing someone else's words without giving credit, intentionally *or* unintentionally). If you quote only sections of a sentence or paragraph, use an ellipsis — three dots (...) indicating that words have been left out of a sentence (Appendix A). If you leave out the last part of the sentence, the first part of the next sentence, or one or more whole sentences or paragraphs, use four dots (....):

> If you quote only sections ... use an ellipsis. ... If you leave. ...

If you insert your own comments within the quote, place brackets around them to distinguish your words from the author's (as discussed in Appendix A):

> "This job [aircraft ground controller] requires exhaustive attention.

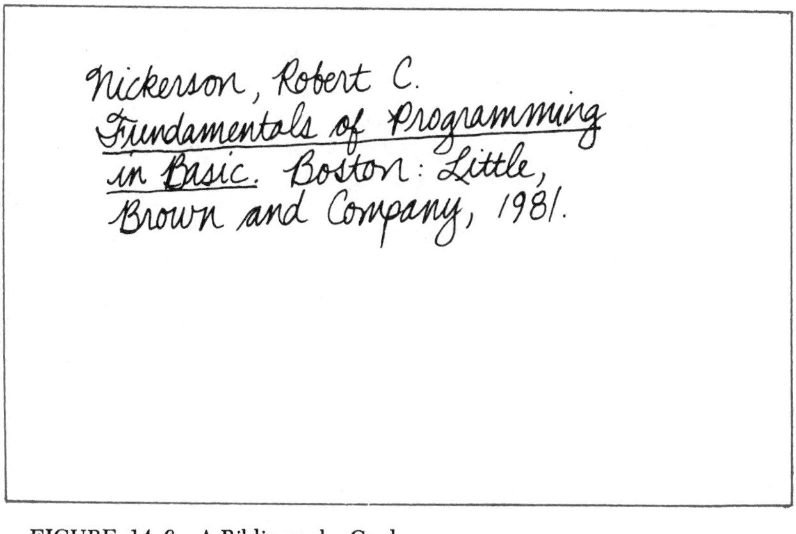

Nickerson, Robert C.
Fundamentals of Programming
in Basic. Boston: Little,
Brown and Company, 1981.

FIGURE 14–6   A Bibliography Card

Nickerson, Robert. <u>Fundamentals</u> II-A-1

p. 34

"The first step in the programming process is understanding the problem to be solved. Understanding the problem involves determining the requirement of the problem and how these requirements can be met."

FIGURE 14–7   Sample Notecard for a Quotation

Avoid using direct quotes too often because your report will then be simply a collection of borrowed words. Instead, synthesize and condense information by paraphrasing and summarizing.

Figure 14–8 shows a paraphrased entry. The researcher condenses the following original in his own words.

> Finally, the programming process is completed by bringing together all the material that describes the program. This is called *documenting* the program, and the result of this activity is the program's *documentation.* Included in the documentation is the program listing and a description of the input and output data. Documentation enables other programmers to understand how the program functions. Often it is necessary to return to the program after a period of time and to make corrections or changes. With adequate documentation, it is much easier to understand a program's operation.[5]

Most of your notes will be paraphrased, but some notecards may have both paraphrases and direct quotations. Paraphrased material needs no quotation marks; however, it *must be documented,* to indicate your debt to a particular source.

---

[5] This passage and those in Figures 14–7 and 14–8 are adapted from *Fundamentals of Programming in BASIC,* by Robert C. Nickerson.

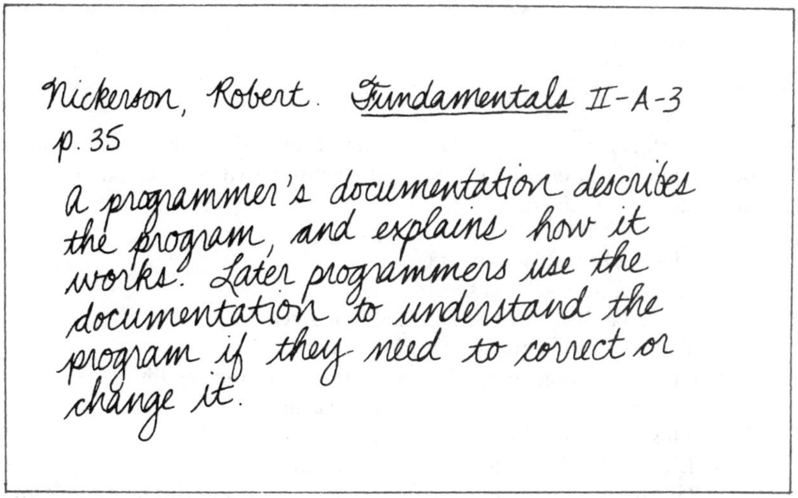

Nickerson, Robert. Fundamentals II-A-3
p. 35

a programmer's documentation describes the program, and explains how it works. Later programmers use the documentation to understand the program if they need to correct or change it.

FIGURE 14–8   Sample Notecard for a Paraphrase

5. Be selective about what and how much you write in your notes. The summarizing techniques in Chapter 6 should help. Follow these guidelines: (a) Preserve the original message when quoting. Don't distort it by omitting vital information. Your aim is not to prove yourself correct, but rather to uncover the facts. If the data disprove your view, do not ignore them. (b) When you get an idea of your own, write it on a notecard immediately. Keep a few cards handy to record observations, questions, or ideas as they pop into your head.

CHAPTER SUMMARY

Research is done to uncover facts, to separate fact from opinion and to arrive at valid and convincing conclusions.

Many sources exist for research data. Primary sources include memory, personal observation, questionnaires, interviews, letters of inquiry, and business or technological records; secondary sources include books, articles, reports, brochures, films, and other publications. The library is your most valuable secondary source; so learn early how to use it. Ask your reference librarian for help with the card catalog, reference works, indexes, government publications, automated searches — or any other problem. Take effective notes.

EXERCISES

1. Each sentence below states a fact or an opinion. Rewrite all statements of opinion as statements of fact. (Remember that a fact can be measured, duplicated, or defined.)

    a. This set of instructions is useless.
    b. My vacation was too short.
    c. The salary for this position is $18,000 yearly.
    d. This desk calculator is priced reasonably.
    e. Our reforesting team planted 12,000 red pine seedlings last week.
    f. Our new delivery trucks get great gas mileage.
    g. This course has been very helpful.
    h. Electric heating is less efficient than oil heating.
    i. This room is much too small for our drafting and duplicating equipment.
    j. Our new company car is the most expensive of all compact sedans.

2. a. Find three articles about the *same* newsworthy event. Read each article, separating fact from opinion. Underline all facts and circle all opinions. Combine the articles and write a summary based exclusively on factual information. (Your instructor may provide the articles.)
    b. Locate a technical article from your field, and underline the facts and circle the opinions on a photocopy that you submit to your instructor.
3. Make a list of periodicals in your library that relate to your field, and arrange these titles in descending order of technicality or in descending order of relevance to *your* work.
4. Assume you belong to a statewide association of TV technicians. Summarize the interview on pages 298–300 in three to four hundred words for publication in the association's monthly newsletter.
5. Choose one situation from your life where research could help you solve a problem or answer an important question. In a memo to your instructor, describe the problem or question and list five sources you would consult to obtain answers.
6. Revise the following questions to make them adequate for inclusion in a questionnaire:

    a. Would a female president do the job as well as a male?
    b. Don't you think that euthanasia is a crime?
    c. Do you oppose increased government spending?
    d. Do you feel that welfare recipients are too lazy to support themselves?
    e. Are teachers responsible for the literacy decline among students?
    f. Aren't humanities studies a waste of time?
    g. Do you prefer Dipsi Cola to other leading brands?

7. *In class:* As a group, choose a familiar campus or community issue and compose a set of interview questions that could be given to people involved in the issue. Also, compose a questionnaire about the same issue.

8. Arrange an interview with a successful person in your field. Make a list of general areas for questioning: job opportunities, chances for promotion, salary range, requirements, outlook for the next decade, working conditions, job satisfaction, and so on. Compose interview questions; conduct the interview; and summarize your findings in a memo to your instructor.

9. Using the library and other sources, locate and record the required information for *ten* of the following items, and fully document each source. (See pages 316–324 for documentation practices.)

    a. the population of your home town
    b. *The New York Times'* headline on the day you were born
    c. the world pole-vaulting record
    d. the operating principle of a diesel engine
    e. the name of the first woman elected to the United States Congress
    f. the exact value of today's American dollar in French francs
    g. the comparative interest rates on an auto loan charged by three banks
    h. the major cause of small-business failures
    i. the top-rated *compact* car in the *world* today
    j. the origin of the word *capitalism*
    k. the distance from Boston to Bombay
    l. the definition of *pheochromocytoma*
    m. an effective plant for keeping insects out of a home garden
    n. the originator of the statement: "There's a sucker born every minute"
    o. the names and addresses of five major United States paper companies
    p. Australian job opportunities in your field
    q. the titles and specific locations of five recent articles on robotics
    r. the half-life of plutonium
    s. a brief description of your intended occupation
    t. the side effects of certain tranquilizers (e.g., Valium or Librium)
    u. the difference between an analog and a digital computer

10. With help from the reference librarian, list the major indexes, reference books, and journals in your specialty. Locate a recent article (or reference-book entry) on a specific topic (e.g., artificial intelligence in medical diagnosis). Write an informative abstract of the article (or photocopy the reference-book entry), and submit it with full documentation (Ch. 15).

11. Consult the appropriate librarian and identify two data-bases you would search for information on the topic in exercise 10.

12. Using *The Monthly Catalog,* find and photocopy a citation for a government publication on a specific topic.

15

Writing and Documenting the Research Report

RESEARCH REPORT DEFINED

A research report records and discusses your findings. Its purpose is to provide the facts that lead to an informed opinion. The information might well provide a basis for decision making, but your job in writing a research report is to *inform*, not to recommend, unless requested. Recommendations are part of the proposal or the analytical report (Chapters 18 and 19). As we will see, research data often make up much of a proposal or an analytical report. Our concern in this chapter, however, is only with the research report.

Chapter 14 discussed information sources and methods for recording data. Here we review and expand that discussion, and offer a plan for completing the research report.

PLANNING THE REPORT

Choice of Topic

Your instructor probably will ask you to choose a topic. Instead of something that merely fulfills an assignment, choose a technical or business topic that interests you, one you would like to learn about. Or ask professional acquaintances if they need research done on a specific topic. If your research report is well written and documented, you can include it as part of your employment dossier, and refer to it in your résumé. The world is full of problems to solve and questions to answer. Make your research a worthwhile contribution.

Focus of Topic

Narrow your topic so you can discuss it fully for your intended audience. To achieve a clear direction, always phrase your topic as a question. Assume, for instance, that you've decided to research certain needs of local businesses. You have to focus on a specific need you can research thoroughly. Let's say you are interested both in the communication needs of local businesses and in your school's communication program. After some hard thinking, you decide on this specific question: How can our university's communication program better serve students and local businesses? Your audience will be faculty, administrators, and executives.

Working Bibliography

When you have selected and narrowed your topic, be sure you can find adequate resources in your library, community, or electronic database service. Make your bibliography early, to avoid choosing a topic only to learn later that not enough sources are available. You might in fact choose your topic after a preliminary search for primary and secondary sources.

Do a quick search of the card catalog, periodical indexes, reference books, and government publications. Record the bibliographic information, as shown in Chapter 14. Many books contain bibliographies that lead you to additional sources. With a current topic, such as 16-bit microcomputers, you might expect to find most information in recent articles. Your bibliography, of course, will grow as you read. Assess possible interview sources and include probable interviews in your working bibliography.

Statement of Purpose

Remember you are not merely collecting views; you are screening and evaluating facts for a definite purpose. Identify your goal and your plan for achieving it. Then, your research will have a direction.

> My purpose is to learn how our communication program can better serve students and area businesses. I will survey local needs, and evaluate our offerings in relation to these needs. To enhance my study, I will include national data on other professional communication programs.

Background List

Make a list of facts you already know about your topic. Brainstorming (see pages 19–21) should produce some good ideas for investigation.

Working Outline

Now that you have identified your definite purpose, you need a road map. By partitioning your topic into subtopics, you emerge with a working outline:

    I. Profile of Communication in Local Businesses
    II. Internal Communication in those Businesses
    III. Frequency and Types of Writing
    IV. Perception of Writing Effectiveness
    V. Improving Service for Business Community and Students

Each of these topics can be further divided (Chapter 9). Your working outline is only tentative; you will change it and add subtopics as your research progresses. Depending on your findings, the outline may shift radically before your final draft. But with a working outline on paper, you can begin gathering information.

GATHERING INFORMATION

Beginning with a General View

Move from general to specific data. Encyclopedias are a good place to begin, because they contain general information. Or, read a book or pamphlet that offers a comprehensive view of your subject before moving to specialized articles in periodicals. Technical dictionaries and newspaper or magazine articles also provide background. Your reference librarian can help you find sources.

Skimming

Before reading any book, look over the table of contents and the index. Check the introduction for an overview or thesis. In long articles, look for headings that may help you locate specific information. Short articles and pamphlets should usually be read fully.

To skim effectively, you have to concentrate. If you feel yourself drifting, take a break.

Selective Note Taking

Resist the temptation to copy or paraphrase every word. As you read, try to develop your own view instead of leaning heavily on the ideas of others. Your finished report should be a combined product of your insights and collected facts you have woven together.

Stay on track by following your outline, but remember that it can be modified. Don't let your outline prevent you from discovery.

Designing Interviews, Questionnaires, or Letters of Inquiry

Your investigation may call for primary research. If interviews, questionnaires, and letters are appropriate, prepare for them well ahead of time.

Direct Observation

Once you have read, questioned, and pondered, take a first-hand look — when practical. If possible, take photos and draw maps, sketches, etc. Save this step for the end of your research, so you know what to look for.

When your information is collected you should have a stack of notecards keyed to your working outline (Chapter 9). Now write up your findings.

WRITING THE FIRST DRAFT

Outline Revision

Throughout your investigation, you have relied on your working outline. By now, you probably have uncovered new material that was not part of your original outline. Before writing your first draft, examine your outline, and make any necessary changes.

Section-by-Section Development

Concentrate on only one report section at a time. Students often find this the most intimidating part of research: pulling together a large body of information in a report. Don't frantically throw everything on the page, simply to get done. Remember that your final report will be the *only* concrete evidence of your labor.

Begin by classifying your notecards in groups according to the section of your outline to which each card is keyed. Next, find a flat surface. Take the notecards for your introduction, and arrange them in order. Now, lay them out in rows, as you would lay out playing cards. Thus armed with your outline, your statement of purpose, your expertise, and your ordered notecards, you are ready to write your first section.

As you move from subsection to subsection, provide commentary and transitions, and document each source. When you complete your introductory section, proceed to the others, weaving ideas together by following the outline (and modifying as needed).

DOCUMENTING THE REPORT

Most research draws on the information and ideas of others. Credit each source of direct quotations, paraphrases, and visuals. Proper documentation satisfies the professional requirements for ethics, efficiency, and authority.

Documentation is a matter of *ethics* in that the originator of an idea deserves to be acknowledged whenever that idea is mentioned. All published material is protected by copyright laws. Failure to acknowledge your source could make you liable to a charge of plagiarism, even if your omission was unintentional.

Documentation is also a matter of *efficiency*. It provides a chain through which our world's printed knowledge can be located. If, for example, you quote a journal article, your reference will enable an interested reader to locate that source easily.

Finally, documentation is a matter of *authority*. In making any claim ("A Mercedes-Benz is a better car than a Ford Granada") you are liable to be challenged with "Says who?" Data on road tests, frequency of repairs, resale value, workmanship, and owner comments can help validate your claim by showing that your opinion is based on *fact*. Your credibility increases in relation to the expert references supporting your claims. For a controversial subject, you may need to cite several authorities (instead of forcing a simplistic conclusion on your material), as in this example:

Opinion is mixed as to whether a marketable quantity of oil exists under Georges Bank. Cape Cod geologist John Blocke feels that extensive reserves are improbable ("Geologist Dampens Hopes" 3). Oil Geologist Donald Marshall is uncertain about the existence of any oil in quantity under Georges Bank ("Offshore Oil Drilling" 2). But the U.S. Interior Department reports that the Atlantic continental shelf may contain 5.5 billion barrels of oil (Kemprecos 8).

Choosing a System of Citation

Documentation practices vary widely. Many disciplines, institutions, and industrial organizations publish their own documentation manuals. Many are listed in the card catalog under "Technical Writing." In writing your paper on food additives, for example, you might have used the American Psychological Association's *Publications Manual* or the American Medical Association's *Advice to Authors,* in response to your readers' requests.

Other specific style manuals include these:

> *Style Guide for Chemists*
> *Geographical Research and Writing*
> *Style Manual for Engineering Authors and Editors*
> *IBM Style Manual*
> *NASA Publications Manual*
> *Guide for Preparation of Air Force Publications*

When a specific format is not stipulated, you can consult one of three general style manuals. *The MLA Handbook for Writers of Research Papers,* by the Modern Language Association, outlines documentation in the humanities. *A Manual of Style,* by the University of Chicago Press, covers documentation in the humanities and related fields, as well as natural sciences. *The Style Manual,* by the U.S. Government Printing Office, covers documentation for government writing.

For a more detailed list and discussion, see John A. Walter, "Style Manuals," *Handbook of Technical Writing Practices,* 2 vols., ed. Jordan, Kleinman, and Shimberg (New York: John Wiley and Sons, Inc., 1971) 2:1267–1273.

This chapter discusses three different systems of documentation: in-text citations, author-year designation, and numerical designation.

In-Text Citations

The Modern Language Association (MLA) has replaced footnotes and endnotes with in-text citations (also called "parenthetical references"). So, instead of placing numbered footnotes (e.g., [1]) in the text and listing source references at the bottom of the page or as endnotes, list your abbreviated references within the text, then provide full documentation on a "Works Cited" page at the end of the report.[1] (The "Works Cited" page replaces the "Bibliography" page.) Let's look at a few examples.

An in-text or parenthetical citation usually includes the author's last name and the page cited, as in: (Barrett 69). Here's what the citation would look like in the report:

> Therefore, the advantages of automation far outweigh the disadvantages (Barrett 69).

Readers needing the full citation for Barrett can turn to "Works Cited," listed alphabetically by author, to get complete publishing information.

Keep parenthetical references brief. For instance, if you mention the author's name in your discussion, don't repeat it in the citation; simply provide the page reference.

For example,

> Barrett claims that the advantages of automation far outweigh its disadvantages (69).

If the work is by a corporate author or if the work is unsigned (i.e., author unknown), use a shortened version of the title or corporate name in your citation, as in, ("Information Systems" 18). Be sure, though, that shortened titles correspond with the entries in "Works Cited" (e.g., "Information Systems for Tomorrow's Office," *Fortune* 19 (Oct. 1982): 18).

Unless your readers request otherwise, use the following format for references on your "Works Cited" page.

Works Cited Form—Books

A reference for a book should contain the following information (as applicable): author, title, editor or translator, edition, volume number, facts about publication (city, publisher, date). Here are some examples:

SINGLE AUTHOR

    Katzan, Harry. Office Automation: A Manager's Guide. New
        York: American Management Association, 1982.

TWO AUTHORS

    Adam, Everett, E. Jr., and Ronald J. Ebert. Production and
        Operations Management. Englewood Cliffs, N.J.: Pren-
        tice-Hall, 1978.

---

[1] Footnotes are now used only when you wish to comment or expand on material in the text (as done here).

THREE OR MORE AUTHORS

Levin, Richard I. et al. <u>Quantitative Approaches to Man-
agement</u>, 5th ed. New York: McGraw-Hill, 1982.

AUTHOR(S) NOT NAMED

<u>Computer Documentation</u>. Boston: Meredith Associates, 1983.

TWO BOOKS WITH THE SAME AUTHOR

Lamont, John W. <u>Biophysics</u>. Boston: Little, Brown, 1984.
---. <u>Diagnostic Techniques</u>. Boston: Little, Brown, 1983.

[When citing more than one work by an author, don't repeat the author's name; simply type three hyphens followed by a period.]

AN EDITOR

Meadows, A. J., ed. <u>The Random House Dictionary of New
Information Technology</u>. New York: Vintage, 1983.

A QUOTE OF A QUOTE

Kline, Thomas. <u>Automated Office Systems</u>. New York: Random House, 1983, p. 97, as cited in John Swenson,
<u>How to Increase White-Collar Productivity</u>. Boston:
Little, Brown, 1983.

Works Cited Form—Periodicals

A reference for an article should contain the following information (as applicable): author, article title, periodical title, volume and/or number, date, and page numbers for the entire article, not for pages cited. Here are some examples:

A MAGAZINE ARTICLE

Main, Jeremy. "The Executive Yearn to Learn." <u>For-
tune</u> 3 May 1982 : 234-239.

[No punctuation separates the magazine title and the date. Note also that the abbreviation *p.* or *pp.* isn't used to designate pages.]

If no author is given, list all other information:

```
"Information Systems for Tomorrow's Office." Fortune
      18 October 1982 : 18-56.
```

A JOURNAL ARTICLE

```
Thackman, John.  "Computer-Assisted Research." The
      American Librarian 51 (1981): 3-9.
```

The "51" represents volume number. No punctuation separates journal title and volume number.

A NEWSPAPER ARTICLE

```
Barrett, Marianne.  "The Decline in White-Collar Pro-
      ductivity." Boston News 15 Jan. 1984, Evening
      ed., sec. 2: 3.
```

When a daily newspaper has more than one edition, cite the edition after the date.

If no author is given, list all other information.

Citing Miscellaneous Items

AN ENCYCLOPEDIA, DICTIONARY, OR OTHER ALPHABETIC REFERENCE WORK

```
"Communication." The Business Reference Book. 1977 ed.
```

If the entry is signed, begin with the author's name.

AN INTERVIEW

```
Jones, Al. President, Al's TV Shop.  Personal Interview.
      Swansea, MA, 2 April 1982.
```

A QUESTIONNAIRE

```
Taylor, Lynne.  Questionnaire sent to 612 Southeastern
      Massachusetts Business Presidents.  14 February 1983.
```

A LECTURE

```
Thompson, Edwin.  "Bureaucratic Follies and Blunders."
      Lecture presented at Southeastern Massachusetts Uni-
      versity.  Dartmouth, 1983.
```

DATABASE SOURCE

Keyes, Langley Carlton. "Profits in Prose." Harvard
    Business Review 39 (1961): 105-112; Latham, New York:
    Bibliographic Retrieval Service, 1984, Accession No.
    611080.

COMPUTER SOFTWARE

Levy, Michael C., William J. Froming, and Marcia Belcher.
    Statmaster: Exploring and Computing Statistics. Com-
    puter Software. Boston: Little, Brown, 1982. IBM
    PC, 48KB, disk.

When documenting software, name the appropriate computer, the kilobytes
(48KB), and the software format (disk).

CORPORATE AUTHOR

The Presidential Task Force on Acid Rain. Acid Rain and
    Corporate Profits. Washington, D.C.: GPO, 1984.

Cite other items (corporate or government pamphlets, reports, dissertations,
and other unpublished work) as follows:

> Author (if known), Title (in quotes), Sponsoring Organization, date,
> page number(s).

In the list of works cited, arrange the entries alphabetically by the author's
last name (as the first item of the entry). When the author's name is unknown,
list the title alphabetically according to its first word (excluding *a, an,* and
the).

Here is the list of works cited by Lynne in her report on pages 325-330.[2]

WORKS CITED

Bennett, James C. "The Communication Needs of Business
    Executives." Readings in Business Communication. Ed.
    Robert D. Gieselman. Urbana, Illinois: Stipes, 1982.

DiGaetani, John L. "The Business of Listening." In Com-
    munication for Management and Business by Norman Sig-
    band, 3rd ed., Glenview, Illinois: Scott, Foresman,
    1982.

Halpern, Jeanne W. "What Should We Be Teaching Students in
    Business Writing?" The Journal of Business Communica-
    tion 18 (1981): 39-53.

---

[2] Normally, of course, this list would appear at the end of the report.

Main, Jeremy. "The Executive Yearn to Learn." Fortune 3
    May 1982: 234-248.

Rochester, Jack B., and John L. DiGaetani. "Managerial Com-
    munication: Total Business Communication for the 1980s."
    The ABCA Bulletin 44 (1981): 9-10.

Sigband, Norman B. Communication for Management and Busi-
    ness, 3rd ed., Glenview, Illinois: Scott, Foresman,
    1982.

Author-Year Designation

In the author-year system, list your references alphabetically, by authors' last names. For a book, give author, date, title (underlined), city of publication, and publisher. For an article, give author, article title (not underlined), periodical title (underlined), volume, number, and pages where the article appears. Capitalize only the first letter of work titles.

Here is a sample author-year reference list that includes two works by the same author and an unpublished work.

REFERENCES

Albey, J. 1984. Modern career choices. San Francisco:
    Hamilton Publishers.

Crashaw, H. et al. 1982. Technology and careers. Boston:
    Little, Brown and Co.

_____. 1983. Careers in the natural sciences. Dallas:
    Bovary Press, Inc.

Donne, M. 1984. Job prospects for college graduates.
    Education Digest 28, no. 2:86-89.

Marsh, A., and Smith, C., eds. 1979. Advice for the job
    seeker. Boston: Arngold Press.

Peters, Claire. 1982. Wage scales for women in civil
    engineering. Master's thesis, Brandon University.

For other typical entries, see the University of Chicago Press *Manual of Style*, 13th ed. or a style manual in your discipline.

In your report text, when you refer to the work, insert the author's name and publication date in parentheses:

> Seventy-five percent of chemistry technicians interviewed expressed a desire for further training (Albey, 1975).

With this system, you can also include page numbers in the text:

> Seventy-five percent of chemistry technicians interviewed expressed a desire for further training (Albey, 1975:114).

When you mention the writer's name in the text of your discussion, do not repeat it within the parentheses:

> Crashaw et al. (1977:93–94) claim that medical technologists "feel challenged by their work."

If citing two works published by the same author(s) in the same year, insert an "a" and a "b" after the dates, both in your reference list and textual citation.

> (Jones, J., 1975a)

As an illustration in actual practice, the sample report on pages 508–516 has an author/year list of references and citations.

Numerical Designation

In the numerical reference system, number references alphabetically or in the order first cited in the text. (Use one arrangement or the other, consistently.) Otherwise the format resembles the author-year system. Here are typical entries for a reference list in order of first citation in the text.

REFERENCES

1. Donne, M. Job prospects for college graduates. Education Digest 28, no. 2 (1984):86–89.

2. Albey, J. Modern career choices. San Francisco: Hamilton Publishers, 1984.

3. Crashaw, H. et al. Careers in the natural sciences. Dallas: Bovary Press, Inc., 1983.

4. Marsh, A. and Smith, C., eds. Advice for the job seeker. Boston: Arngold Press, 1979.

5. Crashaw, H. et al. Technology and careers. Boston: Little, Brown and Co., 1982.

With each reference numbered upon first citation, you then can use the same number in your text whenever you cite that particular work:

> Seventy-five percent of chemistry technicians interviewed expressed a desire for further professional training. (2:83)

Adding page numbers helps clarify the reference.

As an illustration, the report on pages 483–506 has a numbered list of references and citations.

Whichever system of documentation you choose, be consistent throughout your report.

WRITING THE FINAL DRAFT

Below is a sample research report. Study it carefully to see how Lynne has woven her sources, using the new MLA documentation.

AUDIENCE-AND-USE ANALYSIS

Lynne Taylor is preparing the following report as part of her course work in Professional Communication. Because she is writing for a diverse audience, she writes at the lowest level of technicality.

Her primary audience includes her course instructor and the deans of the College of Business and Industry and Arts and Sciences who need her report to make decisions about the feasibility of offering more courses. Her secondary audience consists of the executives she surveyed along with other interested faculty members.

Planning her research, Lynne concluded that her first priority should be to present a profile of communication practices in area companies, to learn if the present course correlated with their needs. Her next step was to determine if an expanded program was warranted, and if so, which courses would best meet the needs of students and area businesses. By citing secondary data, Lynne lends credibility to her findings.

Abiding by the rules of ethical research, Lynne is rigorously impartial, and gives a balanced assessment. She allows no bias on her part to influence her readers' attitudes. She lets the facts speak for themselves, thus allowing her readers to make *educated* judgments. Aware of the fact that the primary purpose of a research report is to *inform*, not to *persuade,* she supplies hard information and presents it in such a way as to hold the interest of her audience. In other words, more than simply presenting bare facts, she consolidates her data to provide coherent and sensible explanations. Lynne does include recommendations, however, to answer the central question she is researching.

To save space, the title page, letter of transmittal, table of contents, and informative abstract are omitted.

HOW CAN SMU'S PROFESSIONAL COMMUNICATION PROGRAM BETTER SERVE STUDENTS AND AREA BUSINESSES?

INTRODUCTION

Description and History

At the beginning of the semester, some students asked why only two courses in professional communication are offered at SMU. Professor Dumont mentioned that proposals for additional courses had been rejected because of inadequate staff, and because data substantiating need had never been collected. He said area businesses should be surveyed, suggesting that the topic would make an excellent research report.

Statement of Purpose

The genesis of this report was the question, "How can SMU's professional communication program better serve students and area businesses?" To answer that question, a questionnaire (see Appendix I) was sent to the presidents of 612 area businesses.

Target Audience

The primary audience is Professor Dumont and the deans of the College of Business and Industry and Arts and Sciences. The secondary audience includes area executives and interested faculty members.

Information Sources

Primary data were collected from the questionnaire responses. Of the 612 questionnaires sent, 103 were returned, for a return rate of 16.83 percent. Secondary sources include various publications.

Limitations

A higher rate of return would have been desirable. But area responses do correlate with responses to other surveys. In fact, secondary sources confirm this report's findings, and demonstrate that the communication needs of area businesses do not differ significantly from those anywhere else in the country.

This report does differ from others, though, since it also attempts to determine whether the communication needs of small companies (fewer than 100 employees) differ from those of large companies (more than 100 employees).

Scope

The report covers two major topics: A profile of communication in local companies, and an assessment of how SMU can better serve the business community.

PROFILE OF COMMUNICATION IN LOCAL BUSINESSES

Internal Communication

In large companies, messages to employees (internal communication) are transmitted through memos 88 percent of the time. In small companies, memos are used 54 percent of the time. Similarly, large companies use supervisors for downward communication 90 percent of the time, while small companies use supervisors 63 percent of the time. The most frequent means for communicating information within small companies is through the president or vice president speaking directly to employees 67 percent of the time.

Frequency of Types of Writing Done by Area Businesses

All presidents state that *letters* are the most common correspondence, with *interoffice memos* ranked second. *Reports* are ranked third in frequency, with *procedures* next.

In either large or small companies, far less unanimity exists about operating manuals, house organs, advertisements, catalogues, handbooks, and manuals. Small companies rarely or never write house organs, manuals, or handbooks. Fourteen large companies (out of 53) write *house organs* often, and eight sometime write them. Only three large companies often write *manuals,* and only four often write *handbooks.*

Most companies surveyed are either manufacturers or banks. Neither banks nor clothing manufacturers have much need for such technical formats as reference manuals and handbooks. No doubt, as the community moves toward high-tech industries, we will see a greater demand for professional communication education.

Responsibility for Writing Assignments within Area Companies

Large companies indicate that managers do 76 percent of all company writing. Vice-presidents do 18 percent, communication specialists 4 percent, and presidents 2 percent.

Conversely, presidents and vice-presidents do 67 percent of the writing in small companies. Managers do 26 percent, and personnel ranging from clerks to plant supervisors do 7 percent.

Because managers in large companies do most of the writing, companies nationally are demanding their managers learn to communicate. In fact, Jack Rochester and John DiGaetani note that many companies, "are sending managers and management trainees to writing courses through continuing education or by private seminar services" (9). Some companies, such as the accounting firm of Coopers and Lybrand, are "so desperate for people who can write effective proposals, they are considering hiring English majors, then teaching them accounting" (Rochester 9). To meet these needs, many universities are developing courses in managerial communication to complement existing courses in Organizational and Professional Communication (Rochester 9).

Perception of Writing Effectiveness within Companies

Although many of the companies surveyed desire more communication courses at SMU, only 2 percent of large companies and 7 percent of small companies rate their writing as poor. Fifty-eight percent of large companies believe their writing is good, compared with 48 percent of small companies. Forty percent of large companies and 38 percent of small companies report their writing as fair.

Guidelines for Written Communication

Despite the many companies who believe their writing is only fair, 85 percent of small companies and 82 of large companies have no writing guidelines. Ninety-eight percent of small companies and 76.5 percent of large companies have no in-house programs to teach employees to communicate. But 75 percent of large companies provide tuition reimbursement to employees. Only 39 percent of small companies offer this incentive.

Consulting Services in Writing and Editing

Only fourteen large companies (26.9 percent) and seven small companies (13.7 percent) have used communication consultants. Correspondingly, 84 percent of large and 87 percent of small companies claim no need for consultants.

Although few businesses use or see the need for communication consultants, 73 percent of large companies and 35 percent of small companies hire writing consultants when marketing new products. The contradiction here may arise because companies see differences between communication consultants and marketing writers. Or perhaps the companies are simply unaware of the services rendered by communication consultants. This latter assumption appears valid, since many companies that characterize their writing as only *fair* see no need for consultants. Or perhaps these companies would prefer that SMU offer seminars to meet their specific needs (see below).

HOW SMU CAN BETTER SERVE THE BUSINESS COMMUNITY

Course Preferences

All companies would like to see more communication courses at SMU. Their first choice is a course in *report writing*. Seventy-one percent of large companies and 48 percent of small companies indicate report writing as the preferred sequel to Business or Technical Communication. This preference correlates with a survey by the *Harvard Business Review*, where 72 percent of respondents (executives and managers) stated report writing is "the best training for business" (Bennett 28). A third survey further confirms these results, with the qualification that "the informational report should receive stronger emphasis than the analytical report" (Bennett 32).

As their second preference, large companies chose *interpersonal communication,* followed by *public speaking.* Among small companies, the order is slightly different. Their second choice is *sales and promotional writing,* followed by *public speaking and interpersonal communication.* Since 73 percent of large companies hire consultants for sales and promotional writing, they evidently do not believe that SMU needs to offer such a course. Since few small companies use marketing consultants, they see a definite need for a course in *sales and promotional writing.*

Few companies, large or small, see the need for courses in writing *procedures* or *abstracts.* This lack of interest in technical communication coincides with the infrequency with which the companies surveyed use technical formats.

Course Content

Both large and small companies rank the importance of topics in a writing program as follows:

1. *organizing information*
2. *editing* for clarity, conciseness, and fluency
3. *summarizing information*
4. *adapting various messages for various audiences*
5. *grammar and punctuation*

The responses coincide with Halpern's. She found that executives consider the following topics most useful: organizing information effectively; polishing drafts of letters, memos, and reports; adapting writing to various audiences; clarifying purpose; and controlling tone in communication (39).

Halpern's results and the results of this survey correlate highly with contemporary rhetorical theory and the emphasis found in many recent texts on business communication. For instance, in his 1982 text, Sigband stresses that "planning, organizing, writing, and editing are intimately associated" (69).

Halpern considers it "notable" that her respondents did not stress the need for "grammar, spelling, and mechanics" (39). Nor did area businesses. Respondents apparently believe that any student in an upper level communication course should have mastered these basic skills.

Although some students do lack basic skills, a course in Professional Communication or Report Writing must reflect the needs of advanced students and the business community — not those of remedial students.

Desired Skills

Sixty-nine percent of large companies and 65 percent of small companies indicate *listening* is the most important or a very important skill. Large companies rank *writing* as second most desirable, followed by *speaking face-to-face, reading,* and *speaking to groups.*

In contrast, small companies rank *reading* as the second most desirable skill, followed by *writing, speaking face-to-face* and *speaking to groups.*

Almost all companies believe that the ability to dictate well is of little importance.

The overall responses differ somewhat from those received by Rochester and DiGaetani. In their survey, *writing* was ranked as most important, followed by *listening, speaking,* and *reading.* The conflicting rankings for writing and listening could be a result of the different groups surveyed. Rochester and DiGaetani questioned "college and university communication teachers and chairpersons" (9), whereas we surveyed business presidents. Each group nonetheless believes that writing and listening are vital skills.

Although the need for writing expertise in business has long been recognized, listening has been neglected. As DiGaetani notes in a recent article in *Business Horizons,* "Listening is a rare skill, the most often used yet least understood and researched of the communicating processes" (622). Recently, though, it is being proclaimed as "the newest area of communication interest" (Rochester and DiGaetani 10). In fact, businesses now consider "listening . . . the key to growth and success" (Rochester and DiGaetani 10), since recent studies show "that most executives spend between 45 and 63 percent of their day listening" (DiGaetani 622). The Sperry Corporation recently has begun a program to improve its employees' listening skills, and advertises the importance of listening with the slogan "When You Know How to Listen, Opportunity Only Has to Knock Once" (*Fortune* 238).

Contract Learning

Fifteen percent of the small companies would participate in a contract learning program whereby SMU students taking a Professional Communication course would work with them. Among large companies, 18 percent are willing to participate. Another 12 percent indicate that agreement would depend upon the specific proposal. Some large companies agree with the concept of contract learning, but note that their companies do little in terms of business and technical communication.

Seminars in Professional Communication

Sixty-one percent of large companies and 40 percent of small ones indicate that SMU should offer Saturday seminars in Professional Communication. These figures reflect a national trend. For example, a recent *Fortune* article, "The Executive Yearn to Learn," notes that companies are increasingly spending much money to send executives and managers to seminars (Main 234).

That a majority of the large area companies want communication seminars also reflects the trend toward insisting that employees know how to communicate. As Sigband stresses, communication is "the lifeblood of every organization" (4). Companies are prone to "hire the best communicator rather than the best accountant . . . engineer . . . (or) manager . . ." (Sigband 3–4). Employees unable to communicate are of little value to an organization.

Too few business and technical majors receive the communication skills they will need. As Rochester and DiGaetani note, "The business world is in desperate need of personnel who can become total communicators" (10).

CONCLUSION

Summary of Findings

Area businesses typically write letters, memos, reports, and procedures. Among large companies, managers do almost all writing; presidents and vice-presidents do most writing among small companies.

Few companies provide guidelines for written communication and even fewer have in-house teaching programs. Yet, 40 percent of large companies and 38 percent of small companies perceive their writing effectiveness as only fair. Many large companies (75 percent) and some small companies (39 percent) do provide tuition reimbursement.

All companies desire more communication courses at SMU. Their first choice after Professional Communication is report writing, followed by interpersonal relations and public speaking. Small companies would also like a course in sales and promotional writing.

Both large and small companies want writing courses to emphasize organizing; editing for clarity, conciseness, and fluency; summarizing; and adapting messages for various audiences. In terms of total communication, area companies indicate listening skills are vital and should be taught along with writing, reading, and speaking.

Finally, many companies indicate that seminars should be offered in Professional Communication.

Recommendations

As a result of these findings and current research in Professional Communication, I offer the following recommendations:

1. All Professional Communication courses should have either informational or analytical reports as their major component. In addition, strong emphasis should be given to various letter formats, memos, and procedures.
2. All Professional Communication courses should involve students in the total communication process — that is, all courses should emphasize and integrate writing, listening, speaking, and reading into the syllabus.
3. Since most writing in large companies is done by managers, SMU should offer Managerial Communication as a sequel to Professional Communication.
4. Seminars in Professional Communication should be offered for SMU's business community.
5. Professional and Technical Communication professors should establish a contract learning program with area businesses.
6. All business and technical majors should take at least one substantive course in Professional Communication.

CHAPTER SUMMARY

Follow these steps in planning your report:

1. Choose a topic that is interesting and sufficiently focused.
2. Make a working bibliography before you proceed.
3. Compose a clear statement of purpose and a working outline.
4. Move from general sources to particular ones, skimming your material.
5. Take selected notes, keyed to your outline, with all sources identified.
6. Plan early for interviews, questionnaires, and inquiry letters.
7. If possible, conclude by having a look for yourself.

Follow these steps in writing your report:

1. Revise your working outline.
2. Follow an introduction-body-conclusion structure.
3. Concentrate on only one section at a time.
4. Credit each data source, using an acceptable documentation format.
5. Write your final draft.

EXERCISES

1. Prepare a research report by completing the following steps. (Your instructor might establish a timetable for completion of each phase.)

Phase One: Preliminary Steps
   a. Choose a topic of *immediate practical importance,* something that concerns you or your community directly. (The list of research possibilities on pages 332–334 should give you some ideas.)
   b. Identify a specific audience and its intended use of your information.
   c. Narrow your topic, and check with your instructor for approval.
   d. Make a working bibliography to ensure sufficient primary and secondary resources. Don't delay this step!
   e. Make a list of things you already know about your topic.
   f. Write a clear statement of purpose and submit it in a proposal memo (pages 425–428) to your instructor.
   g. Make a working outline.

Phase Two: Collecting Data
   a. In your research, move from general to specific; begin with general reference works for an overview.

    b. Skim your material, looking for high points.

    c. Take selective notes. Don't write everything down! Use notecards.

    d. Plan and administer (or distribute) questionnaires, interviews, and letters of inquiry.

    e. Whenever possible, conclude your research with direct observation.

Phase Three: Organizing Your Data and Writing Your Report

    a. Revise and adjust your working outline, as needed.

    b. Compose an audience-and-use analysis like the sample on page 324.

    c. Follow the introduction-body-conclusion format in your report.

    d. Concentrate on only *one* section of your report at a time.

    e. Fully document all sources of information.

    f. Write your final draft.

    g. Proofread carefully and add all needed supplements (title page, letter of transmittal, abstract, summary, appendix, glossary—chapter 12).

Due Dates:

List of possible topics due:

Final topic due:

Proposal memo due:

Working bibliography and working outline due:

Notecards due:

Copies of questionnaires, interview questions, and inquiry letters due:

Revised outline due:

First draft of report due:

Final draft of report with all supplements and full documentation due:

Here are some possible research projects for Phase One:

    a. Because of budget cuts and declining enrollment, your school wants to eliminate some departments. Trace the history of your major department through its stages of planning, formation, staffing and growth as part of your college. Do your findings show a steady growth, periods of decline, or some other trend? What is the outlook for your department within the institution? Write for the student and faculty senates.

    b. Trace the history of your major as an academic specialty. Which institutions offered the first courses or the first major? To what needs were institutions responding in offering this major? Where are the highest-ranked programs now offered? What is the outlook for growth in this discipline?

    c. Identify a modern discovery in your field that dramatically advanced the state of the art (as the discovery of antibiotics revolutionized medical treatment). Establish how the discovery was made and trace its beneficial and negative effects (antibiotics can cause severe allergic reactions and organic complications, and have created resistant strains of organisms).

    d. Survey student opinions about some proposed changes in the curriculum or another controversial campus issue. If possible, compare your findings with national statistics about student attitudes on this issue.

e. Perhaps you attend a school where young and popular faculty often are not granted tenure. Investigate the tenure policy, and prepare a report for anyone concerned. Discover how faculty hiring and tenure decisions are made at your school. What are the primary criteria, and how are they ranked (e.g., research activity, publication, quality of teaching, service to the department and college, community service)? Trace the history of tenure decisions over the last ten years. Do any trends emerge?

f. As much as 30 percent of groundwater in some states is contaminated. To alert readers, find out how the quality of your local groundwater measures up to national standards. Has the quality increased or decreased over the past ten years? What are the major factors affecting local water quality? What is the outlook for the next decade? Can your family feel safe drinking tap water?

g. Trace the modern history of zoning ordinances in your town. How has residential zoning affected the character of the town? How has business zoning affected the character and economy of the town? What is the outlook for future zoning changes?

h. Find out which major department on your campus attracts the most students, and why. Has this department always been so popular? Will it continue to be? Trace its recent history.

i. Discover what qualities most employers look for in a job candidate. Have employers' expectations changed over the last ten years? If so, why?

j. Find out which geographic section of the United States is experiencing the greatest economic and population growth. What are the major reasons for this growth? Trace the recent history of this change.

k. Some observers claim that productivity in American industry is on the decline. Find out whether this claim is valid. If it is, identify the reasons. Include both primary and secondary research data.

l. Assume that you and a business partner have developed a desk-top duplicating machine (or some other product) that can be manufactured at low cost. What are the intricacies involved in applying for and getting a patent for your product? Include both primary and secondary research data.

m. Trace the typical promotional route for people in your specialty. What are the possibilities and requirements for promotion to executive status?

n. What is the present status of women and minorities in your specialty? Assess their hiring and promotional opportunities, relative salaries, and percentages in executive positions. Do data over the past decade suggest a trend toward equal opportunity?

o. What fields will offer the best job opportunities through the next decade? Where will you and your fellow majors fit into this market?

p. What are the prospects for a lifetime career in your field? What percentage of your colleagues now change careers, and why?

q. The federal government is planning at least two national sites for disposal of spent nuclear fuel rods and other high-level radioactive waste. State governments plan many more sites for disposal of low-level nuclear waste and other toxic waste. Find out the location of these sites, and determine whether the site planned closest to your community poses any health or economic threat.

r. Home buyers (especially of older homes) face a frightening array of possible hazards, including chlordane sprayed for termites, wood preservatives, urea formaldehyde foam insulation, radon gas (radioactive) or chemical contamination of well water, lead paint, and asbestos. Research the effects of these hazards for someone you know who is buying a home.

s. Are the schools in your town free of dangers from insulation, asbestos, art supplies, or chemical fumes from cleaning fluids and solvents? Find out, and write a report for the PTA and school committee.

t. Research the anticipated medical effects of full-scale nuclear war. Write for the average reader.

u. Assume you live within fifty miles of a nuclear power plant. In the event of a meltdown, how would your community be affected? Is there an adequate evacuation plan for communities within thirty miles of the plant?

v. Which area of your state has the cleanest air and groundwater, and which has the most polluted? Write for someone who wants the safest place to raise a family.

w. Has acid rain caused any specific damage in your area of the state? Write for a wilderness conservation group.

x. How, specifically, has your school (or some community agency) been affected by budget cuts in recent years?

y. In 1984, the government spent roughly $169 million on civil defense preparedness for nuclear war. Assume that voters will decide whether your state will participate in the preparedness program. In a full-scale nuclear war, would such preparedness make any real difference? Find out.

z. Can peanut butter, black pepper, potatoes, and buttered toast cause cancer? Which of the most common "pure" foods can be carcinogenic? Find out, and write a report for your cafeteria dietician.

SPECIFIC
APPLICATIONS

16

Writing Effective Letters

LETTERS DEFINED

Whereas a report may be compiled by a team of writers and read by many readers, a letter usually is written by one writer for one or more definite readers. Because a letter is more personal than a report, tone is a major ingredient of your message. Because most letters are written to elicit a definite response, you want the reader to be on your side. Because your signature certifies the content of your letter (which may be used later as a legal document), precision is crucial.

PURPOSE OF LETTERS

In any nonmenial job, you will write letters like those listed below. The broad purpose of each is to inform, request, or persuade.

– sales letters designed to create interest in a product or service
– letters of instruction outlining a procedure to be carried out by the reader
– letters of recommendation for friends, fellow workers, or past employees
– letters of transmittal to accompany mailed reports and other documents
– letters to inquire about a product, procedure, or person
– letters to complain about service or products, and to request adjustment
– letters to apply for jobs

You may also need to write letters of response to those letters received by your company.

A full discussion of letters would more than fill a textbook. This chapter, therefore, covers only four common types of letters: the inquiry letter, the claim letter, the adjustment letter, and the letter of application, along with its accompanying résumé. Other useful types — the letter of transmittal and letter reports — are discussed in Chapters 12 and 17. Regardless of type, however, all letters share certain elements.

ELEMENTS OF EFFECTIVE LETTERS

Remember this rule: *never send a letter until you genuinely feel good about signing it;* your signature certifies your approval of the contents. Use the guidelines below for judging the adequacy of letters before you sign them.

Introduction-Body-Conclusion Structure

Structure most letters to include (1) a brief *introduction* paragraph (five lines or fewer) where you identify yourself and your purpose; (2) one or more *body* paragraphs containing the details of your message; (3) a *conclusion* paragraph where you sum up and courteously encourage your reader to act. For readability, keep your paragraphs short (usually fewer than eight lines). If your body section is long, divide it into shorter paragraphs, as shown in Figure 16–1. Notice that this body section is broken down into four specific questions for easy answering. Figure 16–2 shows the response to these questions.

Standard Parts

Letters typically have six parts, in order from top to bottom: heading, inside address, salutation, the text (introduction, body, and conclusion), complimentary close, and signature.

154 Sea Lane
Harwich, MA 02163
July 15, 1984

Land Use Manager
Eastern Paper Company
Waldoboro, Maine 04967

Dear Manager:

Mr. Melvin Blotter, your sales representative, has told me
that Eastern Paper Company makes available certain parcels
of its lakefront property in northern Maine for public lease.
I am interested in a lease, and would appreciate answers to
the following questions:

1. Does your company have any lakefront parcels in
highly remote areas?

2. What is the average size of a leased parcel?

3. How long does a lease remain in effect?

4. What is the yearly leasing fee?

I would welcome any other details you might send along.

Yours truly,

John M. Lannon

John M. Lannon

P.S. I am planning a trip to Maine for August 5-11, and
would be happy to stop by your office any time you
are free.

FIGURE 16-1   A Letter of Inquiry

Eastern Paper Company
Waldoboro, Maine 04697

July 25, 1984

Mr. John M. Lannon
154 Seaweed Lane
Harwich, MA 02163

Dear Mr. Lannon:

This is in answer to your recent inquiry about leasing
lakeside lots in Maine.

We have no lands for lease in what you would classify as
"highly remote areas." The most remote area is on the
northern shore of Deerfoot Lake where presently you would
have to reach the camp lot by boat. Leases on these 30,000
square foot parcels are renewable on a yearly basis each
June. The leasing fee is approximately $250 per year. We
have a limited number of leases available, but you would have
to visit our office to get meaningful information about exact
locations. If you are in the area, you may drop by any weekday morning before 11 o'clock.

The Land Use Regulation Commission in Augusta, Maine,
regulates campsite leasing, and you would need to apply to
this body for a building permit including soil tests, etc.,
before you would be allowed to build.

Thank you for your inquiry.

                                Yours truly,

                                EASTERN PAPER COMPANY

                                A.B. Coolidge
                                A. B. Coolidge
                                Townsite Manager

ABC/de

cc: Mr. Blotter

FIGURE 16–2   A Letter of Response to an Inquiry

Heading

If your stationery has a company letterhead, simply include the date two spaces below the letterhead (Figure 16–2) or at the right or left margin. On blank stationery, include your address and the date (but not your name), as shown in Figure 16–1 and here:

```
Street Address              154 Sea Lane
City, State  Zip Code       Harwich, MA 02163
Month Day, Year             December 24, 1984
```

Avoid abbreviations (but do use the Postal Service's two-letter state abbreviations when addressing the envelope, and in the heading itself when the state has a long name). Depending on the length of the letter, place your heading at least 2 inches below the top of your page and far enough to the left so the longest line abuts your right margin.

Inside Address

Two to six spaces below your heading and 1¼ inches from your left margin is the inside address.

```
Name and Title of Reader (and position)   Dr. Marsha Mello, Dean
Company Name                              Western University
Street Address (if applicable)           Muncie, Indiana 13461
City, State  Zip Code
```

Whenever possible, address your letter to a specifically named reader, using that reader's appropriate title (Attorney, Major, etc.), but don't be redundant by writing "Dr. Marsha A. Mello, Ph.D."; use Dr. or Ph.D., Dr. or M.D. Abbreviate only titles that are routinely abbreviated (Mr., Ms., Dr., etc.). Titles such as Captain are written out in full. Although some women prefer the traditional titles, "Mrs." or "Miss," many prefer "Ms.," since this title does not violate their privacy by declaring their marital status. When unsure of your reader's preference, use "Ms."

Salutation

Your salutation, two spaces below the inside address, is a greeting to the reader. Begin with "Dear," and end with a colon. Include the person's full title. If you don't know the person's name or gender, use his or her position title: "Dear Manager."

```
Dear Ms. Jones:

Dear Professor Smith:

Dear Supervisor:
```

No satisfactory guidelines exist for addressing several people at once. Women rightly resent the term "Gentlemen," and "Ladies and Gentlemen" sounds like the beginning of a speech. "Dear Sir or Madam" sounds formal and old-fashioned; besides, many women dislike being addressed as "Madam." And "To Whom It May Concern" is too vague and impersonal. In such cases, your best bet is to eliminate the salutation completely by using an attention line (page 343).

Letter Text

Begin your letter two spaces below the salutation. For letters that will fill most of the page, use single spacing within paragraphs, and double spacing between. For short letters, double space within paragraphs, and triple space between, to balance the page.

Complimentary Close

Place the complimentary close two spaces below the concluding paragraph, aligned with your heading. Any conventional closing (polite but not overly intimate) will do, but it should parallel the level of formality in your salutation, and should reflect your relationship to the addressee. The following possibilities are ranked in decreasing order of formality.

```
Respectfully,

Yours truly,

Sincerely yours,

Cordially yours,

Sincerely,

Cordially,

Best wishes,

Warmest regards,

Regards,

Best,
```

The complimentary close is followed by a comma.

Signature

Type your full name and title four spaces below your complimentary close. Sign in the space between.

Sincerely yours,

Martha S. Jones

Martha S. Jones
Personnel Manager

Your signature indicates your approval of and responsibility for the letter (even if it's typed by a secretary). If you are writing as a representative of a company or group that bears legal responsibility for the correspondence, type the company's name in full caps two spaces below the complimentary close; place your typed name and title four spaces below the company name, and sign in between.

Yours truly,

HASBROUCK LABORATORIES

Lester Fong

Lester Fong
Research Associate

Specialized Parts

Certain letters require one or more of the following specialized parts.

Attention Line

Use an attention line when writing to an organization and when you want a particular person (whose name you don't know), title, or department to receive your letter.

ATTENTION: Research and Development Division

ATTENTION: Quality Assurance Supervisor

Drop two spaces below the inside address, and place the attention line either flush with the left margin or centered on the page. The attention line replaces your salutation, as shown on page 357.

Subject Line

The subject line forecasts what your letter is about, a good device for getting a busy reader's attention.

SUBJECT: <u>Market Analysis for the Absco Portable Computer</u>

Place the subject line two spaces below the inside address (or attention line), as shown on page 357. Write the subject in caps or underline it.

Typist's Initials

If someone else types your letter, your initials (in caps) and your typist's (in lowercase letters) should appear two spaces below the typed signature, abutting the left margin.

    JJ/pl

Because they repeat the signature block, the writer's initials may be eliminated.

Enclosure Notation

When other documents accompany your letter, add one of the following enclosure notations one space below the typist's initials, abutting the left margin:

    Enclosure

    Enclosures 2

    Encl. 3

If the enclosures are important documents, name them:

    Enclosures: 2 certified checks, 1 set of KBX specifications.

Distribution Notation

If you will distribute copies of your letter to other readers, so indicate, one space below any enclosure notation.

    cc: office file   cc: S. Furlow
         Mr. Blotter   B. Smith

Postscript

A postscript is designed to draw your reader's attention to a point you wish to emphasize. Do not use a postscript for a point you've forgotten in the letter. Rewrite instead. Do use a postscript to add a personal note.

> P.S. You will love the way this model prints documents.

Place the postscript two spaces below any other notation, abutting your left margin. Use the postscript sparingly, if at all.

Appropriate Format

Here is a summary of the format instructions in Chapter 12: Use high quality 20-lb. bond, 8½-by-11 plain white stationery with a minimum fiber content of 25 percent. Type neatly, avoid erasures, and make sure you have clear typewriter keys and a fresh ribbon. Use uniform margins, spacing, and indentation: frame your letter with a 2½-inch top margin and side and bottom margins of 1 to 1¼ inches; single space within paragraphs and double space between; avoid hyphenating at the end of a line.

If your letter extends to more than one page, begin the second page five spaces from the top, with a notation identifying the addressee, the date, and the page number:

> Walter James, June 25, 1981, p. 2

Begin the text of your second page two spaces below this notation. Place at least two lines of your paragraph at the bottom of page one, and at least two lines of your final text on page two.

Your 9½-by-4⅛-inch envelope should be of the same quality as the stationery. Center your reader's address, and single space if it takes three lines or more. Use only accepted abbreviations. Place your own single-spaced address in the upper left corner (Figure 16–3).

Accepted Letter Form

Although several acceptable letter forms exist, and your own company may have its own requirements, we will discuss two most common forms: semi-block, with no indentations (Figure 16–4), and modified block, with the first sentence of each paragraph indented five spaces (Figure 16–5). The letters

```
Marvin Hanley
154 Seaweed Lane
Harwich, MA 02163



                    Dr. Marsha Mello, Dean
                    Western University
                    Muncie, CA 13461
```

FIGURE 16–3   A Sample Envelope

of inquiry and response earlier in this chapter are examples of semiblock and modified block forms, respectively.

Plain English

First, we should distinguish between the *conventions* of letter writing and letterese. The conventions of letter writing concern format, form, letter parts, and general structure — the elements that govern the appearance and organization of your letter. These conventions exist because, over the years, people have agreed on a set of rules for composing attractive, informative, and diplomatic letters. One convention, for example, requires that you type your name (and position) under your signature. This procedure became accepted practice since it was a handy way of clarifying your identity, especially if you wrote in a scrawl. Other conventions (e.g., subject and attention lines) developed because they too were useful.

Letterese, on the other hand, consists of time-worn, stuffy, and overblown phrases some writers think they need to make their communication seem important. Here is a typically overwritten closing sentence:

> Humbly thanking you in anticipation of your kind cooperation, I remain
>
>   Faithfully yours,
>
>   *Marvin Hanley*
>   Marvin Hanley

Although no one speaks this way, some writers lean on such heavy prose instead of simply writing, "We will appreciate your cooperation." Not only is the

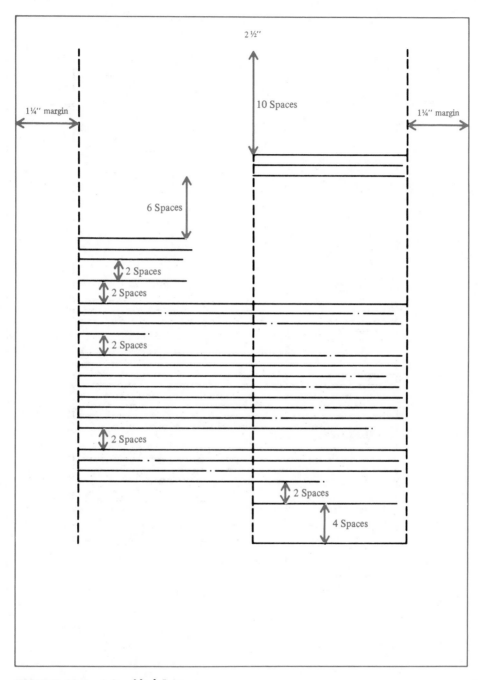

FIGURE 16–4   A Semiblock Letter

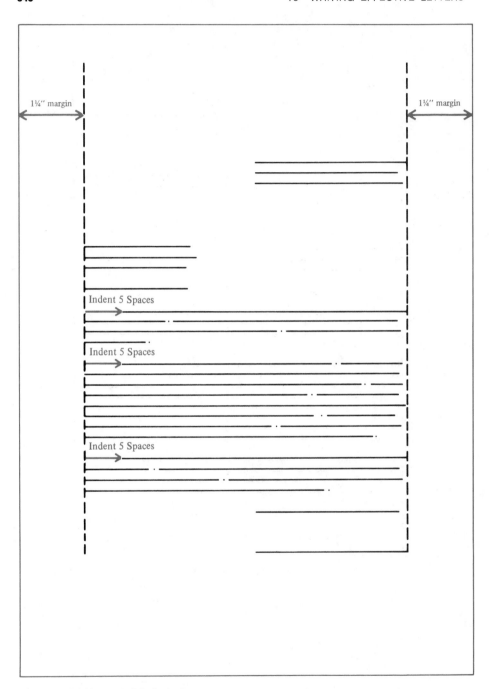

FIGURE 16–5   A Modified-Block Letter

earlier example overworked; it is so clearly exaggerated it sounds insincere. Unfortunately, statements like this and the others listed below are popular because they are easy to use: like TV dinners, effortless but unimpressive. Don't borrow instant phrases.

Here are a few of the many old standards that make letters unimaginative and boring:

Letterese	Translation into Plain English
As per your request	As you requested
Having received your letter, we . . .	We received your letter.
Enclosed please find my résumé.	My résumé is enclosed.
I regret to advise you that I must delay payment.	I must delay payment.
It is imperative that you write at once.	Please write at once.
I am cognizant of the fact that my payment is overdue.	I know that my payment is overdue.
At the earliest possible date	As early as possible
I beg to differ with your estimate.	I disagree with your estimate.
Please be advised that my new address is . . .	My new address is . . .
This writer	I
At the present time	Now
I humbly request that you consider my application.	Please consider my application.
I beg to acknowledge receipt of your check.	I received your check.
In the immediate future	Soon
We are in hopes that you succeed.	Good luck.
In accordance with your request	As you requested
Our situation is such that we cannot immediately pay our complete bill.	We cannot immediately pay our complete bill.
Due to the fact that	Because
Herein enclosed	Enclosed
Please be kind enough to grant me an interview.	May I have an interview?
I wish to express my gratitude.	Thank you.
At this point in time	Now

Be natural: write as you would speak in a classroom. Avoid the temptation to copy a textbook example. Use the samples in this book as models only. Anyone who reads letters regularly can spot a borrowed letter.

Below is a request letter that draws heavily on tired phrases. As you can see, the reader will find it difficult to dig out the meaning buried under the verbiage.

234 Idle Way
Hoboken, N.J. 34567
September 10, 1981

Marvin Mooney
Registrar
Calvin College
Plains, GA 38475

Dear Mr. Mooney:

Pursuant to your notice of September 6, this writer regrets to advise you that my tuition payment will be delayed until September 21, when I receive my scholarship check.

I humbly request you to be cognizant of the fact that this writer's tuition for all five prior semesters has been paid on time. At the present time, my first, and hopefully last, late payment is due to the fact that a computer breakdown in the NDEA offices has occasioned a delay in the processing of all scholarship renewal applications for two weeks. Enclosed please find a copy of a recent NDEA notice to this effect.

I am in hopes that you will be kind enough to grant me an extension of my tuition-due date for this brief period of time. Thanking you in anticipation of your cooperation, I remain

Gratefully yours,

Charles Jones

Charles Jones
Class of '85

Enclosure

The letter below shows the same message in plain English. The message remains unchanged, but it seems more sincere.

Dear Mr. Mooney:

I received your tuition-due notice of September 6 and regret that my payment will be delayed until September 21, when I receive my scholarship check.

Your payment records should show that my tuition bills for all five prior semesters have been paid on time. This first, and I hope last, late payment is the result of a computer breakdown in the NDEA offices, which has delayed the processing of scholarship renewals for two weeks, as explained in the enclosed NDEA notice.

May I have this brief extension of my tuition-due date, without causing you or the college too much inconvenience? Your patience in this difficult time would be a great help.

"You" Perspective

In speaking face-to-face with someone, you unconsciously modify your delivery as you read the listener's signals: a smile, a frown, a raised eyebrow, a nod, etc. Even in a phone conversation a listener provides cues that signal approval, anger, and so on. Writing a letter, however, has one major disadvantage: because you face a blank page, you can too easily write only to please yourself, forgetting that a flesh-and-blood person will be reading your letter.

The "you" perspective concerns your tone; by careful word choice, you show respect for your readers. Put yourself in their place; ask yourself how readers will respond to what you've just written. A letter creates a relationship between writer and reader, and, like the letter's appearance, the words themselves are another basis for that relationship. If you bury readers in letterese and clichés, they are bound to conclude you care little about your writing task or about them. (The second version of the above request letter is written from a "you" perspective, whereas the first is not.)

Even a single word, carelessly chosen, can offend readers. In a letter complaining about the monitor on your new personal computer, for example, you have the choice of saying, "Although the amber screen causes very little eye strain, the character resolution is not sharp enough for lengthy word-processing use" or "The monitor is lousy for word processing." Clearly, "lousy" would be a poor choice here because of the implied insult to the manufacturer or dealer, and because it does not precisely describe the problem (i.e., poor resolution).

Imagine how a parent felt after receiving a school letter noting that her child is a "fortunate deviate." Only after three angry phone calls to school officials did she learn that "fortunate deviate" is educational jargon for "bright child."

Put yourself in the reader's place in the following example: Imagine you manage the complaint department for a large mail-order company. Which of these versions requesting repair, replacement, or refund would you honor quickly and efficiently?

1. I demand that you bums immediately send me a replacement for this faulty desk calculator, and I only hope that it won't be as big a piece of junk as the first! (the belligerent attitude)

2. This new desk calculator does not seem to work, and I wonder if you might kindly consider the possibility of sending me a replacement, if that is all right with you. Please accept my humble thanks in advance. (the pardon-me-for-living attitude)

3. I beg to advise you that I am appalled by the patent paucity of workmanship in this calculator and find it imperative that you refund the full price at the earliest possible date. (the pompous-indignation attitude)

4. After I blew all my bread on this bogus calculator, it blew a fuse and blew my mind. Put your money where your mouth is, and send me a refund. (the pass-me-the-joint attitude)

5. Please send me a replacement or a refund for this defective calculator as soon as possible. (the courteous, confident, and direct attitude)

Statement 5 would most likely achieve results. This statement is neither antagonistic nor apologetic. It is courteous, suggests confidence in the reader's integrity, and makes a direct request in plain English. The earlier versions express only the writer's need to sound off (except 2 where the "you" perspective is carried to a grotesque extreme), and would likely end up in the reader's wastebasket. Choose words carefully.

Clear Purpose

Like any effective writing, good letters do not just "happen." Each is the product of a deliberate process. Most effective letters are *rewritten;* words almost never tumble out for a perfect message on your first try. As you plan, write, and revise, answer these questions:

1. *What purpose do I wish to achieve?* (get a job, file a complaint, ask for advice or information, answer an inquiry, give instructions, ask a favor, share good news, share bad news, etc.).

2. *What facts does my reader need?* (measurements, dates, costs, model numbers, enclosures, other details)

3. *To whom am I writing?* (Do you know the reader's name? Whenever possible, write to a person, not a title.)

4. *What is my relationship to my reader?* (Is the reader a potential employer, an employee, a person doing you a favor, a person whose products are disappointing, an acquaintance, a business associate, a stranger?)

Answer those four questions *before* drafting the letter. Then, after you have written a draft, think about the answers to the next three questions, which pertain most directly to the *effect* of your letter on readers. Will they be encouraged to respond favorably?

5. *How will my reader react to my statements as phrased?* (with anger, hostility, pleasure, confusion, fear, guilt, resistance, satisfaction, etc.).

6. *What impression of me will my reader get from this letter?* (intelligent, courteous, friendly, articulate, pretentious, illiterate, confident, servile)

7. *Am I ready to sign my letter with confidence?* (This bears some thought!)

Don't mail your letter until you have answered each question to your full satisfaction. Revise as often as needed to achieve your purpose.

Appropriate Plan for Good or Bad News

The reaction you anticipate from your reader (question 5, above) should determine the organizational plan of your letter: either a *direct* or *indirect* plan. The direct plan puts the point of major interest right in the first paragraph, followed by the explanation. Use the direct plan when you expect the reader to react favorably or want the reader to know immediately the point of your letter (say, in good-news, routine complaint, inquiry, or application letters — or other routine correspondence).

If you expect your reader to react negatively or to need persuading, use the indirect plan: give the explanation *before* the main point (the refusal of a claim, the admission of a mistake, the request for a pay raise). The indirect plan generally makes readers more tolerant of bad news or more receptive to your argument.

INQUIRY LETTERS

Letters of inquiry may be solicited or unsolicited. You often write the first type as a consumer requesting information about an advertised product. You can expect such a letter to be welcomed. After all, the addressee stands to benefit from your interest. In this case, you can afford to be brief and to the point: "Please send me your brochure on . . ." or some such.

Many of your inquiries will be unsolicited, that is, not in response to an ad, but requesting information for a report or project. Here, you are asking a favor of your addressee, who must take the time to read your letter, consider your request, collect the information, and write a response. Therefore, you need to apologize for any imposition, to express appreciation, and to state a reasonable request clearly and briefly (Long, involved inquiries are unlikely to be answered). Begin your letter with something more cordial and less abrupt than "I need some information."

Before you can ask specific questions, do your homework. Don't expect the respondent to read your mind. A general question ("Please send me all your data on . . .") is likely to be ignored. Only when you know your subject can you focus your questions.

Don't wait until the last minute to write your letter. Write at least three weeks before your report is due, politely indicating the due date in your letter.

Here is a typical inquiry situation: Imagine you are preparing an analytical report on the feasibility of harnessing solar energy for home heating in northern climates. You learn that a private, nonprofit research group has been experimenting in ecologically efficient energy systems. After deciding to write for details, you plan and compose your inquiry, basing it on the questions in the previous section.

Structuring the Inquiry

Like most good communication, a routine inquiry has a distinct introduction, body, and conclusion.

Introduction

Introduce yourself, and state your purpose. Your reader should know who wants the information, and why. Maintain the "you" perspective by opening with a statement that will spark reader interest and goodwill.

Body

Write specific and clearly worded questions that are easy to understand and answer. If you have a number of questions, list them rather than lumping them together in one or two paragraphs. (Lists help readers organize their answers, thereby increasing your chances of getting all the information you want.) Number each question, and separate it from the others, perhaps leaving space for responses right on the page. If you have more than five or six questions, you might place them in an attached questionnaire.

Conclusion

Conclude by explaining how you plan to use the information and, if possible, how your reader might benefit. Offer to send a copy of your finished report. Close with a statement of appreciation; it will encourage your reader to respond.

Revising the Inquiry

Revise your letter until its tone and content measure up to the quality of a letter you would like to receive. When you are fully satisfied with your letter, sign it. Include a stamped, return-addressed envelope.

When completed, your letter might look like the one below (or Figure 16–1).

234 Western Road
Arlington, VT 05620
March 10, 1984

Director of Energy Systems
The Earth Research Institute
Petersham, ME 04619

Dear Director:

While gathering data on home solar heating, I encountered references (in *Scientific American* and elsewhere) to your group's pioneering work. Would you please allow me to benefit from your experience?

As a student at Evergreen College, I am preparing a report on the feasibility of solar energy as a major source of home heating in northern climates within a decade. Your answers to the questions below would help me complete my course project (April 15 deadline).

1. At this stage of development, have you found active or passive solar heating more practical?
2. Do you hope to surpass the 60 percent limit of heating needs supplied by the active system? If so, what efficiency do you expect to achieve, and how soon?
3. What is the estimated cost of building materials for your active system, per cubic foot of living space?
4. What metal do you use in collectors so as to get the highest thermal conductivity at the lowest maintenance costs?

Your answers, along with any recent findings you can share, will enrich a learning experience I will put into practice next summer by building my own solar home. I would be glad to send you a copy of my report, along with the house plans I have designed. Thank you.

Sincerely yours,

Alan Greene

Alan Greene

Sometimes, your questions will be too numerous to be answered by letter or questionnaire. Or, instead of several specific questions, perhaps you have one or two very broad questions that would require lengthy answers. In these cases, you might request an informative interview (if the respondent is nearby). Karen Granger, the following writer, wanted a state representative's "opinions on the EPA's progress" in cleaning up local chemical contamination. Because she anticipated a complex answer to her inquiry, she wisely wrote to request an interview.

> 82 Mountain Street
> New Bedford, MA 02740
> March 8, 1984

The Honorable Roger R. Goyette
Massachusetts House of Representatives
Boston, MA 02202

Dear Representative Goyette:

As a technical writing student at Southeastern Massachusetts University, I am preparing a formal report analyzing the progress the EPA has made cleaning up the PCB's in the New Bedford Harbor.

Throughout my preliminary research, I have encountered your name again and again. Your dedicated work has had a definite impact on this situation, and I hope to benefit from your knowledge.

I was surprised to learn that, although this contamination is considered the most serious anywhere, the EPA has still not moved beyond conducting studies. My study assesses the need for such extensive data-gathering. Your opinion, as I can ascertain from *Standard Times* articles, is that they are definitely moving too slowly.

The EPA refutes that argument by asserting they simply do not yet have the information necessary to begin a clean-up operation.

As both a writer and a New Bedford resident, I am very interested in hearing your opinions on the EPA's progress. Perhaps you can lead me toward information on which to base my report. Would you grant me an interview? With your permission, I will telephone you in a few days to make an appointment to discuss this issue. An interview would be an invaluable aid to me.

I would appreciate your assistance and would be glad to send you a copy of my completed report.

> Very truly yours,
>
> *Karen P. Granger*
> Karen P. Granger

CLAIM LETTERS

Claim letters request adjustments for defective goods or poor services, or they complain about unfair treatment, poor policies, or the like. Claim (or complaint) letters fall into two groups: routine claims and arguable claims. Routine claims follow the direct plan, since the claim is backed by a contract, guarantee, or the company's reputation. In contrast, arguable claims call for an indirect plan, since they are open to interpretation.

Routine Claims

In a routine claim, make your request or state the problem in your introductory paragraph; then explain in the body section. Close courteously, repeating the action you request. Do not thank the company in advance; to do so is presumptuous and unnecessary. If your claim is valid, a reputable firm will honor it.

Make the tone courteous and reasonable. You might well be angry or frustrated, but to insult or berate an innocent reader is offensive and ineffective. Your reader could retaliate by burying your claim at the bottom of the pile. Your goal is not to express dissatisfaction, but to achieve results: a refund, replacement, apology, etc. Avoid a belligerent tone, as well as one that is apologetic or meek. Press your claim objectively, yet firmly, by explaining it clearly, and by stipulating whatever *reasonable* action the firm must take to satisfy you.

Always *explain* the problem in enough detail for your reader to understand the basis for your claim. Explain that your new alarm clock never rings, instead of merely saying it's defective. Identify the faulty item clearly, giving serial and model numbers, and date and place of purchase. Then propose a fair adjustment. Conclude with a firm statement that implies your goodwill and confidence in the reader's integrity.

The following writer does not ask whether the firm will honor his claim; he assumes it will. He asks directly how to return the skis for repair. He uses an attention line to direct his claim to the right department. The subject line, and its reemphasis in the first sentence, makes clear the nature of the claim.

> Box 264-A, Route 5 South
> Pocatello, Idaho 83201
> April 13, 1983

Hart Ski Manufacturing Company, Inc.
630 Pierce Butler Route
P.O. Box 3049
St. Paul, Minnesota 55165

Attention: Consumer Affairs Department

Subject: Delaminated Skis

States problem and action desired	This winter, my Hart skis began to delaminate. I want to take advantage of your lifetime guarantee to have them relaminated.
Provides necessary details	I bought the skis from the Ski House in Hadley, Massachusetts in November 1967. Although I no longer have the sales slip, I did register them with you. The registration number is P9965.
Explains basis for claim	I'm aware that you no longer make metal skis, but, as I recall, your lifetime guarantee on the skis I bought was a major selling point. Only your company and Head were backing their skis so strongly.
Ends courteously, specifically stating desired action	Would you please let me know how to go about returning my delaminated skis for repair?

Yours truly,

Jeffrey P. Ryder

Jeffrey P. Ryder

Although Jeff's is a routine claim, sixteen years separate the purchase from the claim on a lifetime guarantee. (Hart honored the claim without question, and Jeff had his skis relaminated.)

Arguable Claims

Arguable claims may include those for poor service or goods, inaccurate shipments or billings, unresolved insurance settlements, and unsettled warranty disputes. When your direct request has been refused, ignored, or is in some way unusual, you must persuade the firm to grant your claim. Say your parked car is wrecked by a drunk driver, and the insurance company appraises the car at $1,500. But two months earlier, you had the engine rebuilt, a muffler system installed, and the front end repaired. By accepting the $1,500, you would lose $1,000. So you write a claim letter, explaining your circumstances and requesting a fair adjustment.

Use the indirect plan for an arguable claim. Readers are more likely to respond favorably *after* reading your explanation. Begin your introduction with a neutral statement both parties can agree to — but which also serves as the basis for your request (e.g., "Customer goodwill is often an intangible quality, but a quality that brings tangible benefits" or "In the past five years, we've been pleased with the services of Burke, Blair, and Bono").

Once you've established initial agreement, explain the basis for your claim by presenting evidence. The body is usually the longest part of your letter because you must offer a convincing explanation. Include all information needed for a fair evaluation of your claim: date and place of purchase, order or policy number, dates of previous letters or phone calls, and full background of the problem or issue.

Conclude by letting your evidence lead to the requested adjustment. Request a *specific action* (a credit to your account, a replacement, a rebate, etc.). You might not get exactly what you request, but by specifying the action, you give the company an idea of what it must do to satisfy you. Ask confidently.

Notice how the writer below uses a tactful, reasonable tone and the indirect plan to achieve her goal.

> OFFICE SYSTEMS, INC.
> 657 High St.
> Tulsa, Oklahoma
> February 23, 1984

Consumer Affairs Dept.
Hightone Office Supplies
93 Cattle Drive
Houston, TX 24871

ATTENTION: Ms. Dione Dubree

Begins with Neutral Statements Both Parties Can Agree To; Tone is Friendly and Courteous

Hightone Office Supplies has deserved its reputation as the most reliable wholesaler of office supplies in the Southwest. And for eight years, we depended on that reliability. Whenever our customers have needed an order filled quickly, we could always count on Hightone's prompt and accurate service. But for the past three months, Hightone has disappointed us. Let me review the details.

Presents Facts to Substantiate Claim for an Adjustment

On November 29, 1983, we placed an order (#675198) for five cartons of Maxell 5¼" diskettes (catalog #A74–866), eight cartons of Scotch 5¼" diskettes (catalog #A74–892), and three cartons of Epson MX 70/80 Ribbon Cartridges (catalog #A19–556).

Continues to Present the Facts

On December 5, we received our order. But instead of the double-sided, double-density Maxells ordered, we received single-sided Verbatim diskettes. Rather than 5¼" Scotch diskettes, we got 8" diskettes. And the Epson ribbons were blue, not black as we had ordered. We returned the order the same day.

Includes all Relevant Information

On the 5th, we also called John Fitzsimmons at Hightone to explain our problem. He assured us we would have the corrected order by the 12th. Finally, on the 22nd we did receive an order — the original incorrect one — with a note attached stating we would

have to accept it because the packages had been
water damaged while in our possession.

Sticks to the Facts —
Writer Accuses
No One

Our warehouse manager states emphatically that the
packages were in perfect condition when he gave
them to the shipper. Since the packages were in our
possession for only five hours, and since it did not
rain on the 5th (as confirmed by a weather report),
we are certain the packages were not damaged while
in our possession.

Requests a Specific
Adjustment

Therefore, the responsibility for damages rests with
either the shipper or your warehouse staff. What
concerns us is our outstanding bill from Hightone
($1,049.50) for the faulty shipment. We insist that
it be cancelled, and that we receive a statement
documenting the correction. Until this episode, our
transactions with Hightone had been excellent. We
hope they may be so again.

Yours truly,

Sandra Lichstein

Sandra Lichstein
Manager, Accounting
Department

Hightone agreed that Office Systems, Inc., had a valid claim, and adjusted
the bill.

For another illustration of a successful request to settle an arguable claim,
assume you are in the following situation. You have recently bought an expensive stereo, with top-of-the-line speakers, from a franchised dealer in New
Jersey. Three weeks after your purchase, you moved to Wisconsin, and five
weeks later you noticed an increasing distortion of heavy bass sounds in your
speakers. The problem is that your guarantee requires that you return the
equipment to the store where you bought it. So you decide to write to the
store, requesting that they arrange for a local franchised dealer to repair or
replace your speakers. Here is how your letter might read:

534 Hartford Way
Madison, WI 20967
March 20, 1984

Stereo Components, Inc.
143 Main Street
Newark, NJ 10311

ATTENTION: Service Department

SUBJECT: Bass Distortion in Toneway Speakers

Because of your well-deserved reputation, I'm sure you will do everything possible to help me with a perplexing service problem.

On December 10, 1983 I bought a component system (sales receipt #114621) from your outlet. Three weeks later, I moved to Wisconsin, and after eight weeks of total use, I noticed distortion of heavy bass sounds in my speakers.

As a classical music fan, I bought your best speakers (Toneway 305's, #3624 and 3625) because of their extra-wide bass range. However, their distortion of lower ranges of percussion and keyboard sounds is increasing to the point of actual vibration, making my expensive system useless.

My speaker guarantee states that items for repair or replacement must be returned to the *original* retailer. But because we are now hundreds of miles apart, that arrangement would cost me a great deal of time and money — and further delay my use of the stereo. Under these circumstances, could you kindly arrange for a franchised dealer in the Madison area to honor my guarantee directly?

I'm sure my friends and relatives back in New Jersey will appreciate hearing about your help.

Yours truly,

Raymond Fields

Raymond Fields

Notice how the closing sentence encourages the reader to take action, for the sake of retaining customer goodwill.

ADJUSTMENT LETTERS

Most businesses assume that the customer is always right, and grant all reasonable claims. Instead of disputing an arguable claim, a firm generally will make the adjustment, believing correctly that most customers are honest. The resultant goodwill usually outweighs the cost of granting a few suspect claims. (Of course, in the case of unreasonable or absurd claims, the company should refuse the adjustment, as shown on pages 363–364.)

Granting Adjustments

The goal of adjustment letters is to keep customers by retaining their goodwill. In fact, reputable firms welcome claim letters, to help them assess and improve their products and service. But goodwill easily can be lost if the adjustment is granted grudgingly, as in the following examples:

> Although you must have failed to follow our instructions for thinning the paint, we have decided to grant your claim because we stand behind our products.

> We're surprised to hear that you are not fully satisfied with your new Path-beater shoes. The only explanation we can give for the poor fit is that you failed to take your measurements correctly. We do offer a money-back guarantee, provided the shoes show no signs of wear. Simply pack the shoes in their original package, and we'll send you a refund.

Such abuse will drive customers elsewhere. People with legitimate claims do not wish to be told they're being "granted" something rightfully theirs. Nor do they wish to be blamed for the problem. Moreover, the second example offers a weak alternative: a refund. Although the customer might be happy to get his money back, he is still left without shoes, having to search anew. If the writer had offered a replacement, the customer would have received satisfaction *and* shoes, and the company would have retained his goodwill.

In the following rewrite of the Pathbeater example, the writer makes the adjustment gracefully, keeping the customer's needs central through a "you" perspective. The adjustment's positive, helpful tone should please the customer, encouraging future purchases.

> We're sorry to learn that your Pathbeaters don't fit comfortably. Despite our efforts to insure that all our customers get a good fit, we sometimes get "the shoe on the wrong foot." To insure a proper fit, would you kindly fill out the enclosed form. As soon as we receive your measurements, we will send your new Pathbeaters by express mail. We'll also send an address label and return postage for the shoes you have. Simply repack them in the box your new Pathbeaters arrive in.

If a claim warrants an adjustment, grant it willingly and positively. Satisfied customers are loyal customers.

When granting an adjustment, begin with that good news. If your firm is at fault, a sincere apology helps rebuild customer confidence. Following the good news, provide an *adequate* explanation of what went wrong. If this is impossible, explain what your company does to maintain quality control. Without an explanation, you leave the impression that such problems are common or beyond control.

When explaining, never blame employees, as convenient scapegoats. To blame someone in the firm reflects poorly on your firm's operations. Also, never say the problem *never* will recur; mishaps are inevitable.

Refusing Adjustments

Favorable adjustments help a company's reputation. But when the customer has misused the product, or is mistaken about company procedures or services, you must write a refusal, using the indirect plan.

A refusal calls for delicate balance. On the one hand, you must clearly explain why you cannot grant the adjustment. On the other, you must diplomatically tell the reader he is mistaken. Maintain this balance by (1) keeping your tone friendly, (2) using the passive voice to avoid accusations ("The wrong size bolts were used," instead of "You used the wrong size bolts"), and (3) subordinating negative details. Otherwise, you might offend the reader.

Anna Jenkins maintains this balance in the following response to a customer who requested a refund on a 10-speed bicycle.

PEOPLE POWER, INC.
101 Salem Street
Springfield, Illinois 32456

Mrs. Alma Gower
32 Wood Street
Lewistown, IL 32432

Dear Mrs. Gower:

Neutral Statement

You're right. The Windspirit *is* rated as the fastest, most durable 10-speed on the market. In fact, the U.S. team has chosen it for the Olympics.

Explanation

When we advertise the Windspirit as the toughest, most durable 10-speed, we stress it's a racing or cruising bike, built to withstand the long, grueling miles of intense competition. The bike is made of the strongest yet lightest alloys available, and each part is calibrated to within 1/1000 of an inch. That's why we can fully guarantee the Windspirit against defects resulting from the strain of competitive racing.

The Windspirit, though, is not built to withstand the impact of ramp jumps such as those attempted by your son. The rims and front fork would have to be made from a much thicker gauge alloy, thereby increasing weight and decreasing speed. Since we build racing bikes, such a compromise is unacceptable.

Refusal

To insure that buyers are familiar with the Windspirit's limits, in the owner's manual we stress that the bike should be carried over curbings and similar drops, since even an 8-inch drop could damage the front rim. Damage from such drops isn't considered normal wear, so it is not covered by our guarantee.

Helpful Close

Since your son appears to be more interested in a bike able to withstand the impact of high jumps, you could recoup a large part of the Windspirit's price

by advertising it in your local newspaper. Many
novice racers would welcome the chance to buy one
at a reduced price. Or if you prefer having it repaired, you could take it to Jamie's Bike Shop, the
dealer closest to you.

Yours truly,

Anna Jenkins

Anna Jenkins
Manager, Customer Services

Although Mrs. Gower may not like Anna Jenkins' explanation, it is thorough
and reasonable. A less tactful writer might have written a reply like this:

> We guarantee the Windspirit against defects resulting from normal wear.
> No reasonable person's definition of *normal* would include using any 10-
> speed — never mind a precision racing machine like the Windspirit — for
> ramp jumps. That's comparable to using a microscope as a hammer. Ob-
> viously, your claim is baseless, and we must therefore refuse it.

At times we are tempted to write such replies. They're easier, and a lot more
fun than a carefully constructed refusal. They're great for the soul, too, but
awful for goodwill. So if you do lose patience and write such a reply, put it
aside for a day or two. When you return, your desire to retain goodwill will
usually lead to a major rewrite.

Clients sometimes are mistaken about policy or contract terms. Explaining
the situation is important, since you want to retain goodwill, confidence, and
business. Because you're giving bad news, use the indirect plan, as in the
following letter.

Explanation and Neutral Statement	After three years of studying legal documents, answering interrogatories, and submitting estimates, you must be relieved to have this lawsuit resolved. We at Shurter, Goldstein, and Brown are pleased we were instrumental in helping you win this case to your satisfaction.
Implied Refusal	You're correct in your reading of paragraph 4 of our contingency fee agreement. We did agree that our fee would be one-third of the gross recovery. The $856 difference you question stems from paragraph 5, which states that we are also entitled to reasonable expenses and disbursements. Paragraph 4 therefore concerns our fees for labor; paragraph 5 concerns expenses incurred while settling your case. These expenses include long-distance telephone calls, subpoenas, court filings, sheriff's fees, and such. (A break-

down of these expenses is enclosed.) Thus our contingency fee for the $20,000 settlement is $6,000; expenses totaled $856, leaving you a balance of $13,144.

Positive Close

Would you please stop by the office at your convenience to sign the two release forms I mentioned last week. Once they are signed, we can officially close this case, and you will receive the balance owed you by the Clevelands.

RÉSUMÉS AND JOB APPLICATIONS

In today's job market, many applicants compete for few openings. Whether you are applying for your first professional job or changing careers in midlife, you have to wage an effective campaign for marketing your skills. Your résumé and letter of application must stand out among those of your competitors.

Job Prospecting: The Preliminary Step

Before writing a letter and résumé, do some prospecting: study the job market to identify realistically the careers and jobs for which you qualify.

Being Selective

Many new graduates make the mistake of applying for too broad a range of jobs, including many for which they have no real qualifications. Such a shotgun approach is ambitious, but may decrease your chances — and can be most discouraging.

Concentrate on openings for which you are best suited. Begin by evaluating your skills, aptitudes, and general goals:

- What skills have you acquired in school, in jobs, from hobbies and interests?
- What can you do well: write, draw, speak other languages, organize, lead, instruct, socialize, sell, solve problems, think creatively?
- Are you good at making decisions?
- Are you better at original thinking or following directions?
- Are you looking for security, excitement, money, travel, power, prestige, or something else?

Besides helping focus your search, answers to the above questions will be handy when you write your résumé.

If your definite goal is to join some major corporation, consult the following library resources for information about prospective employers.

– *Moody's Industrial Index* or Standard and Poor's *Register of Corporations*, for data on plant locations, major subsidiaries, products, executive officers, and corporate assets.

– *The Business Index*, for articles about particular companies in publications such as *Wall Street Journal, Fortune, Forbes, The New York Times, Business Week*, and *Office Administration and Automation*.

– The yearly issue of *Fortune* Magazine that lists "The Fortune 500" companies.

– Annual reports, for data on a company's assets, innovations, recent performance, and prospects. (Many libraries collect annual reports in a separate file cabinet. Ask your librarian.)

Once you have a general direction, you can begin the actual job search.

Launching Your Search

Launch your job campaign early. Don't wait for the job to come to you. Take the initiative by observing these suggestions:

1. As early as your sophomore year, begin scanning the want ads; major newspapers publish an entire Help Wanted section in Sunday editions. Here you will find descriptions, salaries, and qualifications for countless jobs.

2. Ask your reference librarian to point out occupational handbooks, government publications, newsletters, and magazines or journals in your field.

3. Visit your college placement service; here, openings are posted; interviews are scheduled; and counselors have advice about job hunting.

4. Speak with someone working in your field for an inside view and practical advice.

5. Sign up at your placement office for interviews with company representatives who visit the campus.

6. Seek the advice of faculty in your major who do outside consulting or who have worked in business, industry, or government.

7. Look for a summer job in your field; this experience may count more than your education.

8. Establish contacts; don't be afraid to ask for advice. Make a list of names, addresses, and phone numbers of people willing to help.

9. Many professional organizations invite student memberships (at reduced fees). Such affiliations can generate excellent contacts, and they look good on the résumé. If you do join a professional organization, try to attend meetings of the local chapter.

By taking the above steps well before your senior year, you may learn that certain courses make you more marketable. In many fields, some knowledge of computers is desirable; certain nursing positions require counseling experience, and so on. Tailor your final semesters to these requirements (taking introductory computer courses, taking counseling courses and doing volunteer work for a community service organization, or the like).

If you're changing jobs or careers, employers will be more interested in what you've accomplished *since* college. Be prepared to show how your work experience is relevant to the new job. Capitalize on the contacts you've made over the years. Collect on favors owed.

Whether you're a beginner or a veteran, you might register with an employment agency. Of course, a fee is payable after you're hired, but in many cases the employer pays the fee. Ask about fee arrangements *before* you sign up.

Being Realistic

When you do apply for jobs, be realistic. If the advertised requirements include years of experience or a master's degree, look elsewhere. Sometimes, however, the gap between your qualifications and those required might not be great. In this case, an enthusiastic, well-written letter, along with alternative qualifications (related volunteer work or outside interests) might land you the job. Rely on your good judgment and the advice of placement counselors and faculty members. Don't hesitate to ask for advice! Besides applying for advertised openings, you might write to organizations that have not advertised. Both solicited and unsolicited letters are discussed later.

Once you have a picture of where you fit into the job market, you will set out to answer the big question asked by all employers: "What do you have to offer?" Your answer must be a highly polished presentation of yourself, your education, work history, interests, and special skills — your résumé.

The Résumé

The résumé is a summary of your experience and qualifications. Written before your application letter, it provides background information to support your letter. In turn, the letter will emphasize specific parts of your résumé, and will discuss how your background is suited to that particular job. The résumé gets you the interview, not the job.

Most employers spend less than 60 seconds scanning a résumé. They look for an obvious and persuasive answer to this question: What can you do for us? Any employer expects a résumé:

1. to look good (conservative and tasteful, on high-quality paper),

2. to read easily (headings, typeface, spacing, and punctuation that provide clear signals), and

3. to provide information the employer needs to make an interviewing decision.

Most employers discard immediately résumés that are mechanically flawed, cluttered, sketchy, or hard to follow. So don't leave readers guessing or annoyed; make your résumé perfect.

Organize your information within these categories:

- name and address
- career objectives
- education
- work experience
- personal data
- interests, activities, awards, and skills
- references

Select and organize your material to best show what you have to offer. Don't just list *everything;* be selective. (We're talking about *communicating* instead of merely delivering information.) Don't abbreviate, because not all readers may know the referent. Use punctuation to clarify and emphasize, not to be "artsy" or "unique." Try to limit your résumé to one page.[1]

Begin your résumé at least one month before your job search. You will need that much time to do a first-class job. Your final version can be duplicated for various targets (but each letter has to be original).

Name and Address

Under the first heading, include your full name and mailing address (if different), and phone number (many interview invitations and job offers are made by phone). If your school address and phone number differ from your summer address and number, include both, indicating dates you can be reached at each.

Career Objectives[2]

From your collected information (steps 1–9, listed earlier), you should have a clear idea of the *specific* jobs for which you *realistically* qualify. Resist the

---

[1] Of course, if you are changing jobs or careers, or if your résumé looks cramped, you might need a second page.

[2] Be prepared to have different versions for different résumés.

impulse to be all things to all people. The key to a successful résumé is the image of *you* it projects — disciplined and purposeful, yet flexible. State both immediate and long-range goals, including plans for continuing your education:[3]

> My immediate goal is to join the intensive-care nursing staff of an urban teaching hospital. Through on-the-job experience and part-time graduate study in crisis treatment and life-support systems, I hope eventually to supervise an intensive-care unit and instruct student nurses.

Your statement of career objectives should show you have a clear sense of purpose and have given serious thought to your future. *Do not* borrow a trite statement of objectives from a placement office brochure.

Educational Background

If your education is more impressive than your work experience, place it first. Begin with your most recent school and work backward, listing degrees, diplomas, and schools *beyond* high school (unless prestige, program, or your achievement warrant its inclusion). List the courses that have directly prepared you for your career. If your class rank is in the upper 30 percent, mention it. Include any schools attended or courses completed while you were in military service. If you financed part or all of your education by working, say so, indicating the percentage of your contribution.

Work Experience

If you have solid work experience, place it before your education. Beginning with your most recent job and working backward, list and clearly identify each job, giving dates and names of employers. Tell whether the job was full time, part time (hours weekly), or seasonal. Tell exactly what you did in each job, indicating promotions. If the job was major (and related to this one), describe it in detail; otherwise, describe it briefly. Include any military experience. If you have no real experience, show you have potential by emphasizing your preparation, and by writing an enthusiastic letter.

Do not use complete sentences in your job descriptions; they take up room best left for other items. But do use action verbs throughout (*supervised, developed, built, taught, opened, managed, trained, solved, planned, directed,* and so on). Such verbs emphasize your vitality, and help you stand out.

---

[3] To save space, you can omit your statement of career objectives from the résumé and include it in your letter instead.

Personal Data

Title VII of the Civil Rights Act of 1964 (amended in 1972 and 1978) has made it illegal for an employer to discriminate on the basis of sex, religion, color, age, or national origin. Therefore, you aren't required to provide this information or a photograph. But if you believe that any of this information could advance your prospects for the job, by all means include it.

Personal Interests, Activities, Awards, and Skills

List hobbies, sports, and other pastimes; memberships in teams and organizations; offices held; and any recognition for outstanding performance. Include dates and types of volunteer work. These items give employers a profile of such important traits as variety of interests, creative use of leisure time, concern for personal growth and community welfare, team spirit, ability to work within a group, and leadership qualities. A history of volunteer work suggests you give freely of your time without concern for material gain. Employers know that persons who seek well-rounded lives are likely to take an active interest in their jobs. Who you are away from work helps define who you will be on the job.

Be selective in this section. List only items that show the qualities employers seek.

References

Your list of references names four or five people *who have agreed* to write strong, positive assessments. Often a reference letter is the key to getting an employer to want to meet you, so choose your references carefully.

Select references who can speak with authority about your ability and character. Avoid members of your family and close friends not in your field. Choose instead among professors, previous employers, and community figures who know you well enough to write concretely on your behalf. In asking for a reference, keep these points in mind:

1. *A mediocre letter of reference is more damaging than no letter at all.* So don't merely ask, "Could you please act as one of my references?" This question leaves the person little chance to say no. He or she might not know you well or may not be impressed by your work but, instead of refusing, might write a watery letter that will do more harm than good. Instead, make an explicit request: "Do you feel you know me and my work well enough to write me a strong letter? If so, would you act as one of my references?" This second version gives people the option to decline gracefully. Otherwise, it elicits a firm commitment to a positive letter.

2. *Letters are time-consuming.* References have no time to write individual

letters to every prospective employer. So ask for only one letter, with no salutation. Your reference keeps a copy; you keep the original for your personal dossier (so you can reproduce it as necessary); and a copy goes to the placement office for your placement dossier. Because the law permits you to read all material in your dossier, this arrangement provides you with your own copy of your credentials.[4] (The dossier is discussed later in this chapter.)

If the people you select as references live far away, you may wish to make your request by letter, like this one:

> Dear Mr. Knight:
>
> From September 1977 to August 1979, I worked for you at Teo's Restaurant as waiter, cashier, and then assistant manager. Because I enjoyed my work, I decided to study for a career in the hospitality field.
>
> In three months I will graduate from San Jose City College with an A.A. degree in Hotel and Restaurant Management. Next month I will begin my job search. Do you feel you know me and my work well enough to write me a strong letter of recommendation? If so, would you be one of my references?
>
> To save your time, please omit the salutation from the letter. If you could send me the original, I will forward a copy to my college placement office.
>
> To update you on my recent activities, I've enclosed my résumé. I appreciate your help.

Opinion generally is divided about whether the names and addresses of references should be included in a résumé. If saving space is important, simply state, "References available on request," keeping your résumé only one page long, but, if your other résumé items take up more than one page, you probably should include names and addresses of references. (An employer might recognize a name, and thus notice *your* name among the crowd of applicants.) If you are changing careers, a full listing of references is especially important.

Composing the Résumé

With data collected and references lined up, you are ready to compose your actual résumé. Imagine you're a twenty-four-year-old student about to graduate from a community college with an A.A. degree in Hotel and Restaurant Management. Before college, you worked at related jobs for over three years.

[4] Under some circumstances you may — and may wish to — waive the right to examine your recommendations. Some applicants, especially those applying to professional schools as in medicine and law, do waive this right. They do so in concession, one supposes, to a general feeling that a letter writer who is assured of confidentiality is more likely to provide a balanced, objective, and reliable assessment of a candidate. In your own case, you might seek the advice of your major advisor or a career counselor.

You now seek a junior management position with a nationwide hospitality chain while you continue your education, part time. You have spent two weeks compiling and selecting information for your résumé, and getting commitments from four references. Figure 16–6 shows your résumé. Notice that this résumé mentions nothing about salary. Wait until this matter comes up in your interview, or later. And because James Purdy is not entering a beauty contest, he includes no photograph. The information, specific but concise, describes what he has to offer, and requires only a moment to read. There are no statements of self-praise like "I did such a terrific job I was promoted to manager." The facts speak for themselves.

As a final check, look at your résumé as a personnel director might, and analyze your presentation for weaknesses. Below is Personnel Director Mary Smith's assessment of Purdy, in a memo to colleagues.

JANUARY 20, 1981

TO: Hiring Committee

FROM: Mary Smith,
              Personnel Director

SUBJECT: Follow-up on James Purdy's
              Application (copy enclosed)

I recommend we pursue James Purdy's candidacy. This applicant shows strong purpose and responsible planning. His background provides detailed and specific support for his stated plans.

The fact that he financed his own education yet achieved a high cumulative average indicates he is both a diligent and capable worker. His courses, along with experience in sales, food service, and hospitality suggest his career choice is based on sound knowledge of the hospitality field and an obvious talent for direct customer contact.

His history of job promotions indicates that the quality of his work has impressed employers repeatedly. His skiing ability and interest in cooking and sailing make him a strong candidate for one of our northern resorts. His activities and awards suggest he works well with others (including youngsters), is respected by his peers, has leadership qualities, and is willing to volunteer his time and talent. Finally, his foreign-language skill could be an asset to our customer relations division.

Overall, James Purdy seems a well-rounded person who knows what he wants and who can offer youth, enthusiasm, and experience. He promises to be a responsible employee who will continue to improve personally and professionally.

For a résumé composed by a person with extensive experience, see Figure 16–7. In that example, notice that work experience is placed before education, and that names and addresses of references are included.

James David Purdy

203 Elmwood Avenue
San Jose, California 90462
Tel.: 214-316-2419

Professional Objective	To work in customer relations for a hospitality chain, to continue my education part time, and eventually to assume market management responsibilities.
Education 1979–1981	*San Jose City College, San Jose, California* Associate of Arts Degree in Hotel/Restaurant Management, June 1981. Grade point average: 3.25 of a possible 4.00. All college expenses financed by scholarship and part-time job (20 hours weekly).

Employment

1979–1981	*Peek-A-Boo Lodge, San Jose, California* Began as desk clerk and am now desk manager (part time) of this 200-unit resort. Responsible for scheduling custodial and room service staff, convention planning, and customer relations.
1977–1979	*Teo's Restaurant, Pensacola, Florida* Began as waiter, advanced to cashier, and finally to assistant manager. Responsible for weekly payroll, banquet arrangements, and supervised dining room and lounge staff. Left to enroll in college.
1976–1977	*Encyclopedia Britannica, Inc., San Jose, California* Sales representative (part time). Received top bonus twice.
1975–1976	*White's Family Inn, San Luis Obispo, California* Worked as bus boy, then waiter (part time).
Personal	*Awards* Captain of basketball team, 1976; Lions' Club Scholarship, 1979.
	Special Skills Speak French fluently; expert skier.
	Activities High School basketball and track teams (3 years); college student senate (2 years); Innkeepers' Club – prepared and served monthly dinners at the college (2 years).
	Interests Skiing, cooking, sailing, oil painting, and backpacking.
References	Available from: Placement Office, San Jose City College, San Jose, CA 90462

FIGURE 16–6   A Résumé for an Entry-Level Candidate

Peter Arthur Profitt
14 Cherokee Road
Tucson, Arizona 85703
Telephone: Home: 602-516-1234
            Office: 602-567-5000

Qualifications and Career Objectives

Comptroller, designer of data processing systems, international sales, manager of large foreign office, manager of accounting firm, budget officer.

My immediate goal is to continue my career in fiscal/budget management, in a position with major challenges and responsibilities. Continuing my formal education part time, I plan to fit myself for top executive responsibilities.

Work Experience

1972-present   Comptroller, Datronics, Phoenix, Arizona — Oversee formation of fiscal policies of Datronics; develop appropriate operational procedures; maintain overall coordination of daily business activities, including, for example: (1) supervise development and operation of an accounting system including payrolls, operation and capital equipment budgets, and R&D funds; (2) advise president in forming company policies, plans, and procedures; (3) oversee receipt and control of operational revenues and expenditures; (4) prepare annual budgets and long-range fiscal policy for directors' approval.

1965-1972   Assistant Manager, then Manager, Financial Operations, Abernathy's, New York — (1) supervise maintenance of operations accounts; (2) supervise expenditure and receipt of funds (under vice-president for operations); (3) develop forms/procedures for accounting, purchasing, cost systems, and computerizing of entire financial operation; (4) establish and supervise inventory control system; (5) assist in preparation of budgets, financial data, and reports for immediate and long-term use.

1960-1965   Payroll Manager, Milene's Boston — (1) Supervise payroll department, including preparation of all branch store payrolls, deductions, etc.; (2) responsible for state/federal payroll audits; (3) direct issuance of U.S. Savings Bonds; (4) assist in budget estimates of employee costs and promotions; (5) prepare periodic payroll reports.

1958-1960   Assistant in Fiscal Management, Milene's, Boston — (1) supervise three employees in preparation of departmental payrolls and maintenance of relevant personnel records and files; (2) interview and recommend applicants for clerical employment; (3) code and computer index file material.

Note: At both Abernathy's and Datronics I established the training programs for accounting and computer personnel — programs still being carried on.

FIGURE 16–7   A Résumé for a Person with Extensive Experience

2

Education Background

M.B.A. candidate, University of Tucson, 1974

Related graduate-level courses: Cases in Personnel Management, Advanced Cost Accounting, Statistical Analysis of Business Trends, Computerized Payroll Systems Development, and others

B.S., Accounting, Northeastern University, Boston, 1958 – graduated *cum laude*

A.S., Business Administration, Mass. Bay Community College, Watertown, 1956

Certificate, Proficiency in French, WSAFI, Stuttgart, Germany, 1953

Certificate, Data Processing Specialist, U.S. Air Force Base, Omaha, 1951

Personal Interests, Activities, Awards, and Special Skills

Interests: Native American archaeology, skiing, chamber music (I am first violin in an amateur group), gourmet cooking, whitewater canoeing, French and German literature.

Activities: 1968 Class Agent, Northeastern University; Rotary Club chapter president (1 year), Framingham, Mass.; United World Federalists chapter treasurer (3 years), Beverly Mass.,; Beverly Hospital Fund chairman (4 years); American Field Service chapter president (3 years), Tucson; United Fund (Commercial) chairman (2 years), Tucson; Sierra Club member (10 years).

Awards: Young Executive of the Year, Beverly Chamber of Commerce, 1964; Record Fund Raising Award, United Fund, Tucson, 1975; Alumni Fund Awards (for highest total), Northeastern University (2 years).

Special Skills: Written and oral fluency – French and German; conversational Spanish; computer operations; successful training programs.

References

Mrs. James Stirling Fell, President Datronics, 1142 Arroyo Grande, Phoenix, AZ 85903

Dr. Walter J. Enos, Vice-President (Operations), Milene's, Box 1000, Boston, MA 02114

Mr. Albert Fresco, President, Abernathy's, 500 Fifth Avenue, New York, NY 10014

Mr. Peter S. Pence, Chairman, United Fund, 5 Union Place, Tucson, AZ 02103

FIGURE 16–7   (*Continued*)

When fully satisfied with your résumé, consider having your model printed. For about $50 you can obtain a better-looking copy than a typewriter could produce. This one prototype, in turn, will yield as many copies as you need. For neat copies, use a photocopying machine or offset printing; *never* send out carbon, thermofax, or mimeographed copies.

A final suggestion: avoid using a résumé-preparing service. Although professionals can use your raw materials to produce an impressive résumé, employers often recognize the source by its style, and could conclude that you are incapable of communicating on your own.

Now, with your résumé prepared, you are ready to plan and compose the job application letter.

The Job Application Letter

Your Image

Your job application letter is one of the most important pieces you will ever write. Depending on its quality, your letter either will open doors or be a waste of time. Although it expands on your résumé, you must emphasize personal qualities and qualifications in a convincing way. Here you project an image of your personality. In your résumé, you merely present raw facts; in your application letter, you discuss these facts. The tone and insight you bring to your discussion suggest a good deal about who you are. The letter is your chance to explain how you see yourself fitting into the organization. Your purpose is to interpret your résumé and show an employer how valuable you will be.

People are uncomfortable talking about themselves in letters to strangers. They often feel they can say little without seeming conceited, but the most essential ingredient in your letter is confidence. If you don't believe in yourself, who will? Your letter is a sales letter: it markets your greatest commodity — *you*. To be effective, it *must* stand out among other applications — without being kinky or cute.

Targets

Unlike the résumé, the letter should never be photocopied. Although you can base letters to different employers on the same model — with appropriate changes — type each letter freshly.

The immediate purpose of your letter is to secure an interview. Therefore, the letter itself should make the reader want to meet you. Make your statements engaging, precise, and *original*. Borrowed phrases from textbook examples and "letterese" will not do the job.

Sometimes you will apply for jobs advertised in print or by word of mouth (solicited applications). At other times you will write prospecting letters to organizations that have not advertised but might need someone like you (unsolicited applications). Either letter should be tailored to the situation.

The Solicited Letter

Imagine you are James Purdy. In *Innkeeper's Monthly,* you read this advertisement and decide to apply:

RESORT MANAGEMENT OPENINGS

Liberty International, Inc. is accepting applications for several junior management positions at our new Lake Geneva Resort. Applicants must have three years practical experience, along with formal training in all areas of hotel/restaurant management. Please apply by June 1, 1984 to:

Elmer Borden
Personnel Director
Liberty International, Inc.
Lansdowne, Pennsylvania 24135

Now, plan and compose your letter, using the questions on pages 352–353 as a guide.

Introduction. Create a confident tone by directly stating your reason for writing. Name the exact job, and remember you are talking *to* someone; use "you" often (especially important here, where you must talk about yourself without seeming conceited). If you can, establish a connection by mentioning the name of a mutual acquaintance; say you learn that your professor of nutrition, Dr. H. V. Garlid, is a former colleague of Elmer Borden; mention of his name — with permission — will call attention to your letter. Finally, after referring your reader to your enclosed résumé, discuss your qualifications.

Body. Concentrate on what you, specifically, can bring to *this* job. Don't come across as a jack-of-all-trades. Avoid flattery ("I am greatly impressed by your remarkable company"). Be specific. Replace "much experience," "many courses," or "increased sales" with "three years of experience," "five courses," or "a 35-percent increase in sales between June and October 1983." Always support your claims with *evidence,* and show how your qualifications will benefit this employer. Create a dynamic tone by using *active* rather than *passive* voice, as well as action verbs:

Weak	Increased responsibilities were steadily given to me.
Stronger	I steadily assumed increasing responsibilities.

Trim the fat from your sentences:

Flabby	I have always been a person who enjoys a challenge.
Lean	I enjoy a challenge.

Project confidence with your language:

Unsure	It is my opinion that I will be a successful manager because. . . .
Certain	I will be a successful manager because. . . .

Never be vague:

Vague	I am familiar with the 1022 interactive database management system, and RUNOFF, the text-processing system.
Definite	As a lab grader for one semester, I kept grading records on the 1022 database management system, and composed lab procedures on the RUNOFF text-processing system.

Also, avoid "letterese." Write as you speak in the classroom. Describe how you will fit in, and remember that an enthusiastic tone can go a long way. Your attitude can be as important as your background. Show a clear sense of purpose.

Conclusion. Restate your interest in the job; emphasize your flexibility and willingness to retrain (if necessary); and review other important personal qualities. If your reader is nearby, request an interview; otherwise, request a phone call, stating times you can be reached. Your conclusion should leave your reader with the impression you are more than a name on a page; you are worth knowing.

Revision. **Never** settle for a first draft — or even a second or third! Perhaps by your fourth you will be approaching the best possible answers to our guide questions. And because this letter is your model for letters serving a variety of circumstances, it must be your best effort. When you are sure your letter has quality content, an appropriate tone, and an impeccable format, sign it. If you still feel unsure, revise again.

Purdy's letter to Elmer Borden was the product of careful attention to these principles. After several revisions, he finally signed the letter in Figure 16–8.

203 Elmwood Avenue
San Jose, California 10462
April 22, 1981

Mr. Elmer Borden
Personnel Director
Liberty International, Inc.
Lansdowne, Pennsylvania 24135

Dear Mr. Borden:

Please consider my application for a junior management position at your Lake Geneva resort. I will graduate from San Jose City College on May 30 with an Associate of Arts degree in Hotel/Restaurant Management. Dr. H. V. Garlid, my nutrition professor, has told me of his experience as a consultant for Liberty International, which further encouraged me to apply.

For two years I worked as a part-time desk clerk and now am the desk manager at a 200-unit resort. This experience, combined with earlier customer contact work in a variety of situations (see enclosed résumé) has given me a clear and practical understanding of customers' needs and expectations. As an amateur chef, I know of the effort, attention, and patience required to prepare fine food. Moreover, my skiing and sailing background might well be assets to your resort's recreation program.

I have confidence in my managerial ability and am determined to succeed in the hospitality field. My experience and education have strengthened the most vital skills any management professional can have: to work well with others and to respond creatively to changes, crises, and added responsibilities.

I would very much like to discuss how my placement with Liberty International would benefit us both. Would you please phone me any weekday after 4 p.m. at 214-316-2419? I hope to hear from you soon.

Yours truly,

James David Purdy

James David Purdy

Enclosure

FIGURE 16–8   A Solicited Job Application Letter

He wisely emphasized practical experience because his background is varied and impressive. An applicant with less experience would emphasize education instead, discussing courses and activities.

The Unsolicited Letter

Ambitious job seekers do not limit their search to advertised openings. The unsolicited, or "prospecting," letter is a good way to uncover other possibilities. Such letters have advantages and disadvantages.

Disadvantages. The unsolicited approach has two drawbacks: (1) You might waste time writing to organizations that have no openings. (2) Because you don't know what the opening is — if there is one — you cannot tailor your letter to the specific requirements.

Advantages. This cold-canvassing approach does have one big advantage: for an advertised opening you will compete with legions of qualified applicants, whereas your unsolicited letter might arrive just when an opening has materialized. If it does, your application will receive immediate attention and you just might get the job! Even when there is no immediate opening, the company may file an impressive application until an opening does occur. Or the application may be passed along to a company that has an opening.

There are often good reasons for going beyond the Help Wanted columns. Unsolicited letters generally are a sound investment if your targets are well chosen and your expectations are realistic.

Reader Interest. Because your unsolicited letter is unexpected, attract your reader's attention early, and make him or her want to read on. Don't begin: "I am writing to inquire about the possibility of obtaining a position with your company." By now, your reader is asleep. If you can't establish a connection through a mutual acquaintance, use a forceful opening like this:

> Does your hotel chain have a place for a junior manager with a college degree in hospitality management, a proven commitment to quality service, and customer relations experience that extends far beyond mere textbook learning? If so, please consider my application for a position.

Unlike the usual, time-worn, and plastic openings, this approach gets through to your reader.

The Prototype

Many of your letters, solicited or unsolicited, can be versions of one model, or prototype. So your prototype must represent you in the best possible light.

Take plenty of time to compose the model letter and résumé. Employers will regard the quality of your application as an indication of the quality of work you will do. Businesses spend much money and time projecting favorable images. The image you project, in turn, must measure up to their standards.

SUPPORTING YOUR APPLICATION

Your Dossier

Your dossier is a folder containing your credentials: college transcript, letters of recommendation, and any other items (such as a notice of a scholarship award or letter of commendation) that document your achievements. In your letter and résumé, you talk about yourself; in your dossier, others talk about you. An employer impressed by what you say about yourself will want to read what others think, and will request a copy of your dossier. By collecting recommendations in one folder, you spare your references from writing the same letter over and over.

Your college placement office will keep your dossier on file and send copies to employers who request them. Always keep your own copy. Then, if an employer requests your dossier, you can make a photocopy and mail it, advising your reader that the placement office copy is on the way. This is not needless repetition! Most employers establish a specific timetable for (1) advertising an opening, (2) reading letters and résumés, (3) requesting and reviewing dossiers, (4) interviewing, and (5) making offers. Obviously, if your letter and résumé do not arrive until the screening process is at step 3, you are out of luck. The same is true if your dossier arrives when the screening process is at the end of step 4. Timing is crucial. Too often, dossier requests from employers sit and gather dust in some "incoming" box in the placement office. Weeks can pass before your dossier is mailed. The only loser in this case is you.

Interviews

An employer impressed by your credentials will phone or write to invite you for an interview. You are one of the finalists. But now you must show the interviewer(s) you are as impressive in person as you seemed on paper — and the best person for the job.

You may meet with one interviewer, a group, or several groups in succession. You may be interviewed alone or with several candidates at once. Interviews can last one hour or less, a full day, or even several consecutive days. The character of the interview can range from a pleasant, informal chat to grueling quiz sessions. Some interviewers may antagonize you deliberately to observe your reaction ("Whatever made you imagine that you could fill this position?"). If that happens, suppress your annoyance; look the interviewer straight in the eye, and answer calmly and confidently.

Your library has books with more detailed interview advice than we have space for here; because confidence is essential, prepare by studying techniques of being interviewed. Better yet, get practice. Take as many interviews as you can. Your confidence and competence will increase with each.

Prepare for the interview by learning about the company (its products or services, history, prospects, branch locations) in trade journals and industrial registries or indexes. If time permits, request company brochures and annual reports. Prepare specific answers to the obvious questions:

> Why do you wish to work here?
> What do you consider to be your strongest quality?
> What do you see as your biggest weakness?
> Where would you like to be in ten years?

Plan informative and direct answers to questions about your background, training, experience, and salary requirements. Project a sense of purpose. Prepare your own list of questions about the job and the organization; you will be invited to ask questions, and the questions you ask can be as revealing as the answers you give.

If you've prepared, your interview should be a stimulating experience where you and the employer can learn much about each other. The purpose of the interview is to confirm the impressions an employer has from your application. Be yourself. You have specific skills and a unique personality to offer. Busy people are taking the time to speak with you because they recognize your worth. Knowing this, you can enter the interview with confidence.

Know the exact time and location of the interview. Come dressed as if you already work for the company. Maintain eye contact most of the time; if you stare at your shoes, the interviewer will not be impressed. Relax in your chair, but don't slouch. Don't smoke, even if invited. Don't pretend to know more than you do; if you can't answer a question, say so. Avoid abrupt yes or no answers, as well as life stories. Make your answers concrete but to the point.

Don't be afraid to allow silence. An interviewer simply may *stop* talking, just to observe your reaction to silence. If you really have nothing more to say or ask, don't feel compelled to speak; caught off-guard, you're likely to blather. Instead, prepare for silences so you can feel comfortable if they occur. Let the interviewer make the next move.

When your interviewer hints the meeting is ending (perhaps by checking his or her watch), take the cue. Restate your interest in the job; ask when you can expect further word; thank your interviewer; and leave.

The Follow-up Letter

Within a few days of your interview, jog the employer's memory with a letter of thanks that restates your interest. Here is Purdy's follow-up to his interview with Elmer Borden.

> Thank you again for your hospitality during my visit to your Lansdowne offices. The trip and scenery were delightful.
>
> After meeting you and your colleagues, and touring the resort, I remain convinced that I could be a productive member of your staff.

The Letter of Acceptance

If all goes well, you will receive a job offer by phone or letter. If by phone, request a written offer, and respond with a formal letter of acceptance. This letter may serve as part of your contract; so be sure to spell out the terms of the offer you are accepting! Here is Purdy's letter of acceptance.

> I am happy to accept your offer of a position as assistant recreation supervisor at Liberty International's Lake Geneva Resort, with a starting yearly salary of $18,500.
>
> As you requested, I will phone Ms. Druid in your personnel office for final instructions on reporting date, physical examination, and employee orientation.
>
> I look forward to a long and satisfying career with Liberty International.

The Letter of Refusal

You may find yourself in the enviable position of having to refuse offers. Even if you refuse by phone, write a prompt and cordial letter of refusal, explaining your reasons, and leaving the door open for future possibilities. You may find later that you dislike the job you accepted, and wish to explore old contacts. Here is how Purdy handled a refusal.

> Although I was impressed by the challenge and efficiency of your company, I must decline your offer of a position as assistant desk manager of your Beirut hotel. I have taken a position with Liberty International because it will allow me to complete the requirements for my B.S. degree in hospitality management full time.
>
> If any later openings should materialize, however, I would again appreciate your considering me as a candidate.

Thank you for your interest and courtesy, and best wishes for continued success.

CHAPTER SUMMARY

The letter is a more personal form of communication than the report; its tone must connect with the reader. Never mail a letter until you feel good about signing it.

Any effective letter has an appropriate format, form, and plan, is written in conversational language with a "you" perspective, and expresses a clear purpose. Three common types of letters are:

- the inquiry letter, requesting information
- the claim letter, registering dissatisfaction and requesting adjustment
- the adjustment letter, either granting or refusing a claim

Follow these steps in writing your job application:

1. Spend time job prospecting.
2. Begin your résumé and letter early, and plan on many revisions.
3. Compose your résumé as an inventory of your qualifications. Make a perfect model from which you can run off photocopies.
4. Compose your letter, emphasizing the major features from your résumé. Project confidence without being pompous. Tailor your letter to fit the job and the application situation (solicited or unsolicited). Write a fresh letter (based on your prototype) for each application.

Follow these steps in supporting your application:

1. Compile your dossier, and retain control of its distribution.
2. Prepare well for interviews, and take as many as you can, to sharpen your skills.
3. Write follow-up letters after completing interviews.
4. Request job offers in writing, and respond with a letter of acceptance.
5. For offers you refuse, send refusal letters that leave doors open.

REVISION CHECKLIST FOR LETTERS

Use this checklist to refine the content, arrangement, and style of your letters.

Content

1. Does the letter contain all standard parts (heading, inside address, salutation, letter text, complimentary close, signature)?

2. Does it contain all needed specialized parts (attention and/or subject line, typist's initials, enclosure notation, distribution notation, postscript)?

3. Have you given the reader all necessary information?

4. Have you identified the name and position of your reader?

Arrangement

1. Does the letter have an introduction, body, and conclusion?

2. Does a good-news letter follow the direct plan? The bad-news letter, the indirect?

3. Is the format appropriate (good paper, neat typing, uniform margins and spacing)?

4. Does the letter follow an accepted form (semiblock or modified block)?

Style

1. Is the letter phrased in conversational language (free from "letterese")?

2. Does the letter reflect a "you" perspective?

3. Is the opening paragraph designed to interest the reader?

4. Does the tone reflect your relationship with your reader?

5. Is the reader likely to derive a favorable impression from this letter?

6. Is the closing paragraph designed to encourage the reader to act?

7. Are all sentences clear, concise, and fluent?

8. Is the letter in correct English (Appendix A)?

9. Does the letter's appearance enhance the writer's image?

Now list those elements of the letter that need improvement.

EXERCISES

1. Bring to class a copy of a business letter addressed to you or a friend. Compare letters. Choose the most and least effective.

2. Write and mail an unsolicited letter of inquiry about the topic you have investigated, or will investigate, for an analytical report or research assignment. Your letter may request brochures, pamphlets, or other informative literature, or it may ask specific questions ("What chemicals are used to clean algae, barnacles, and other marine vegetation from the cooling system's filters?" "Are these chemicals then discharged into the sea?"). Submit a copy of your letter and your addressee's response to your instructor.

3. a. As a student in a state college, you learn your governor and legisla-

ture have cut next year's operating budget for all state colleges by 20 percent. This cut will cause the firing of many young and popular faculty members, a drastic reduction in student admissions, reduction in financial aid, cancellation of new programs, and erosion of college morale and quality of instruction. Write a complaint letter to your governor or your legislative representative, expressing your strong disapproval and justifying a major adjustment in the proposed budget.

b. Write a complaint letter to a politician about some issue affecting your school or community.

c. Write a complaint letter to an appropriate school official about a campus problem.

4. Besides providing data about one's background, what does a letter of application say about an applicant? Explain in a short essay titled "Reading Between the Lines."

5. Write a 500-word essay explaining your reasons for applying to a specific college for transfer or for graduate or professional school admission. Be sure your essay covers two general areas: (1) what you can bring to this school by way of attitude, background, and talent and (2) what you expect to gain from this school in personal and professional growth.

6. Write a letter applying for a part-time or summer job. Choose an organization that can offer you experience related to your career goal. Identify the exact hours and calendar period during which you are free to work.

7. a. Identify the job you hope to have in a few years. Using newspaper, library (see your reference librarian for assistance), placement office, and personal sources, write your own full description of the job: duties, responsibilities, work hours, salary range, requirements for promotion, highest promotion possible, unemployment rate in the field, employment outlook for the next decade, need for further education (advanced degrees, special training, etc.), employment rate in terms of geography, optimum age bracket (as in football, does one fade around thirty-five?).

b. Using the above sources and your good judgment, construct a profile of the ideal employee for this job. If you were the personnel director screening applicants, what specific qualifications would you require (education, experience, age, physical ability, appearance, special skills, personality traits, attitude, outside interests, and so on)? Try to locate an actual newspaper describing job responsibilities and qualifications. Better yet, assume you're a personnel officer, and compose an ad for the job.

c. Assess your own credentials against the ideal-employee profile. Review the plans you have made to prepare for this job: specific courses, special training, work experience, etc. Assume you have completed your preparation. How do you measure up to the requirements in b? Are your goals realistic? If not, why not? What alternative plan should you formulate?

d. Using your list from part c as raw material, construct your résumé. Revise until it is perfect.

e. Write a letter of application for the job described in part a. Revise until you feel good about signing it.

f. Write a follow-up letter thanking your fictional employer for your interview and restating your interest.

g. Write a letter accepting a job offer from this same employer.

h. Write a letter graciously declining this job offer.

i. Submit each of these items, in order, to your instructor.

Note: Use the sample letters in this chapter for guidance but don't borrow specific expressions.

8. Assume the following advertisement has appeared in your school newspaper:

STUDENT CONSULTANT WANTED

The office of Dean of Students invites applications for the position of student consultant to the Dean for the upcoming academic year. Duties will include (1) meeting with fellow students as individuals and groups to discuss issues, opinions, questions, complaints, and recommendations regarding all areas of college policy, (2) presenting oral and written reports to the Dean of Students on a regular basis, and (3) attending various college planning sessions in the role of student spokesperson. Time commitment: 15 hours weekly both semesters. Salary: $2000.

Candidates should be full-time students with at least one year of student experience at this college. The ideal applicant will be skilled in report writing and oral communication, will work well in groups, and will demonstrate a firm commitment to our college. Application deadline: May 15.

a. Compose a résumé and a letter of application for this position.

b. Split your class into screening, interview, and hiring committees.

c. Exchange your group's letters and résumés with another group.

d. As an individual committee member, read and evaluate each of the six applications you have received. Rank each application, privately, on paper, according to the criteria discussed in this chapter before discussing them with the colleagues in your group. *Note:* While screening applicants, you will be competing for selection by another committee who is reviewing your own application.

e. As a committee, select the three strongest applications, and invite the applicants for interviews. Interview each selected applicant for ten minutes, after you have prepared a list of standardized questions.

f. On the basis of these interviews, rank your preferences privately, on paper, giving specific reasons for your final choice.

g. Compare your conclusion with those of your colleagues and choose the winning candidate.

h. As a committee, compose a memo to your instructor, giving specific reasons for your final recommendations.

9. Most of the following sentences need to be overhauled before being included in a letter. Identify the weakness in each statement, and revise as needed.

   a. Pursuant to your ad, I am writing to apply for the position of junior accountant.
   b. I need all the information you have about methane-powered engines.
   c. You idiots have sent me a faulty disk drive!
   d. It is imperative that you let me know of your decision by January 15.
   e. You are bound to be impressed by my credentials.
   f. I could do wonders for your company.
   g. I humbly request your kind consideration of my application for the position of junior engineer.
   h. If you are looking for a winner, your search is over!
   i. I have become cognizant of your experiments and wish to ask your advice about the following procedure.
   j. You will find the following instructions easy enough for an ape to follow.
   k. I would love to work for your wonderful company.
   l. As per your request I am sending the county map.
   m. I am in hopes that you will call soon.
   n. We beg to differ with your interpretation of this leasing clause.
   o. I am impressed by the high salaries paid for this kind of work.

10. Write and mail an inquiry letter about an item or service you've seen advertised. Ask no fewer than six questions, and provide whatever explanations you think necessary to help the reader answer your questions fully. Write as a prospective customer, not as a student. Turn in the advertisement with your letter. Your instructor might also ask for a copy of the reply.

11. As secretary of your college outing club, you are to write an inquiry to the Department of Lands and Forests in Ontario, Canada, requesting information on canoeing the Winisk River in Northern Ontario. Specifically, you want to know if permits are needed for running the river, for campsites, for campfires, and for fishing. You also need information on buying contour maps of the area, taking your equipment across the border, exchanging currency, and finding reliable guides. Since you're unfamiliar with the area, you want to know about the wildlife you might encounter. Should you (can you) take a gun? What kinds of fish can you expect to catch, and can you count on catching enough fish for meals? What are the possibilities of re-supplying along your route? What about weather and insects? When is the best time to travel the river to avoid ice and insect swarms? Are there any brochures or guidebooks to the rapids and the best approaches to them? Once you get to Winisk (on Hudson's Bay), is there transportation for you and your canoes back to your starting point (600 miles away)? If so, to whom should you write? Organize your questions under specific headings to make your respondent's job easier. Address your inquiry to Art Rivers, Department of Lands and Forests, 112 North Way, Toronto, Canada POH 328.

12. As manager of Heritage Forms word processing center, you've received increasing complaints from your employees about working conditions.

Their chairs give them backaches and stiff necks; their eyes hurt from monitor glare; their legs ache and go numb; their work stations are arranged inefficiently. In addition, they suffer from dizzy spells, and often get migraine headaches. As a result, morale is low; sick days used up. You're unsure of what to do. You've already asked the sales reps for your new office systems for help, but they insisted that your employees simply need more time to adjust. You've rejected that excuse because the employees have had six months to "adjust." Their complaints are legitimate.

Yesterday, you read an article by Professor Jane Sloane in *Office Administration and Automation* on ergonomics, the study of ways to improve working conditions by adapting work stations to workers' needs and comforts. Write an inquiry letter to Professor Sloane, Department of Human Resources, Yorkshire College, Wichita KS 43198. State your purpose and explain your problem (adding whatever details you need). After your explanation, ask a minimum of five questions that will generate specific answers to help you eliminate employee discomfort. You might also ask about other sources of information on ergonomics. Since you're asking a big favor from Professor Sloane, offer something in return (e.g., a detailed analysis of how ergonomics improved working conditions for your employees).

13. Write a letter of complaint about some problem you've had with goods or services. Be sure to state your case clearly and objectively, and to request a specific adjustment.

14. Last October 25, you bought an internal frame backpack (the Mountaineer, Item #51–6131) from Maple Mountain Outfitters, a direct-mail outfitter specializing in expedition equipment. The pack costs $159.95, plus $5.65 for shipping. You've been pleased with other gear bought from Maple Mountain (their Staysnug down sleeping bag and Maple Mountain two-person lightweight tent, for instance), but the backpack is worthless. Since you take good care of your equipment, you were looking forward to years of use from the pack. But on your third trip, a two-week hike along the Sawtooths in Idaho, you had lots of problems. The zippers on two of the outside pockets and on the main compartment came unsewed; the ice-axe loop ripped off the first time your axe snagged, and worst of all, on your fifth day out, one of the carrying straps ripped out — an unpleasant experience when you're lugging 70 pounds of gear up a 60-degree slope. You endured the trip by promising yourself revenge on whoever constructed the pack. That shoddy pack turned a beautiful trip into a monumental headache. With the trip over, you have decided you would be satisfied with a new pack — definitely not the same model — and an apology. You will accept a Maple Mountain Expedition Pack as a replacement. It costs $40 more, but you figure the company should pay the difference to make up for the problems you've had. Write a diplomatic letter, explaining the situation and requesting the replacement. The company's address is 10 Mountain Way, Jim's Creek WA 83190.

15. Assume you're a consumer affairs representative at Maple Mountain Outfitters. Respond to the previous case (#14) with the following facts.

Eight people have had similar problems with the Mountaineer. Lab tests show that the thread used in those bags was defective, and Maple Mountain has sued the manufacturer. From information you've received, the manufacturer knew the thread was defective, but figured it would hold up better than it did. Mention that you're checking past orders so that everyone who bought a Mountaineer will be notified. Your aim is satisfied customers; the Maple Mountain Expedition Pack has been shipped today at no extra cost. Since you will need the Mountaineer pack as evidence, ask that it be returned in the new pack's box. Indicate that you've also enclosed a shipping label and a check for $5.65 to cover shipping charges. Offer a sincere apology for the inconvenience, and show you are aware of the hardships that can occur with defective packs. Send your reply to Ray Winston, 31 Southway St., Cody WY 64551.

16. As purchasing agent for Greyfox Electronics, you decided to save money by ordering invoice forms, letterheads, and envelopes from a new vendor, Express Forms. The shipment arrived, but the company's name is misspelled: Greyfox is spelled Greyfor. Since your purchase orders have Greyfox Electronics prominently displayed, the vendor is responsible. The problem is compounded because the shipment arrived two weeks late, and you're almost out of invoices. You need corrected invoices within seven working days — or your great money-saving idea may turn into a disaster, which certainly won't help your chances for promotion. Write Express Forms, explaining the problem and the urgency. Mention that repeat business depends on their meeting the deadline. Express Forms, 82 Ranger Road, Smithville AR 51001.

17. Nine months ago, you bought a Big Sam Coffee Brewer (Style 341) from Bradfords, a local department store. The brewer's heating element stopped working last week; Bradfords closed for business three months ago. Write the manufacturer: ABC Corp., 91 Ventura Way, Perkins, OH 41628. You want the brewer replaced or repaired, and you have fifteen months left on the warranty. (Enclose a copy of your sales slip.) You've shipped the brewer to them by UPS. Use attention and subject lines.

18. Assume you're a consumer affairs representative for the ABC Corp. (see #17). Write Marvin Molloy that his Big Sam Coffee Brewer has been repaired and shipped. He should receive it in a week. Explain that the heating element broke because the brewer's base had been immersed in water. Such immersion causes the heating element to warp and eventually crack. Enclose a brochure that gives instructions for cleaning and maintaining the unit. Also enclose a coupon for a pound of Big Sam Coffee. Marvin's address is 5 Circle Road, Tampa, FL 31987.

19. As director of Consumer Affairs, you've received an adjustment request from Brian Maxwell. Two years ago, he bought a pair of top-of-the line Gannon speakers. Both speakers, he claims, are badly distorting bass sounds, and he states that his local dealer refuses to honor the three-year warranty. After checking, you find that the dealer refused because someone had obviously tampered with the speakers. Two lead wires had been re-

spliced; one of the booster magnets was missing; and the top insulation also was missing from one of the speaker cabinets. Your warranty specifically states that if speakers are removed from the cabinet or subjected to tampering in any way, the warranty is void. You must refuse the adjustment; however, because Maxwell bought the speakers from a factory-authorized dealer, he is entitled to a 30-percent discount on repairs. Write the refusal, offering this alternative. His address: 691 Concord Street, Biloxi, MS 71690.

20. Steve Barnes has sent you his waterbed mattress, and demanded a refund. His claim: it leaks like a sieve. A quick check confirms his complaint: one side of the rubber mattress is full of tiny holes. Barnes himself has inadvertently provided the answer. In his letter, he mentioned that the mattress hadn't been mistreated; only he and his two Siamese cats slept on it. The cats undoubtedly caused the holes by clawing the mattress. So the mattress is not flawed. (A flawed mattress splits, usually at a crease or a seam.) You must refuse his claim and inform him that his mattress can't be repaired. Write the letter, including suggestions (say, a thicker mattress pad) so the problem will not recur. Be tactful; don't insult Barnes' pride or intelligence. His address: 66 Stone Road, Middleton, MA 02674.

21. As superintendent of a large warehouse, you've just received a return on a six-month-old shipment of nuts, bolts, and other fasteners from a hardware retailer who wants a credit on the return. Inspection of the goods shows they've been water damaged, and many of the fasteners are rusted. Since you're not responsible for the damage, you can't give the dealer credit. The cartons in which the goods were shipped aren't water damaged, so the shipper doesn't appear responsible either. Write Glen Harper, Harper Hardware, 100 East Elm St., Trenton, NJ 31267. Explain the situation and suggest how he could sell the merchandise to recoup some of his money.

22. Luke Harrington wants a $400 refund for four dead Douglas Fir trees your workers planted in his yard two years ago. Since you guarantee your transplants for three years, he wants his money returned. After checking Harrington's contract, you recall his problem: you wouldn't guarantee the five Douglas Firs he ordered because he wanted them planted in a wet, marshy area, and Douglas Firs need well-drained soil. A check of Harrington's lot confirms that four trees planted in the wet area have died of root rot. Write him, reminding him of the contract, and refusing the adjustment. As you did two years ago, suggest that he plant balsam firs in the wet area. Although balsam needles are slightly darker than the Douglas Firs', both trees have the shape he wants. The balsams would retain the symmetry of his tree line. Harrington's address: 921 Daisy Lane, Churchill, MO 61516.

23. Art Kurt has mailed back a red shirt he bought through your catalogue three months ago. Mr. Kurt claims that the shirt is defective, and he wants a $67.25 adjustment to his credit card. Your textile technologist has discovered that the shirt was washed at least twice with a harsh detergent, and that detergent residue remained in the fibers, causing further breakdown. The care label stipulates that the shirt must be washed by hand with a mild detergent. Refuse the adjustment, but don't accuse; instead, explain.

Mention you're having a sale this week on shirts that have the look and feel of silk, cost only 1/3 the price, and can be machine washed. His address: 391 Beacon St., Vergennes, VT 96701.

24. Three months ago, you hired the firm of Peabody, Maxwell, and Swain to do a title search and represent you at the closing on a four-acre industrial lot in Waterbury, Connecticut. Today, you received a $3,200 bill for the firm's services. The amount is outrageous. You've never paid more than $1,500 for such services. Since the title was clear, the search couldn't have taken more than three hours. The remaining work, including showing up at the closing, couldn't have taken more than four hours. Besides, law clerks would have done most of the work. In effect, you've been charged more than $450 per hour for work done by clerks. Write the firm. Request an adjustment (choose your amount; you can hire excellent lawyers for $150 per hour), an hourly breakdown of the bill, and an explanation. Since you plan to build a plant on that lot within the year, you will need the services of a local firm — but certainly not one that charges such exorbitant fees. Address your letter to William Swain, Esq., 132 Main St., Waterbury, CT.

25. Last month, you hired Interstate Roofing to repair the factory roof. According to the agreement, they were to put down four layers of heavy-duty tarpaper, then spread 1/4 inch of hot tar over the entire roof. Interstate's crew showed up on a cloudy day. Your maintenance supervisor asked the crew not to begin the job because rain was forecast. Interstate's crew chief decided to work anyway, since he'd lose time moving the crew to another job. By 11:00 a.m. it was drizzling; when the crew finished at 3:00 p.m., it was raining steadily. You were skeptical. The job was supposed to take 2½ days, and they finished in less than one — in the rain no less.

Two weeks later, when the tar was dry enough to walk on, you checked the roof. As you had suspected, it was a mess. The tar was full of moisture bubbles, some over a foot square. And the tar was spread so thinly, the edges of the tar paper were curling. On Monday, May 15, you called Wayne Driscoll, Interstate's head manager, to complain. He promised to send someone to check the roof. No one showed up. You made two more calls, on May 18 and 23. Each time, Driscoll apologized and promised to send someone. Finally, on May 27, the crew chief who had done the job showed up. He looked at the roof and claimed it was fine. Furious, you called Driscoll again. He was out of town and would call when he returned. The call never came. Today, June 2, you received a bill for $2,658 from Interstate. You've had it; you want the roof repaired correctly or you won't pay a cent. Write Driscoll, reviewing the events and explaining that it's in his best interest for him to do the job correctly. His address: 321 Elm Street, Elmira, NY 12101.

26. Two weeks ago, Jane Swan, your third-shift supervisor, called to say the plant had no heat. You immediately called Acme Heating and requested emergency service. Acme sent a repairperson, and by 3:00 a.m., the heating system was running. A day later, the system again shut down; so you called Acme, and again, they got the system running. When the system shut down

a third time, you called Harry Foreman at Acme to find out why they hadn't repaired the system the first or second time. He explained that the problem was excessive condensation in the fuel tank. As a result, water was getting into the fuel lines, causing the system to fail. The only solution was to drain the tank to get rid of the water. You approved this plan. Today you received Acme's bill for $648. The breakdown shows $225 for three emergency calls. Since the same problem occurred three times, and since Acme didn't explain until the final occurrence, you don't believe you should have to pay for two of those calls. Besides, you're one of Acme's biggest customers, buying at least 12,000 gallons of fuel oil each winter. Request an adjustment on your bill. The address: Box 229, Hillcrest Ave., Warwick, RI 05131.

27. As head purchasing agent for Greatlink, you were asked to buy an 8-by-10-foot oriental rug for the redecorated foyer of the president's suite. The president stipulated that the rug have a red background. You found a beautiful Bakhtiari rug on sale for $1,560. The president was pleased and had had the rug set in the foyer two days ago. Yesterday, you received a frantic call from the president's secretary. The rug had to go back — at once. The chairman of the board, an authority on oriental rugs, noticed the rug is from Iran — and he refuses to have an Iranian rug in the building. He wants it replaced with a Kirshehr, a Turkish rug. You call the rug dealer and explain the problem. They can't accept a return on a used rug (especially a sale item), and suggest you write their head office. Write the head office and request a trade. Explain that Greatlink will pay the $125 to have the rug cleaned, and that you'll gladly pay the difference between the sale price of $1,560 and the regular price for a Kirshehr. The address: Oriental Rugs, Inc., 612 Madison Avenue, New York, NY 10122.

17

Writing Short Reports

SHORT REPORTS DEFINED

Reports present ideas and facts to decision makers. Primitive people used reports to decide where the best hunting was; generals used them to determine their enemies' strengths and weaknesses; immigrants flocked to America after hearing reports about "this land of opportunity." In all such cases, *decisions* were made based on reports. Today we often rely on weather reports, credit reports, consumer reports, monetary reports, congressional reports, and world reports before we make such decisions as going to the beach, investing in money market accounts, or going to the Bahamas on vacation. If we have made decisions based on someone's inaccurate report, we become more discerning about whose reports we take seriously.

In the professional world, decision makers rely on reports. For every long report, countless short reports lead to *informed* decisions on matters as diverse as the most comfortable office chairs to buy or the best recruit to hire for management training.

PURPOSE OF SHORT REPORTS

The short report's purpose is to communicate objectively and precisely in one of these formats: the letter report, the memorandum, the prepared-form report, or a variety of other forms that fit into none of the first three categories and that we call "miscellaneous."

On the job, personnel must communicate with speed and precision. Your success may well depend on your skill in sharing useful information with colleagues and superiors. Here are some of the kinds of reports you might write on any workday:

- a request for assistance on a project
- a requisition for parts and equipment
- a proposal outlining reasons and suggesting plans for a new project
- a set of instructions
- a cost estimate for planning, materials, and labor on a new project
- a report of your progress on a specific assignment
- an hourly or daily account of your work activities
- a voucher detailing your travel expenses
- a report of your inspection of a site, item, or process
- a statement of reasons for equipment or project failure
- a record of a meeting's minutes
- a report of your survey to select the best prices, materials, or equipment

Whether you report your data in a letter, memo, on a prepared form, or in some miscellaneous form will depend on your purpose and audience. Quite possibly, the same information you cast in a memo to a superior will be incorporated in a letter to a client.

TYPICAL LETTER REPORTS

Letter reports typically go to people outside the organization. For the introductory and closing elements of your letter, follow the standard letter format discussed in Chapter 16. The two format additions in a letter report are (1) a "subject" heading, placed two spaces below the salutation, and (2) other headings, as needed, to segment your letter into specific areas (see Frymore's letter report, pages 396–397). Be sure your letter report keeps the personal, "you," perspective.

Letter reports serve a variety of purposes. Here we show letter reports designed for two broad purposes: to *inform* and to *recommend*.

The Informational Report

The informational letter report typically provides readers with information about a firm's products, services, operations, or about anything they need to know.

The following *informational* report is a letter report to Wilmington's fire marshal. While inspecting Frymore's plant, the marshal found its fire evacuation procedures haphazard, and gave the firm one month to establish uniform and sensible evacuation procedures. In responding, the writer numbers items and underlines job titles to delineate clearly everyone's responsibilities. The fire marshal thus can see at a glance who is responsible for various procedures. Employees receiving the same report (in memo form) can quickly determine what they should do and where they should exit the building.

FRYMORE'S
111 Riddle Lane
Wilmington, Vermont 03212

July 25, 1983

Mr. Jack Pine
Fire Marshal
32 Allumette Lane
Wilmington, Vermont 93212

Dear Mr. Pine:

SUBJECT: FRYMORE'S FIRE EVACUATION PROCEDURES

Here is the report you requested on our evacuation procedures. We have followed your suggestions by clearly delineating responsibilities for all personnel during fire evacuation. This report and the attached diagrams [not included here to save space] have been circulated to all employees and are also posted on all department bulletin boards. As instructed, we have also held two fire drills using the new procedures and were able to evacuate all personnel within five minutes. With practice, we *will* achieve the required 3½-minute evacuation.

Specific procedures are as follows:

1. The <u>Packaging Manager</u> will stand in front of his department near the flow tracks to stop employees from running and pushing. He will direct employees out the employee entrance or through the main office, depending on the location of the fire.

2. The Processing Manager will be responsible for the proper shutdown of all equipment before leaving the building. Press operators and collette operators will perform emergency shut-down of fryers and extruders.

3. The Casing Manager will go to the junction at the packaging line and kitchen lane to keep traffic moving toward exits.

4. During the first shift, the Quality Control Manager will direct traffic at the entrance of the cafeteria hallway to control running and keep employees from going to the locker rooms. On the second and third shifts, the Shift Manager will cover the area.

5. In the carton room, the forklift operator on each shift will direct employees out of the building through the west exit.

6. In the shipping area, the Shift Supervisor will direct people out of the building through the loading dock.

7. The Plant Services Manager will evacuate office personnel through the front door and away from the building.

8. Maintenance employees who are part of the plant's fire fighting department will report to the Maintenance Manager to help locate and contain the fire until the Fire Department arrives.

9. Once the fire is located, the Production Manager or a designated employee will go to the front of the building to direct the Fire Department to the fire.

10. All employees will remain outside the building until told by the Fire Marshal they can safely return.

We hope these procedures are satisfactory since we seek the utmost safety for our employees. Should you have any questions or further suggestions, please call me at 927-8315, ext. 332. We look forward to your next inspection in September.

Yours truly,

Christina Alvarez

Christina Alvarez
Executive Officer

The Recommendation Report

Beyond merely providing information, the writer of the *recommendation* report interprets data, draws conclusions, and offers recommendations, usually in response to a specific reader request. The following recommendation report is cast as a letter. It differs from Frymore's report in that headings are included. (The itemized-list format — most appropriate for the diverse contents of Frymore's report — precluded the need for headings.) This writer uses technical language (liquidity ratios, debt-to-asset ratios, etc.) because he knows his reader is familiar with such terms. Following a direct plan, the report begins with a brief introduction to the topic and the report's purpose, then gives its recommendation, an overview, and the topics of analysis. Note also that the writer's tone is formal, befitting the relationship between investment counselor and client. Although he recommends buying the stock, he remains impartial, letting the facts prove his point.

INVESTMENT COUNSELING SERVICES
6098 Cloverdale Road
Washington, D.C. 20645

May 15, 1984

Mrs. Josephine Osgood
4552 South Chelsea Lane
Bethesda, Maryland 20814

Dear Mrs. Osgood:

SUBJECT: THE FEASIBILITY OF INVESTMENT IN WANG LABS

We are pleased to submit the following report you requested on the feasibility of a common stock investment in Wang Labs.

Recommendation

Wang Labs' liquidity ratio, profit margin, and debt-to-asset ratio are sound, thereby making Wang's stock an excellent investment. We recommend you invest at least $10,000 of your husband's estate in Wang's Class B stock. Although this stock restricts your voting privileges to one vote for every ten shares held, it does entitle you to an additional $.025 per share dividend. Wang is a leader in the rapidly growing computer industry; your investment therefore would not only be low-risk, but also would provide an excellent return for your retirement.

Overview of Wang Laboratories

Wang Laboratories has been in business for over thirty years, successfully competing with IBM, Digital Equipment Corporation, and Hewlett Packard Corporation, to name a few. Wang's share of the market has remained steady for the past four years. Total services and sales ranked 457th in the industrial United States, with orders increasing in the last year from 700 million to over 1 billion dollars. Net income placed Wang 280th in the country, and 11th on return to its investors.

Types of Common Stock

Investors are offered two types of common stock, both of which are listed on the American Stock Exchange. The assigned par value of both classes is $.50 per share. Class B stock entitles the investor to an additional $.025 per share dividend, but it restricts voting privileges to one vote for every ten shares held. Class C stock is not entitled to the additional dividend, but does carry full voting rights.

Liquidity Ratio

The liquidity ratio shows that working capital has increased from $2.57 of current assets to $1.00 of current liabilities in 1980, and increased from $3.56 of current assets to $1.00 of current liabilities in 1981. The current ratios are illustrated on the enclosed Summary of Financial Ratios for Wang Labs [not included here]. The two activity ratios show that the company has a receivables turnover of once every 111 days. This turnover keeps the inventory from becoming obsolete. Inventory is carried on a First in/First out basis that understates net income in periods of inflation.

Profit Margin

The margin for profit on sales is 9 percent, and it has been steady for the last three years. Earnings in 1971 were $.09 per share; in 1981 they were $1.36 per share. Included in these ten-year earnings is a two-for-one stock split issued on October 31, 1980.

Debt-to-Asset Ratio

Wang converted over $200 million of debentures into common stock in 1981. This action dropped debt-to-asset ratio from 1.33 to 1 in 1979 to 8:10 in 1980, and to 3:5 in 1981. This breaks down to three dollars of debt for every five dollars of assets. Assets alone would cover the debt in case of a liquidation.

In sum, Wang is a sound investment that would add strength and diversity to your portfolio. Should you decide to invest in Wang's stock, please call me at your convenience.

Sincerely,

INVESTMENT COUNSELING SERVICES

Harold Bloomstein

Harold Bloomstein
Investment Counselor

TYPICAL MEMORANDUM REPORTS

Memos are the major form of written communication in an organization. Unlike conversations, memos leave a "paper trail" so directives, inquiries, instructions, requests, recommendations, etc., can be used for future reference.

The standard memo (Figure 17–1) has a heading that names the organization and identifies the sender, recipient, subject (often in caps or underlined for emphasis), and date. (As with company letterheads, placement of these items differs slightly among firms.) A memo report includes topic headings for easier reading and better organization. (Refer to memo reports throughout this chapter.)

When your memo report is longer than one page, on following pages list the recipient's name, the date, and the page number (Mr. Roberts, 4/13/84, page 2). Place this information three spaces from the top of the page. Begin your text three spaces below.

Direct-plan memos simply end after the final point; no closing remarks or summary are necessary. Memos don't require a complimentary close and signature. If authentication is needed, initial the "FROM" line after, or close to, your name. When you're distributing the memo to a number of people, place a distribution notation at the end of the memo listing recipients (e.g., Copies: C. Black, J. Capilona, G. Hopkins, P. Maxwell).

Memo reports cover any topic important to a firm's operations. The most common types include informational, recommendation (cast as letter reports earlier), justification, progress, periodic, and survey reports. Examples of each kind follow.

Justification Reports

Justification reports (a type of recommendation report) are in a unique class in that they often are initiated by the writer rather than authorized or requested by the readers. Writers use such reports to suggest and justify changes in policy or procedures. Justification reports overlap with proposals (Chapter 18), but usually require less persuasion, since the "justification" should be obvious. Such reports therefore typically begin rather than end with recommendations, as proposals typically do. A justification report provides an excellent chance for junior staff to show their initiative and be recognized by superiors. Such reports answer this key question for readers: Why should we?

Justification reports should include specific benefits such as savings in time, increased productivity, better performance, or increased profits. The benefits need not always be tangible, however; they could include such intangibles as customer goodwill or improved employee morale.

Typically, justification reports follow a version of this format:

1. Statement of Purpose or Problem: In one or two sentences, make your recommendation and state the possible benefits.

2. Cost and Savings (or Advantages): Don't *explain* savings or advantages

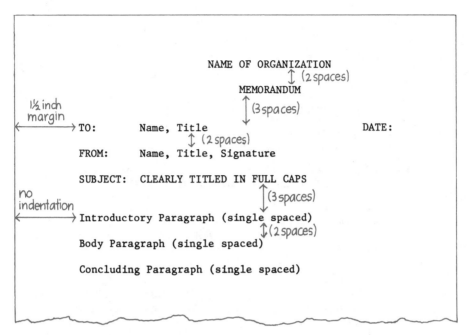

FIGURE 17–1   Standard Memo Format

here; just state what the firm will gain. Save the explanation for the discussion section.

3. Methods or Procedures: Briefly explain how your suggestion can be implemented. For instance, if people need release time for training workshops, state how many people are involved, how much release time is required, and how long the sessions will last.

4. Conclusion: List your conclusions here, but save the details for your discussion.

5. Discussion: Provide details, explanations, and describe how you arrived at your conclusions.

The format is flexible, so alter it when necessary to meet the goals of your report.

The following writer uses the preceding format, but he adds a section describing the facilities since he knows that his audience, Cecile Michno, is unfamiliar with the details of the service. So that both sides of the issue are presented, he includes a section briefly explaining possible disadvantages. Note that he presents all information impartially, concisely, and in a format that guides the reader through the report. By itemizing costs, possible savings, and benefits, he demonstrates his knowledge of the topic.

AQUAFORM

MEMORANDUM

DATE: Nov. 23, 1984

TO: Cecile Michno
             Executive Vice President

FROM: Kurt Porschen
             Assistant Budget Director

SUBJECT: CUTTING TRAVEL COSTS BY USING AT&T'S
          PICTUREPHONE MEETING SERVICE

Purpose

AquaForm can save at least $13,500 annually by using AT&T's Picturephone Meeting Service for its conferences. I recommend that we use teleconferences rather than flying our staff to Houston quarterly.

Cost and Savings

Ninety-minute quarterly teleconferences would cost $10,200 as compared with the $23,500 we spent last year sending four Chicago district managers to Houston every quarter. This represents a yearly savings of $13,300.

Procedure

At a prearranged time, Chicago conferees would go to AT&T's Picturephone Meeting Service, located at 3221 Lakeshore Drive, a 20-minute drive from our main office. Houston staff would attend the teleconference at 693 North Drive, a 15-minute drive from the Houston office.

Description of Facilities at a Picturephone Service Location

Conference participants gather in specially equipped conference rooms at different sites. The room is equipped with incoming and outgoing camera monitor units. The conferees, wearing individual microphones, sit at a table facing a large wall monitor where they view the conferees at the other location. Whenever someone speaks, the camera automatically focuses on that speaker.

Other equipment in the conference room includes a hard-copy machine for sending or receiving documents during the conference; an easel with its own multipurpose camera for displaying and projecting text, graphics, illustrations, charts, etc.; a video cassette recorder for recording meetings for future reference; an encryption (code-scrambling) terminal to insure the meeting's privacy; and an audio add-on telephone that allows people to join the meeting via telephone.

Possible Disadvantages

Some experts contend that a teleconference is not as effective as an actual meeting since conferees don't get a sense of other conferees' "presence."

Others claim that some participants get stage fright when they see themselves on a monitor.

When I polled our managers, none saw either contention as a problem. Since they now videotape their training programs, our managers are accustomed to seeing themselves on TV monitors. Moreover, AT&T allows groups to familiarize themselves with the conference room and its equipment before actually holding a conference.

Conclusions

AquaForm would derive at least six benefits from teleconferences:

1. We would save at least $13,300 annually.
2. Managers wouldn't lose the two days of work a quarter they lose now by flying to Houston.
3. Managers wouldn't suffer the strain and fatigue of traveling.
4. Meetings would be more productive.
5. Assistant managers would be able to attend the conferences.
6. We would establish a more efficient information flow.

Discussion of Conclusions

1. For the past four years, we've flown an average of four district managers to Houston for quarterly conferences. Yearly costs are as follows:

	1981	1982	1983	1984
Yearly costs:	$20,780	$24,620	$26,500	$23,500

Average yearly cost: $23,850

Breakdown of last year's costs:

Round trip plane fare: 4 managers x 4 trips	$12,480
Lodging	1,200
Meals and gratuities	1,600
Taxi fares	396
Entertainment expenses	1,870
Miscellaneous expenses	560
Estimated lost-time salaries	5,394
TOTAL	$23,500

AT&T charges $2,550 for a 90-minute teleconference between Chicago and Houston:

4 conferences @ $2,550 = $10,200

ESTIMATED SAVINGS FROM TELECONFERENCES = $13,300

2. At present, managers lose an average of two working days a quarter, traveling to Houston. Claire Golembewski from accounting estimates these at $5,394 lost-time salaries. Since at least four managers travel to the conferences each quarter, we lose at least eight manager-days per

quarter, or 32 days yearly. By using the Picturephone Service, we would recover these 32 work days, thereby increasing our white-collar productivity by 4 percent annually.

3. All managers, but especially Andy Gray and Marie LaPointe (who both turned sixty this year), often complain about the strain and annoyance of traveling to these conferences. In fact, Marie, our most productive district manager, prefers early retirement to continued travel to Houston. Teleconferences would end travel fatigue. Of the seven district managers, only one would prefer to continue flying to Houston.

4. Meetings would be more productive. Since teleconference time is limited to 30, 60, 90, or 120 minutes, meetings are usually more productive, efficient, and informative because conferees digress less and follow their agendas more closely.

5. Five years ago, we stopped sending assistant managers to Houston conferences because of the expense. There is no limit to the number of conferees at a Picturephone meeting; therefore, assistant managers could once again attend, for valuable, yet inexpensive, training.

6. The Picturephone Service would help establish a more efficient information flow. Now, we are restricted to our schedule of quarterly conferences because of the expense and strain of travel. With the meeting service, we could conceivably hold shorter conferences more often. A 30-minute conference costs $930, an hour's conference, $1,740. When polled, all managers agreed that shorter, more frequent meetings would be quite helpful, especially during the planning stages of new marketing campaigns.

Progress Reports

Large organizations depend on progress (or status) reports to keep track of activities, problems, and progress on various projects. Daily progress reports are vital in a business that assigns several work crews to various projects. Management uses progress reports to evaluate the project and its supervisor, and to decide how to allocate funds. Progress reports typically are informational rather than recommendation reports.

Often, a progress report is one of a series. Together, the project proposal (Chapter 18), progress reports (the number varies with the scope and length of the project), and the final report (Chapter 19) provide a record and history of the project.

To give management the answers it needs, project reports must, at a minimum, answer these questions:

1. How much has been accomplished since the last report?
2. Is the project on schedule?

3. If not, what went wrong?
   a. How was the problem corrected?
   b. How long will it take to get back on schedule?
4. What else needs to be done?
5. What is the next step?
6. Are there any unexpected developments (other than schedule problems)?
7. When do you anticipate completion? Or, on a long project, when do you anticipate completion of the next phase?

If the report is part of a series, you might refer to prior problems or developments, but the bulk of the report must concern itself with the project's current status.

Many organizations have forms for organizing progress reports, so no one format is best. But all reports in a series should follow the same organization. The following report illustrates how one writer organized her report. Like all examples in this text, you should use it as a guide; don't follow it slavishly. The format and organization you choose must fit *your* purpose, audience, and situation.

MEMORANDUM

DATE: May 28, 1982

FROM: Banking Operations, Group Training Managers

TO: P. J. Stone
             Senior Vice President

SUBJECT: Progress Report: Training Equipment for New Operations
             Building

WORK COMPLETED

Our training group has met twice since our May 12 report, to answer the questions you posed in your May 16 memo. In our first meeting, we identified the various types of training we anticipate.

Types of Training
– Divisional Surveys
– Loan Officer Work Experience
– Divisional Systems Training
– Divisional Clerical Training (On-going)
– Divisional Clerical Training (New Employees)
– Divisional Management Training (Seminars)
– Special/New Equipment Training
– Divisional Orientations (New Employees)

In our second meeting, we considered various areas for the training room.

Training Room

The time required for each type of training varies according to each division's needs. However, the frequency of training necessitates our having a training room available daily. The large training room in the Corporate Education area (10th floor) would be ideal. Before submitting our next report, we need your confirmation that this room can be assigned to us.

To support the training programs, we purchased the following equipment:

Equipment	Training Purpose
– Audioviewer (with record, delay, and syncpulse capability)	Surveys, work experience, systems, clerical, management, orientation
– 16mm Projector	Clerical, management
– Video Cassette Recorder and Monitor	Work experience, clerical, management, orientation
– CRT	Work experience, systems, clerical, management, orientation
– Mini/Micro Computer, for Computer-Assisted Instruction	Work experience, clerical, management, systems, orientation
– Slide Projector	Management, orientation, seminars
– Tape Recorder (Wallensack)	Work experience, clerical, management, orientation

This equipment will allow us the flexibility to administer training in a variety of modes, ranging from programmed and learner-controlled instruction, to group seminars and workshops.

WORK REMAINING

To support the training, we should furnish the room appropriately. Since the types of training will vary, the furniture should work in a flexible environment. Outlined below are our anticipated furnishing needs. The project is on schedule, and with your approval of the following specifications, we will be ready to begin training as soon as the new room is ready.

Furnishings

– Tables and chairs that can be set up in many configurations. This would allow for individual, small- or large-group training, and large seminars.
– Portable partitions or room dividers. These would provide study space for training with programmed instruction, as well as allow for simultaneous training.
– Built-in storage space for audio-visual equipment and training supplies. Ideally, this storage space should be multi-purpose, providing work or display surfaces.

– A flexible lighting system. This is important for audio-visual presentations and individualized study.

– Independent temperature control. This will ensure that the training room remains comfortable regardless of group size and equipment used.

As soon as we receive your approval for this plan, we will proceed to the next step: sending out bids for room dividers, and having plans drawn for the built-in storage space.

cc: R. S. Pike, SVP
      G. T. Bailey, SVP
      L. L. Hunt, SVP

As you work on a longer report or proposal, your writing instructor might require a progress report. In the following memo, Karen Granger reports her progress on her term project: an evaluation of the Environmental Protection Agency's effectiveness in cleaning a severely contaminated harbor.

PROGRESS REPORT

TO: Dr. John Lannon

FROM: Karen P. Granger

DATE: April 17, 1984

SUBJECT: An Analysis of the Effectiveness of the EPA's 1983 *Remedial Action Master Plan*

WORK COMPLETED TO DATE

February 23:   Began general research on the PCB contamination of the New Bedford Harbor.

March 8:   Decided to analyze the <u>Remedial Action Master Plan</u> (RAMP), to determine whether residents are being "studied to death" by the EPA.

March 9–19:   Drew a map of the harbor to show areas of contamination. Obtained the RAMP from Pat Shay of the EPA.

               Interviewed Representative Goyette briefly over the phone; made an appointment to interview Goyette and Sharon Dean on April 13.

               Interviewed Priscilla Chapman, Executive Director of the New England Sierra Club, briefly over the phone.

March 24:   Obtained <u>Public Comments on the New Bedford RAMP</u>, a collection of reactions to the RAMP.

April 13:   Interviewed Goyette and Dean; searched Goyette's files for information. Also searched the files of Gerry Sotolongo, New Bedford Project Officer, EPA.

WORK IN PROGRESS

Contacting by telephone the people who commented on the RAMP.

WORK TO BE COMPLETED

April 25:   Finish contacting commentators on the RAMP.

April 26:   Interview an EPA representative about the complaints that
                 the commentators raised on the RAMP.

DATE FOR COMPLETION: May 3, 1984

COMPLICATIONS

The issue of PCB contamination is very complicated and emotionally dis-
tressing. The more I uncovered, the more difficult I found it to remain
impartial in my research and analysis. As a New Bedford resident, I expected to find that we are indeed being studied to death; because my
research seems to support my initial impression, I am not sure that I have
remained impartial.

Lastly, the people I want to talk to are very busy. They have been polite,
but they do not have the time to find the answers for my questions. I am
having trouble getting them to elaborate on their comments on the RAMP.
Everyone I have spoken with, however, has been interested and encouraging,
if not informative.

Periodic Reports

The periodic activity report is similar to the progress report in that it summarizes work activities over a specified period. But unlike the progress report,
which summarizes *project* accomplishments, periodic reports summarize daily,
weekly, or monthly general work activities. Many manufacturers requiring periodic reports have prepared forms, since most of their tasks are quantifiable (e.g.,
Units produced: 672). But the nature of most white-collar jobs doesn't lend
itself to prepared-form reports. So you will likely have to develop your own
activity report, as Fran DeWitt did in the following monthly report.

DeWitt's report answers her boss's primary question: "What did you accomplish last month?" Her response is detailed and informative; otherwise, it
would be useless. Notice that this firm's memo format differs slightly from
others we've seen.

Date   Subject
June 2, 1983   Coordinating Training
                                             Meetings
From
F. C. DeWitt

To
N. Morgan, Assistant Vice President

For the past month, I've been working on a cooperative project with the Banking Administration Institute, Computron Corporation, and several banks. The purpose of our task force has been to develop training programs, specific to banking, appropriate for computer-assisted instruction (CAI).

The project focuses on three major areas: Banking Principles, Proof/Encoding Operator Training, and Productivity Skills for Management.

I will be hosting two meetings for this task force. On June 16, we will discuss Proof/Encoding Training, and on June 17, Productivity Training. The objective for the Proof/Encoding meeting is to compare ideas, information, and current training packages available on this topic. We will then design a training course.

The objective of the meeting on Productivity is to discuss productivity in the banking industry (specifically Banking Operations). The discussion will include instances where computer-assisted instruction is appropriate for teaching productivity skills. Computron will also discuss computer applications currently used to teach productivity. A representative from FPR's Corporate Management Training department will attend.

On June 10 I will attend a meeting in Washington, D.C., to design a course in basic banking principles for high-level clericals/supervisory-level employees. We will also discuss the feasibility of adapting this course to CAI. This type of training, not currently available through Corporate Education, would meet a definite supervisor/management need in the division.

My involvement in these meetings serves two main purposes. First, structured discussions with training personnel in the banking industry provide an exchange of ideas, methods, and experiences. This involvement expedites development of our training programs because it saves me time on research. Second, microcomputers will continue to impact future training. With a working knowledge of these systems and their applications, I will be able to assist my group in designing programs specific to our needs.

Survey Reports

Brief survey reports often are used to examine conditions that affect an organization (consumer preferences, available markets, etc.). The following memo, from the research director for a Midwest grain distributor, gives clear and specific information directly. Notice the absence of background information. (An explanation of how and where these data were obtained would not be significant for this writer's purpose.) To simplify the interpretation of data, the writer arranged the body section in a classification table.

ACME GRAIN WHOLESALERS INC.

MEMORANDUM

April 15, 1984

TO: Charles Jones, Manager,
              Marketing Division

FROM: Margaret Spaulding,
              Research Director

SUBJECT: FOOD-GRAIN CONSUMPTION IN U.S., 1969–1972

Here are the data you requested on April 7 as part of your division's annual marketing survey.

U.S. PER-CAPITA CONSUMPTION OF FOOD GRAINS
IN POUNDS, 1969–1972

	1969	*1970*	*1971*	*1972*
Corn Products:				
Cornmeal and other	15.8	15.8	15.8	15.8
Corn syrup and sugar	20.3	20.8	21.4	21.7
Oat Food Products	3.2	3.2	3.2	3.2
Barley Food Products	1.2	1.2	1.2	1.2
Wheat:				
Flour	112.0	110.0	110.0	111.0
Breakfast cereals	2.9	2.9	2.9	2.9
Rye, Flour	1.2	1.2	1.2	1.2
Rice, Milled	8.3	6.7	7.7	7.0

If you have any questions or require additional information, please call Ms. Smith at 316.

cc: Mr. C. B. Schultz, Vice-President in Charge of Marketing

PREPARED-FORM REPORTS

To streamline communications and keep track of data, many companies use prepared forms for short reports. Such forms are doubly useful:

1. A prepared form provides clear guidelines for recording data. If you complete the form correctly, you are sure to satisfy the readers' needs.

2. A prepared form standardizes data reported from various sources. Identical classes of data are recorded in identical order. This fixed format allows for rapid processing and tabulation of data.

Since countless forms are prepared for countless purposes, no examples are reproduced here. However, the sample questionnaire in Chapter 14 (pages 303–305) and the accident report in Figure 10–3 are good examples of prepared forms: Their data can be reviewed easily and tabulated rapidly.

The one drawback of some prepared forms is limited space for recording data. You may need to attach your own statement explaining certain items on the form.

MISCELLANEOUS REPORTS

Miscellaneous reports, a "catch-all" category, are reports for assorted purposes that don't fit our previous classifications. Reporting minutes of a meeting, for instance, follows fairly standard conventions. Other topics and data may require you to invent a suitable format. When creating a format for miscellaneous reports, organize the information to best answer the questions readers are likely to ask. Use topic headings to guide them.

Minutes

Minutes are the official records of organizational and committee meetings. Copies of minutes are distributed to all members and concerned superiors as a way of keeping track of proceedings. The person appointed secretary records the minutes.

Minutes are filed as part of an official record, so they must be precise, clear, highly informative, and free of the writer's personal commentary ("As usual, Ms. Jones disagreed with the committee") or judgmental words ("good," "poor," "irrelevant," etc.). When you record minutes, be sure to answer these questions:

- Which group held the meeting?
- When and where was it held?
- What was its purpose?
- Who chaired the meeting?
- Who else was present?
- Were the minutes of the last meeting approved (or disapproved)?
- Was anything resolved?

In addition to answering the previous questions, summarize the points made during the group's discussion of each agenda item. Name the person who makes the motion, and the person who seconds it. Record the results of votes on each motion offered, along with a description of the motion itself. If the group votes to support a proposal, include a description of the proposal.

BRUNELL'S DEPARTMENT STORES

MEMORANDUM

DATE: October 12, 1984

TO: Personnel Managers

FROM: Harold Tweeksbury

SUBJECT: MINUTES OF PERSONNEL MANAGERS' MEETING,
            OCTOBER 5, 1984

<u>Members Present</u>
Harold Tweeksbury, Jeannine Boisvert, Sheila DaCruz, Manny Gomes, Ted
Washington, Denise Walsh, Cora Parks, Cliff Walsh, Joyce Capizolo, Percy
Blunt

<u>Agenda</u>
1. The meeting was called to order on Wednesday, October 5, at 10 A.M.
   by Cora Parks.
2. The minutes of the September meeting were approved unanimously.
3. The first order of new business was to approve the following policies for
   the Christmas season:

   a. Temporary employees should list their ID numbers in the upper left
      corner of their receipt envelopes, to help verification. Discount Cler-
      ical Assistant Managers will be responsible for seeing that this pro-
      cedure is followed.

   b. When temporary employees turn in their envelopes, personnel from
      Discount Clerical should spotcheck them for completeness and leg-
      ibility. Incomplete or illegible envelopes should be corrected, com-
      pleted, and, if necessary, rewritten. <u>Envelopes should not be sealed.</u>

4. Jeannine Boisvert made a motion that we also hold one-day training
   workshops for temporary employees to teach them our policies and procedures. Denise Walsh seconded the motion. Joyce Capizolo disagreed,
   saying that on-the-job training (OJT) was enough. The motion for the
   training session carried 6-3. The first workshop, which Jeannine agreed
   to arrange, will be held October 25.
5. Percy Blunt from Sales Audit asked that transmittal sheets be *stapled*
   to batched receipt envelopes. Last year, he and his staff spent hours
   straightening mixups. Cora Parks made a motion of Percy's request; it
   was seconded by Manny Gomes and carried unanimously. Percy will
   send a memo to all staff this week explaining the correct procedure.
6. Manny Gomes made a motion, not seconded, that next year's Christmas
   party be held at Chez Le Monde instead of the Manito. Harold Tweeksbury then made a motion, seconded by Denise Walsh, that a questionnaire

be sent to all staff to learn their preferences. Chez Le Monde would be one of the choices. The motion carried unanimously.
7. Joyce Capizolo requested that temporary employees be sent a memo explaining the temporary employee discount procedure. The request was changed into a motion and seconded by Cliff Walsh. The motion passed by a 7-2 vote.
8. The meeting was adjourned by Cora Parks at 11:55 A.M.

Besides being purely informational (as in the minutes above), miscellaneous reports can offer recommendations (as in the sample that follows).

Preliminary Marketing and Research Report

Miscellaneous reports can deal with any aspect of an organization's operations. The following preliminary report was written to explore ways to market Format Training Corporation's new Strategic Planning Game (a computer-assisted package for management training). Since this report served as an agenda for brainstorming meetings, the writer generated broad topics to stimulate discussion and to answer her audience's main question: What has to be done before we can market this product?

FORMAT TRAINING CORPORATION

MEMORANDUM TO: FTC Management Systems

FROM: Sandra Collins

RE: MARKETING OUR STRATEGIC PLANNING GAME

DATE: OCTOBER 5, 1984

I thought it might be useful to outline ways we can develop a marketing strategy for our Strategic Planning Game. Although collectively we have a lot of experience marketing our consulting services, marketing a product calls for different approaches. Here are some broad areas of marketing research I'd like you to consider before our meeting on October 9.

1. Identify High-Potential Market Segments

   Example: Identify which types or classifications of industries (chemical, manufacturing, financial, etc.) represent the greatest demand for our Strategic Planning Game.

2. Develop and Identify High-Need Topic and Program Areas

   Example: List, in order of preference, the specific management skills and programs that are in greatest demand — and for which employee levels.

3. <u>Generate a Data Base of Executives Responsible for Training and De-</u><u>velopment Decisions</u>

<u>Example:</u>   Generate a list of training executives in high-potential market segments who have or plan to use outside suppliers for management training.

4. <u>Determine How Current Training Needs Are Being Met</u>

<u>Example:</u>   Identify what percentage of current managerial training is done in-house, and what percentage is done by outside suppliers, schools, etc. We might be able to tailor our program to meet these needs.

5. <u>Identify General Perceptions and Attitudes toward Using Games for</u><u>Management Training</u>

<u>Example:</u>   Given the current successes of games in computer-assisted training, is there a substantial market segment we can fill?

These are just a few of the research areas that might help us. Try to expand on this list before the meeting.

CHAPTER SUMMARY

Reports present ideas and facts to people who use them to make informed policy decisions. The short report's purpose is to communicate concisely. Depending on your subject, your reader's needs, and your company's policy, you might record your data in memo form, letter form, on a prepared form, or in a miscellaneous form.

Letter reports provide information to people outside the organization. They are typically informational or recommendation reports.

Memorandum reports, the most common form of in-house communication, follow a fixed format. They make liberal use of headings and itemized lists to guide readers through the report.

Unlike most recommendation reports, *justification reports*, written to suggest changes in policy or procedures, are typically initiated by the writer rather than authorized by someone else. Justification reports list recommendations, benefits, and conclusions before providing discussion, details, and the means used to arrive at the conclusions.

Progress reports are informational reports that help supervisors keep track of activities, problems, and progress on various projects. Whereas progress reports summarize project accomplishments, *periodic reports* summarize daily, weekly, or monthly general work routines.

For ease in standardizing and tabulating data, prepared-form reports can be useful.

Miscellaneous reports follow no specific conventions, since their data are so variable that no conventions can serve as adequate guidelines. Organize the information in such reports to best answer the questions readers are likely to ask.

REVISION CHECKLIST FOR SHORT REPORTS

Use this checklist as a guide to refining and revising your short reports.

1. Have you chosen the best form of short report for your specific purpose?
2. Does the memo have a complete heading (name of organization, name and title of sender and recipient, identification of subject, date)?
3. Have you single-spaced within paragraphs and double-spaced between?
4. Have you used headings, charts, or tables whenever they are needed?
5. If more than one reader is receiving copies, does the memo include a distribution notation (cc) to identify other readers?
6. Does the letter report follow a standard letter format, and does it include a subject heading and any needed internal headings?
7. Have you organized your information for emphasis?
8. Does your justification report include a clear statement of purpose and specific benefits?
9. Does your progress report answer these questions: What has been accomplished since the last report? Is the project on schedule? When do you anticipate completion?
10. Do your minutes summarize key points, record vote results, and describe proposals passed?
11. Are all sentences clear, concise, and fluent?

EXERCISES

1. We all would like to see certain changes in our school's policies or procedures, whether they are changes in our major, school regulations, social activities, grading policies, registration procedures, or the like. Find some area of your school that needs obvious changes, and write a justification report to your instructor. Explain why the change is necessary, and the benefits to be gained. Follow the format on pages 400–401.

2. You would like to see some changes in this course to better reflect your career plans. Perhaps you feel too much emphasis is placed on letters,

too little on reports. Or maybe there's too much lecturing and not enough discussion. Write a justification report to your instructor, justifying the reasons for the changes you propose. Remember, you must illustrate specific benefits for you and your classmates resulting from your plan. Do *not* try to justify spending less time doing course work.

3. Assume you manage a retail store that employs 75 people. Yesterday, you received a memo from Sheila Hawkes, the Chief Executive Officer, requesting suggestions for cutting costs. For some time, you've felt that the 18 percent employee discount is too generous. Besides, although store policy forbids the practice, you know that some employees misuse the discount by making purchases for friends. You've never said anything about the problem because the chain's founder and Chairman of the Board, Sam Hughes, has always claimed the discount increases employee loyalty. Moreover, he argues that the stores benefit by not having to train new employees. He may be right, for although wages are comparable with other retail stores, employee turnover in the chain is only 5 percent; the average is 16 percent. Despite his arguments, you believe you can justify a 12 percent discount. (Other area stores allow 10 percent.) You have no way of knowing if employee turnover will increase, but you think not, since employees have an average of eight years with the store.

Here are the facts you've gathered to justify this policy change. In the past twelve months, employee purchases averaged an astounding $1,292 (before the 18 percent discount) for a total of $96,900. The discount therefore cost your store alone $17,442. A 12 percent discount would have saved $5,814. More telling, though, is the pattern of large purchases. Twenty-eight employees bought televisions; 38 bought microwave ovens; 35 bought 35mm cameras; 52 purchased small computers, and 36 made large furniture purchases. In all, $82,871 was spent for large items, a sum far in excess of typical consumer spending patterns. Thus, not only is the 18 percent discount high, but because it is so high, some employees apparently abuse it by making large purchases for friends. Write a justification report for Sheila Hawkes. Be sure to discuss and counter the issue of increased turnover.

4. Think of an idea you would like to see implemented in your job. Write a justification memo, persuading your audience that your idea is worthwhile.

5. As manager of office services for the past three months, you've noticed various departments stockpiling their own inventories of consumable office supplies (paper, envelopes, pencils, etc.). Because you worked in purchasing before your promotion to office services, you know that office supply costs in the company have increased by nearly 50 percent a year for the last three years. You're convinced that hoarding and random ordering from assorted vendors by various departments is largely responsible for these increases. In fact, inventory is so diffuse, no one has any idea of what consumable supplies are on hand companywide. To decrease consumable office supply expenses and to increase control over inventory, you'd like to develop a central office supply inventory. Inventories of capitalized items such as microcomputers, typewriters, and office furniture would be excluded, as

would any forms, stationery, or supplies unique to a department. Once departments use their present inventories, all supplies would be requisitioned through Office Services. Office Services could therefore keep track of inventory, and place bulk orders through Purchasing. You estimate the company could save at least $8,000, or 16 percent of current expenses. Write your justification memo to Carol Holt, General Manager.

6. Yesterday, your supervisor circulated a memo requesting ideas for modernizing Bains and Bains' image. You'd like to begin with their correspondence. Mr. Bains, Sr. still closes his letters with "Your obedient servant," and "I remain humbly yours." Harold, Jr. isn't much better. He uses so much letterese and so many clichés, you have difficulty finding any content that's actually his. As a result, most other employees do the same. Write a memo to your supervisor, Jonathan Hartwell. Although Hartwell is your primary audience, you can be sure that Harold, Jr. is your secondary audience. (Turn to Chapters 4 and 16 for examples of letterese, clichés, and overblown language you would find in Bains and Bains' correspondence — use your imagination here: if phrasing is pompous, stuffy, or wordy, they're probably using it. Use these examples as actual excerpts from company correspondence, and compare them with plain English.) In a recommendation report, persuade Hartwell (and Jr.) that using plain English would be an excellent way for the firm to modernize its image.

7. Identify a dangerous or inconvenient area or situation on campus or in your community (endless cafeteria lines, a poorly lit intersection, slippery stairs, a poorly adjusted traffic light, etc.). Observe the problem for several hours during a peak-use period. Write a justification report to a *specifically identified* decision maker, describing the problem, noting your first-hand observations, making recommendations, and encouraging reader support or action.

8. Assume you received a $10,000 scholarship, $2,500 yearly. The only stipulation for receiving installments is that you send the scholarship committee a yearly progress report on your education, including courses, grades, school activities, and cumulative average. Write the report, following the guidelines on pages 404–405.

9. In a memo to your instructor, outline your progress to date on your term project. Describe your accomplishments, plans for further work, and any problems or setbacks. Conclude your memo with a specific completion date.

10. Make up a periodic form that will allow you to keep a one-week record of your major daily activities. During that week, record the time spent in each activity. When your prepared-form report is completed, review the data and write an interpretation with conclusions and recommendations for better budgeting your time.

11. Assume you've begun your career. Your boss wants weekly memo reports on projects and accomplishments. In short, what did you do last week to earn your salary? Write your weekly report. Unless you have worked in the field, this assignment requires research. Check your job title in *The*

Dictionary of Occupational Titles. You might also want to check the *Occupational Outlook Handbook.* Both can be found in the library's reference section. If possible, interview someone in your field. Write a realistic report of a typical week's activities. Or, do the same for an actual part-time job.

12. Keep accurate minutes for one class (preferably one with debate or discussion). Submit the minutes in memo form to your instructor.

13. Conduct a brief survey (e.g., of comparative interest rates from various banks on a car loan, comparative tax and property evaluation rates in three local towns, or comparative prices among local retailers for a certain item). Arrange your data and report your findings to your instructor in a memo that closes with specific conclusions and recommendations for making the most economical choice.

14. Although the campus store (or cafeteria) is a convenient place to buy small items (or food), you believe students are overcharged for the convenience. To prove your point, you decide to do a comparative analysis of five items at the campus store (or cafeteria) and two local stores (or restaurants). Be sure to compare like sizes, weights, brands, etc. Conduct your survey, analyze your findings, and submit the results in memo form to the student senate. Your instructor might want you to include recommendations.

15. *Informational Report Topics* (Choose *one* for a memo or letter.)
   a. You are legal consultant to the leadership of a large auto-workers' union. Before negotiating its next contract, the union needs to know what effects robotics technology will have on assembly-line auto workers within ten years. Do the research and write the report.
   b. You are a consulting engineer to an island community of 200 families suffering a severe water shortage. Some islanders have raised the possibility of producing drinking water from salt water (desalination). Write a report for the town council, summarizing the process and describing instances where desalination has been used successfully or unsuccessfully. Would desalination be economically feasible for a community this size?
   c. You are health officer in a town less than one mile from a massive radar installation. Citizens are concerned about the effects of microwave radiation. Do they need to worry? Find the facts and write your report.
   d. You are health officer in a town where a power-company easement allows high-voltage lines to run immediately adjacent to the elementary school playground. Are the children endangered by ionic disturbances? Parents and school committee want to know. Find out, and write your report.
   e. Dream up a scenario of your own, where information would make a real difference.

16. *Recommendation Report Topics* (Choose *one* for a memo or letter.)
   a. You are an investment broker for a major firm. A longtime client

calls to ask your opinion. He's thinking of investing in a company that is fast becoming a leader in fiber-optics communication links. "Should I invest in this technology?" your client wants to know. Find out, and give him your informed opinion in a short report.

b. The buildings in the condominium complex you manage have been invaded by carpenter ants. Can the ants be eliminated by the use of any insecticide *proven* non-toxic to humans or pets? (Many dwellers have small children and pets.) Find out, and write a report making recommendations to the maintenance supervisor.

c. The "coffee generation" wants to know about the properties of caffeine and the chemicals used on coffee beans. What are the effects of these substances on the human body? Write your report, making specific recommendations about precautions coffee-drinkers can take.

d. As a consulting dietician to the school cafeterias in Blandville, you've been asked by the school board to report on the most dangerous chemical additives in foods. Concerned parents want to be sure that foods containing these additives be eliminated from school menus, insofar as possible. Write your report, making general recommendations about modifying school menus.

e. Dream up a scenario of your own, where information and recommendations would make a real difference. (Perhaps the question could be one you've always wanted answered.)

18

Writing
a Proposal

DEFINITION AND PURPOSE

A proposal is an offer to do something or a suggestion for action. The general purpose of a proposal is to *persuade* readers to improve conditions, authorize work on a project, accept a service or product (for payment), or otherwise support a plan for solving a problem or doing a job.

Your own proposal may be a letter to your school board to suggest changes in the English curriculum; it may be a memo to your firm's vice-president to request funding for a training program for new employees; or it may be a 1,000-page document to the Defense Department to bid for a missile contract (competing with proposals from other firms). You might write the proposal alone or as part of a team. It might take hours or months.

Whether in science, business, industry, government, or education, proposals are written for decision makers: managers, executives, directors, clients, trustees, board members, community leaders, and the like. Inside or outside your organization, these are the people who decide if your suggestions are worthwhile, if your project will ever get off the ground, if your service or product is useful. In fact, if your job depends on funding from outside sources, proposals might well be your most important writing activity. To be successful, a proposal must be convincing.

THE PROPOSAL PROCESS

The basic proposal process can be summarized simply: someone offers a plan for something that needs to be done. In business and government, this process has three phases:

1. Client X needs a service or product.
2. Firms A, B, C, etc., propose ways to meet the need.
3. Client X awards the job to the firm offering the best proposal.

The complexity of events within each phase will of course depend on the situation. Here is a typical situation:

Assume you manage a mining engineering firm in Tulsa, Oklahoma. On Wednesday, February 19, you spot this announcement in the *Commerce Business Daily*:[1]

> **R — Development of Alternative Solutions to Acid Mine Water Contamination from Abandoned Lead and Zinc Mines** to Tar Creek, Neosho River, Ground Lake, and the Boone and Roubidoux aquifers in northeastern Oklahoma. This will include assessment of environmental impacts of mine drainage followed by development and evaluation of alternate solutions to alleviate acid mine drainage in receiving streams. An optional portion of the contract, to be bid upon as an add-on and awarded at the discretion of the OWRB, will be to prepare an Environmental Impact Assessment for each of three alternative solutions as selected by the OWRB. The project is expected to take 6 months to accomplish with anticipated completion date of September 30, 1984. The projected effort for the required task is 30 person-months. Request for proposals will be issued and copies provided to interested sources upon receipt of a written request. Proposals are due March 1, 1984. (044)

> Oklahoma Water Resources Board
> P.O. Box 53585
> 1000 Northeast 10th Street
> Oklahoma City, OK 73152
> (405) 271-2541

[1] This daily publication lists the government's latest needs for *services* (salvage, engineering, maintenance, etc.) and for *supplies, equipment, and materials* (guided missiles, engine parts, building materials, etc.). The *Commerce Business Daily* is an essential reference tool for anyone whose firm seeks government contracts.

The Grantsmanship Center News, published six times a year, contains a wealth of information about federal grants for nonprofit organizations in areas such as rural development, media, science, energy and environment, and education. This publication also lists notices of training programs in proposal writing, and offers books on proposal writing and evaluation, along with bibliographies of essential publications about the proposal process.

Your firm has the personnel, experience, and time to do the job, so you decide to compete for the contract. Because the March 1 deadline is fast approaching, you write immediately for a copy of the request for proposal (RFP). The RFP will contain the guidelines for developing the proposal — guidelines for spelling out your plan to solve the problem (methods, timetables, costs, etc.).

You receive a copy of the RFP on February 21 — only one week before deadline. You get right to work, with the two staff engineers you have appointed to your proposal team. Since the credentials of your staff could affect client acceptance of the proposal, you ask all members to update their résumés (for inclusion in an appendix to the proposal).

Several other firms will be competing for this contract. The Oklahoma Water Resources Board will award the contract to the firm submitting the best proposal; board members will use the following criteria (and perhaps others) to review various proposals:

- understanding of the client's needs described in the RFP
- soundness of the firm's technical approach
- quality of project organization and management
- ability to complete the job by the deadline
- ability to control costs
- specialized experience of the firm in this type of work
- qualifications of the staff to be assigned to the project
- the firm's track record for similar projects

A client's specific evaluation criteria often are listed (in order of importance or according to a point scale) in the RFP. Although these criteria may vary, every client expects a proposal that is *clear, informative,* and *realistic.*

Some clients will hold a preproposal conference for the competing firms. During this briefing, the various firms are told of the client's needs, expectations, specific start-up and completion dates, criteria for evaluation, and any other details that might serve as guidelines for the competing firms.

The sample situation above is only one among countless possibilities. You might encounter proposal situations that are quite different, but in every case the process will be similar: you will propose a plan to fill a need. And in most cases, the merit of your proposed plan will be judged *solely* in terms of what you have *on paper.* This reality is summed up in the following advice from a government publication that solicits proposals.

> [Proposal] reviewers will base their conclusions only on information contained in the proposal. It cannot be assumed that readers are acquainted with the firm or key individuals or any referred-to experiments.[2]

Make any proposal able to stand alone in meaning and persuasiveness.

[2] *Small Business Innovative Research Program* (Washington, D.C.: U.S. Department of Defense, 1983), p. 9.

TYPES OF PROPOSALS

Despite their variety, proposals can be classified in three ways: according to *origin, audience,* or *intention.* Based on its origin, a proposal is either *solicited* or *unsolicited* — that is, requested by someone or initiated on your own because you have recognized a need. Business and government proposals are most often solicited and originate from a customer's request (as shown in the sample situation on page 421).

Based on its audience, a proposal may be *internal* or *external* — written for members of your organization or for clients and funding agencies. (The situation on page 421 calls for an external proposal.)

Based on its intention, a proposal may be a *planning, research,* or *sales* proposal. These last categories by no means account for all variations among proposals. In fact, certain proposals may fall under all three categories, but these are the types you will most likely have to write. Each type is discussed and illustrated below.

The Planning Proposal

A planning proposal suggests ways to solve a problem or to bring about improvement. It might be a request for funding to expand the campus newspaper, an architectural plan for new facilities at a ski area, or a plan to develop energy alternatives to fossil fuels. In every case, the successful planning proposal answers this central question for readers:

– What are the benefits of following your suggestions?

The planning proposal that follows is external and solicited. Disappointed with earlier, in-house, writing workshops, the XYZ Corporation has contracted a team of communication consultants to design results-oriented workshops. The writers here need to persuade their reader that their realistic methods are likely to achieve better results than prior approaches, which were much too broad. In their proposal, addressed to the company's training officer, the consultants offer concrete and specific solutions to clearly identified problems.

After a brief introduction summarizing the problem, our writers develop their proposal under two major headings, "Assessment of Needs" and "Proposed Plan," to give the audience a clear forecast of the contents.

Under "Proposed Plan," subheadings offer an even more specific forecast. The "Limitations" section shows that our writers are careful to promise no more than they can realistically deliver. At XYZ, upper-management resistance seems to underlie most other communications problems. Because this ultimate problem apparently has gone unrecognized, the final head, "A Related Problem," is inserted for emphasis.

Because this proposal is external, it is cast as a letter. Notice, however, that the complimentary closing ("Best wishes"), and word choice ("thanks;" "what we're doing on our end;" "Jack and Terry," etc.) create an informal, familiar tone. Such a tone is appropriate in this particular external document because the writers and reader have spent many hours in planning conferences, luncheons, and phone conversations.

Like any document that gains reader acceptance, this one is the product of careful decisions about content, organization, and style.

January 15, 1984

Mary Wilson
Senior Education and Training Officer
XYZ Corporation
69 North Charles Boulevard
Cambridge, MA 02139

Dear Mary:

Thanks for sending along the outline for your technical writing workshop. Understandably, such an ambitious plan for eight hours of contact time would not likely produce noticeable results. Here's what we're doing on our end to design an approach that should be realistic and gratifying.

ASSESSMENT OF NEEDS

After conferring with technicians in both Jack's and Terry's groups, and analyzing their writing samples, we identified the following limited hierarchy of common needs:

– improving readability
– achieving precise diction
– summarizing information
– organizing a set of procedures
– formulating various memo reports
– analyzing audiences for upward communication
– writing persuasive bids for transfer or promotion
– writing persuasive suggestions

PROPOSED PLAN

Based on the above needs, we have limited our instruction package to eight carefully selected and readily achievable goals.

Course Outline
Our eight two-hour sessions are structured as follows:
1. achieving sentence clarity
2. achieving sentence conciseness
3. achieving fluency and precise diction
4. writing summaries and abstracts

5. outlining procedures and manuals
6. editing procedures and manuals
7. formulating various reports for various purposes
8. analyzing the audience and writing persuasively

Classroom Format

The first three meetings will be lecture-intensive. So students can apply the material covered in our sessions, we will assign weekly exercises to be done at home and corrected collectively in class. The remaining five weeks will combine lecture and exercises with group editing of work-related documents. Our intent throughout is to remain flexible enough (within the course outline) so we can respond to emerging needs.

Limitations

Given our limited contact time, we cannot realistically expect to turn out a batch of polished technical communicators. However, by course's end, our students will have begun to appreciate the act of writing as a deliberate process of deliberate decisions.

A Related Problem

Several participants have expressed fear that upper management will consider as trivial any communication that is not appropriately inflated, opaque, and jargon-ridden. If such claims are indeed accurate, some "consciousness-raising" might be in order for those who consider muddy prose more substantial than plain English.

If you have any suggestions for refining this plan, please let us know.

Best wishes,

Carl Winston
Carl Winston

Mae Barrett
Mae Barrett

The Research Proposal

Research (or grant) proposals request approval (and often funding) for a research project. A chemistry professor, for instance, might address a research proposal to the Environmental Protection Agency for funds to identify toxic contaminants in local groundwater. Research proposals are solicited by many governments and private agencies: Department of Health and Human Services, National Science Foundation, National Institutes of Health, Department of Agriculture, Carnegie Foundation, Fulbright-Hays Foundation, and others. Each granting agency has its own requirements for the format and content of proposals, but any successful research proposal answers these central questions:

- Why is this project worthwhile?
- What qualifies you to undertake the project?
- What are its chances of success?

In college, you might submit proposals for independent study, field study, or a thesis project. Here is the title of a research proposal submitted to the thesis committee in a geology department:

A PROPOSAL FOR A MASTER'S THESIS PROJECT
TO INVESTIGATE THE TERTIARY GEOLOGY
OF THE ST. MARIES RIVER DRAINAGE
FROM ST. MARIES TO CLARKIA, IDAHO

A technical writing student might submit an informal proposal requesting the instructor's approval of a term project (which, in turn, may be a formal proposal). The introduction of the following proposal describes the background of the problem and justifies the need for the study. The body outlines the scope, method, and sources for the proposed investigation. The conclusion describes the goal of the investigation and encourages reader support. The proposal is convincing because it answers questions about *what, why, how, when,* and *where.* Because this proposal is internal, it is cast informally as a memo.

March 16, 1984

TO: Dr. John Lannon

FROM: T. Sorrells Dewoody

SUBJECT: A PROPOSAL FOR DETERMINING THE FEASIBILITY
           OF MARKETING DEAD WESTERN WHITE PINE

Introduction

Over the past four decades, huge losses of western white pine have occurred in the Northern Rockies, primarily attributable to white pine blister rust and the attack of the mountain pine beetle. Estimated annual mortality is 318 million board feet. Because of the low natural resistance of white pine to blister rust, this high mortality rate is expected to continue indefinitely.

If white pine is not harvested while the tree is dying or soon after death, the wood begins to dry and check (warp and crack). The sapwood is discolored by blue stain, a fungus carried by the mountain pine beetle. If the white pine continues to stand after death, heart cracks develop. These factors work together to cause degradation of the lumber (as graded by the Western Wood Product Inspectors) and a consequent loss in value.

Statement of Problem

White pine mortality results in a reduction in value of white pine stumpage, since the commercial lumber market will not accept it. The major implications of this problem are two: first, in the face of rising demand for wood, vast amounts of timber are not being used; second, dead trees are left to accumulate in the woods, where they are rapidly becoming a major fire hazard here in northern Idaho and elsewhere.

Proposed Solution

One possible solution to the problem of white pine mortality and waste is to search for markets other than the conventional lumber market. The last few years have seen a burst of popularity and a growing demand for weathered barn boards and wormy pine for interior paneling. Some firms around the country are marketing defective wood as specialty products. (*Note:* These firms call the wood from which their products come "distressed." This term will hereafter be used to refer to dead and defective white pine.) There is a good possibility that distressed white pine might find a place in such a market.

Scope

To determine the feasibility of developing a market for distressed white pine, I plan to pursue six areas of inquiry:

1. What products are presently being produced from dead wood, and what are the approximate costs of production?
2. How much demand is there for distressed wood products?
3. Can distressed white pine meet this demand as well as other species meet it?
4. Is there room in the market for distressed white pine?
5. What are the projected costs of retrieving and milling distressed white pine?
6. What prices for the products can the market bear?

Methods

My primary data sources will include consultations with Dr. James Hill, Professor of Wood Utilization, and Dr. Sven Bergman, Forest Economist — both members of the College of Forestry, Wildlife, and Range. I will also inspect decks of dead white pine at several locations, and visit a processing mill to evaluate it as a possible base of operations. I will round out my primary research with a letter and telephone survey of processors and wholesalers of distressed material.

Secondary sources will include selected publications on the uses of dead timber, and a review of an ongoing study by Dr. Hill concerning the uses of dead white pine.

My Qualifications

I have been following Dr. Hill's study on dead white pine for two years. In June of this year I will receive my B.S. in forest management. I am familiar with wood milling processes and have had firsthand experience at logging. My association with Dr. Hill and Dr. Bergman creates the opportunity for an in-depth feasibility study.

Conclusion

Clearly, something should be done to reduce the vast accumulations of dead white pine in our forests. The land on which they stand is among the most productive forest land in northern Idaho. By addressing the six areas of inquiry mentioned earlier, I can determine the feasibility of directing capital and labor to the production of distressed white pine products. With your approval I will begin my research at once.

The Sales Proposal

A sales proposal is addressed to clients and offers a service or product. Sales proposals may be solicited or unsolicited. If they are solicited, several firms may be competing with proposals of their own. Because sales proposals are addressed to readers outside your organization, they are cast as letters (if they are brief). But long sales proposals, like long reports, are formal documents with supplements (cover letter, title page, table of contents, etc.).

The sales proposal, a major marketing tool in business and industry, will be successful if it answers this question:

> How will you serve our needs better than your competitors?
> *or*
> Why should we hire you over someone else?

The following solicited proposal offers a service. Because the writer is competing with other firms, the body of his proposal explains specifically *why* his machinery is best for the job, *how* the job can best be completed, *what* his qualifications are for getting the job done, and *how much* the job will cost. He will be legally bound by his estimate. So he points out possible causes of increased costs, to protect himself. In your sales proposals, *don't underestimate costs by failing to account for all variables* — a sure way to lose money.

The writer's introduction describes the subject and purpose of the proposal. The conclusion reinforces the confident tone throughout and encourages reader acceptance by ending with — thus emphasizing — two key words: "economically" and "efficiently."

James A. Landmover
Route #1
Buhl, Idaho 83331
Tel. 208-634-5267

March 28, 1984

Greg Haver
Star Route
Bliss, Idaho 83314

SUBJECT: PROPOSAL TO DIG A TRENCH AND MOVE
BOULDERS AT SITE TEN MILES WEST OF BLISS

Dear Mr. Haver:

I've inspected your property and would be happy to undertake the land-scaping project necessary for the development of your farm.

The backhoe I use cuts a span 3 feet wide and can dig as deep as 18 feet — more than an adequate depth for the mainline pipe you wish to lay. Because this backhoe is on tracks rather than tires, and is hydraulically operated, it is particularly efficient in moving rocks. I have more than twelve years of experience with backhoe work and have completed many jobs similar to this one.

After examining the huge boulders that block access to your property, I am convinced they can be moved only if I dig out underneath and exert upward pressure with the hydraulic ram while you push forward on the boulders with your D-9 Caterpillar. With this method, we can move enough rock to enable you to farm that now inaccessible tract. Because of its power, my larger backhoe will save you both time and money in the long run.

The job should take 12 to 15 hours, unless we encounter subsurface ledge formations. My fee is $60 an hour. The fact that I provide my own dynamite crew at no extra charge should be an advantage to you since you have so much rock to be moved.

Please phone me at any time for more information. I'm sure we can do the job economically and efficiently.

Yours truly,

James A. Landmover

James A. Landmover

The above three categories for proposals (planning, research, and sales) are neither exhaustive nor mutually exclusive. A research proposal, for example, may request funds for a study that will lead to a planning proposal. The architectural proposal partially shown on pages 430–435 is a combined planning and sales proposal: if clients accept the writer's preliminary plan, they will hire the firm to design the new ski lodge facilities.

ELEMENTS OF A GOOD PROPOSAL

Readers will evaluate your proposal according to how clearly you answer their questions about *what, why, how, when,* and *how much.* An effective proposal is clear, informative, and realistic and conforms to the following guidelines.

Appropriate Format and Supplements

Short proposals can be memos or letters, depending on whether they are internal or external. Longer memos or letters, however, are more easily read if headings are used. Either format may include appendixes (pages 250–253) for support material (maps, blueprints, specifications, calculations, etc.) that would interrupt the text.

Although a short proposal should serve many of your purposes, some projects call for a long proposal. And different readers inside and outside your organization will be interested in different parts of your proposal: some need only a summary; others already know about the problem and will want only your plan; others want all the details. Or perhaps a soliciting agency will specify a certain format and supplements. Such cases may call for a long proposal that, like the formal report, has all needed supplements (cover letter, title page, informative abstract, etc.). Especially important to nontechnical readers (often the decision makers) of a long proposal is the abstract, containing (1) a statement of the problem and causes, (2) proposed solution(s), and (3) an assessment of the plan's feasibility. (See the sample proposal on pages 443–458.)

Remember that decision makers are busy people who dislike fog. Begin with a title that is absolutely clear about the intent of your proposal:

Foggy   A PROPOSAL FOR FACILITIES
               AT SUGARPLUM SKI AREA

What kinds of facilities are being planned — lifts, lodge, parking? What is being proposed — new construction, remodeling, design?

Revised   A PRELIMINARY DESIGN PROPOSAL
              FOR THE NEW LODGE FACILITIES
              AT SUGARPLUM SKI AREA

Don't write "Recommended Improvements" when you mean "Recommended Wastewater Treatment." A concrete and specific title signals a concrete and specific proposal.

Design your proposal to reflect your attention to detail. A hastily typed and assembled proposal suggests to readers the writer's careless attitude toward the project in general. Avoid, however, any "decorations," such as flashy covers, colored paper, or catchy titles. Keep layout, typeface, and bindings conservative and tasteful.

A Focused Subject and a Worthwhile Purpose

Do not try to solve all the world's problems in a single proposal. A fatal mistake, say, in a research proposal, is to begin writing too soon — before you have zeroed in on your subject and purpose. Focus on *one* specific research question you can answer exhaustively, and make your approach original enough to get the reader's attention and support.

The same precise focus applies to planning or sales proposals. Readers want specific suggestions for filling specific needs. By spelling out your subject and purpose, show them immediately you understand their problem.

> *Subject*
> To develop a year-round destination resort, the Sugarplum Corporation wishes to expand its ski area facilities. As part of its expansion program, Sugarplum needs hotel rooms, a day care center, and more space for its cafeteria, bar, ski-and-rental shop, ski school, ski patrol room, and parking.

> *Purpose*
> Our architectural firm offers the following preliminary design proposal for the lodge facilities. The design is based on our evaluation of site information, the master plan, and a list of required facilities and their relationships, as well as the projected circulation around and through the required spaces. Our building philosophy and design priorities also take into account local building restrictions and possibilities of future expansion.

Notice that the focus here is limited to the *preliminary design phase*. The actual plans for construction, decor, landscaping, etc., will be the subject of later proposals. This, then, might be called a *preproposal*. If accepted, it will lead to more specific proposals.

Identification of Related Problems

Do not underestimate the complexity of the project. Identify *all* problems readers themselves might not recognize. Remember that only problems that have been *fully* and *clearly* defined can possibly be solved. Here is how the architectural proposal treats one such problem:

The Sugarplum Corporation has expressed a desire for glass walls on the north, east, and south exposures of the main lodge. Although improving the view, glass walls would increase heat loss through thermal conductivity. Concentrated areas of glass (triple-glazed) should be limited to southerly exposures for maximum use of solar energy. For further energy efficiency, exterior walls should be insulated to a value of R-19 or better, and the doors built in an airlock configuration to minimize infiltration.

Realistic Methods

Resist the temptation to propose easy answers to hard questions. Be conservative. Propose only those methods that have a good chance of success. If a certain solution is the best available — but still leaves doubt as to its effectiveness — let the readers know; otherwise, you or your firm could later be held liable for the project's failure. Avoid overstatement. Here is how the architectural proposal treats a questionable solution:

> The subsoil near the lodge is sand, with a two-foot cover of humus and no rock outcrops. Some places on the higher mountain have soil depths of over 10 feet. Where the bedrock is close to the surface, however, considerable runoff occurs, causing erosion around the drainage ditches and the parking lot (see the drainage map in Appendix B). A system of French drains, tile beds, and drainage pipes coupled with a landscaping program *can eliminate most, if not all, of the drainage and erosion.*

Notice how the writer qualifies his suggested solution (in the part we italicized here) to avoid promising more than he can deliver.

Concrete and Specific Information

Vagueness is a fatal flaw in a proposal. Before you can *persuade* readers, you must *inform* them; therefore, you need to *show* as well as *tell*. (Review pages 64–65.) Instead of writing "The mountain is covered with mixed vegetation" (thus allowing for many interpretations), write, "The mountain's vegetation consists primarily of tamarack, cedar, and sub-alpine fir, with a ground cover of bear grass and alder bush."

Here are the specifications that will serve as design guidelines for the ski patrol suite; notice their concreteness:

Ski Patrol Suite

Square Footage Requirements: 900 sq. ft. total: first aid room = 400 sq. ft. (to include a 30-sq.-ft. restroom); dressing room/lockers = 500 sq. ft.

Occupants: ski patrol persons; injured skiers and their friends.

Special Equipment: ski, boot, and pole storage; lockers; boot-drying racks; good heater; storage for toboggans and first aid equipment; two cots; coat racks; blackboard; desk; exterior ski racks.

Special Considerations: first aid room should be pleasant and comfortable; ski patrol room should be functional; cots should be well sheltered from drafts and doors; floor surface in the patrol room should be durable; first aid room should have easy slope, snow-cat, and ambulance access.

Location: close to the parking lot.

A less competent writer might have said:

> The ski patrol suite will have to be large enough to accommodate the various personnel, patrons, and equipment, and will have to be located in an easily accessible area.

Notice how abstract words ("large," "accessible") and general words ("personnel," "patrons," "equipment") provide no useful design guidelines until they are supported with concrete and specific details. To avoid any misunderstanding, a proposal must elicit *one* interpretation only.

Visuals

If they enhance your proposal, use visuals, properly introduced and discussed (Chapter 13).

FIGURE 1   Bubble Diagram of the Administrative Complex

Spatial Relationships

The design of the administrative complex should facilitate circulation and communication among various departments, as well as allowing for centralized management. See Figure 1.

Appropriate Level of Technicality

A single proposal might address a diverse audience. A research proposal might be read by experts in the field, who then advise the granting agency whether to accept or reject it. Planning and sales proposals might be read by colleagues, superiors, and clients (who often are laypersons). Informed and expert readers will be most interested in the technical details of the project. Lay readers will be more interested in the projected results, but they will need an explanation of technical details as well. Learn all you can about the needs, concerns, and biases of your audience.

Unless your proposal gives all readers what they need, it is not likely to move anyone to action. This is where supplements are useful (especially abstracts, glossaries, and appendixes). Let your knowledge of the audience guide your decisions about supplements. Who is your secondary audience (pages 14–16)? Who else will be evaluating your proposal?

If the primary audience is expert or informed, keep the proposal itself technical. For uninformed secondary readers (if any), provide an informative abstract, a glossary, and appendixes that explain specialized information. If the primary audience has no expertise and the secondary audience does, follow this pattern: write the proposal itself for laypersons, and provide appendixes containing the technical details (formulas, specifications, calculations, etc.) the informed readers will use to evaluate your plan.

A Tone that Connects with Readers

The successful proposal is the one that is *accepted* by its audience, so make it reader-oriented. Keep your tone confident, encouraging, and diplomatic. Show readers you believe in your plan; urge them to act; anticipate how they will react to your suggestions. Do not come across with a superior attitude.

Superior   The Sugarplum Corporation *has apparently ignored* the *obvious* problems of drainage and erosion, snow accumulation, and hazardous circulation patterns. Moreover, the corporation *should pay close attention* to the following design restrictions: building codes, energy conservation, the limited use of stairs, and anticipated snow loads on the structures.

Although *what* is said above is worthwhile, *how* it is said alienates readers. The italicized words create a superior and insulting tone.

On the other hand, don't come across as a milquetoast:

> Weak — *It would seem that there is a possibility of* problems with drainage and erosion, snow accumulation, and hazardous circulation patterns. Also, *it might not be a bad idea to give some consideration to* the following design restrictions. . . .

The italicized words make the writer seem unsure, a tone not likely to inspire confidence.

This next version has a tone that is confident, encouraging, and diplomatic:

> Revised — Design considerations must address the problems of drainage and erosion as well as snow accumulation and hazardous circulation patterns. Moreover, any design must abide by these restrictions: building codes, energy conservation, the limited use of stairs, and anticipated snow loads on the structures.

COMPOSING A PROPOSAL

Like all informative writing, proposals have an introduction-body-conclusion arrangement. Depending on project complexity, each section contains some or all of the subsections listed in the following general outline:

I. INTRODUCTION
    A. Subject and Purpose (or Objective)
    B. Statement of Problem
    C. Background
    D. Need
    E. Qualifications of Personnel
    F. Data Sources
    G. Limitations
    H. Scope

II. BODY
    A. Methods
    B. Timetable
    C. Materials and Equipment

    D. Personnel
    E. Available Facilities
    F. Needed Facilities
    G. Cost
    H. Expected Results
     I. Feasibility

  III. CONCLUSION
    A. Summary of Key Points
    B. Request for Action

These subsection headings can be rearranged, combined, divided, or deleted as needed. Although not every proposal contains all subsections, each major section must answer certain reader questions, as illustrated below.

Introduction

The introduction answers all the following questions — or all those that apply to the situation:

    – What problem are you proposing to solve?
    – In general, what solution are you proposing?
    – Why are you proposing it?
    – What are the benefits?
    – What are your qualifications for this project?

From the outset, your goal is to sell your idea, to convince readers the job needs doing and you (or someone else) are the one to do it. If your introduction is long-winded, evasive, or vague, readers may not read on. Make it concise, specific, and clear.

Spell out the problem to make it absolutely clear to the audience — and to show you understand it fully. Explain why the problem should be solved or the project undertaken. Identify any sources of data. In a research or sales proposal, state your qualifications for doing the job. If your plan has limitations, explain them. Finally, define the scope of your plan by enumerating the specific subsections to be discussed in the body section.

Here is the introduction for a planning proposal titled "A Proposal for Solving the Noise Problem in the University Library." Jill Sanders, a library work-study student, addresses her unsolicited proposal to the chief librarian and the administrative staff. Because this proposal is unsolicited, it must first make the problem vivid through details that arouse concern and interest. So this introduction is longer than it would be in a solicited proposal, where readers would already agree about the severity of the problem.

INTRODUCTION

Subject

During the October 1983 Convocation at Margate University, many students and faculty members complained about noise in the library. Shortly afterward, certain areas were designated for "quiet study," but the library continues to receive complaints about noise. To create a scholarly atmosphere, the library staff should take immediate action to decrease noise.

Purpose

This proposal examines the noise problem from the viewpoint of students, faculty, and library staff. It then offers a plan to make certain areas of the library quiet enough for serious study and research.

Sources

The data for this proposal come from a university-wide questionnaire, interviews with students, faculty, and library staff, inquiry letters to other college libraries, and my own observations for three years as a library staff member.

Statement of Problem

This subsection examines the severity and causes of the noise.

Severity

Since the 1983 Convocation, the library's fourth and fifth floors have been reserved for quiet study, but students hold group-study sessions at the large tables and disturb others working alone. The constant use of computer terminals on both floors adds to the noise, especially when students working in pairs discuss the program. Moreover, people often chat as they enter or leave study areas.

On the second and third floors, designed for reference, staff members help patrons locate materials, causing constant shuffling of people and books, as well as loud conversation. At the computer service desk on the third floor, conferences between students and instructors create more noise.

The most frequently voiced complaint from the faculty members interviewed concerned the second floor, where people using the Reference and Government Documents services converse loudly. Students complain about the lack of a quiet spot to study, especially in the evening, when even the "quiet" floors are as noisy as the dorms.

Over 80 percent of the respondents (530 undergraduates, 30 faculty, 22 graduate students) to a university-wide questionnaire (Appendix A) insisted that excessive noise discourages them from using the library as often as they would otherwise. Of the student respondents, 430 cited quiet study as their primary reason for wishing to use the library.

The library staff recognizes the problem but has insufficient personnel to cope with it. Because all staff members have assigned tasks, they have no time to monitor noise in their sections.

Causes

Respondents complained specifically about the following causes of noise (in descending order of frequency):

1. Loud study groups that often lapse into social discussions.
2. A general disrespect for the library, with the attitude of some students characterized as "rude," "inconsiderate," or "immature."
3. The constant clicking of computer terminals on all five floors, and of typewriters on the first three.
4. Vacuuming by the evening custodians.

All complaints centered on the lack of enforcement by the library staff.

Because the day staff works on the first three floors, quiet-study rules are not enforced on the fourth and fifth floors. Work-study students on these floors have no authority to enforce rules not enforced by the regular staff. Small, black-and-white "Quiet Please" signs posted on all floors apparently go unnoticed, and the evening security guard provides no added deterrent.

Needs

Excessive noise in the library is keeping patrons away. By addressing this problem immediately, we can help restore the library's credibility and utility as a campus resource. We must reduce noise on the lower floors and eliminate it from the quiet-study floors.

Scope

The proposed plan includes a detailed assessment of methods, costs and materials, personnel requirements, feasibility, and expected results.

Body

The body section will receive the most attention from readers. It answers all the following questions that are applicable:

- How will it be done?
- When will it be done?
- What materials, methods, and personnel will it take?
- What facilities are available?
- How long will it take?
- How much will it cost, and why?
- What results can we expect?
- How do we know it will work?
- Who will do it?

Here you spell out your plan in enough detail for readers to evaluate its soundness. If this section is vague, your proposal stands no chance of being accepted. Besides being clear, be sure your plan is realistic and promises no more than you can deliver. The main goal of this section is to prove your plan will work.

PROPOSED PLAN

The following plan takes into account the needs and wishes of our campus community, as well as the available facilities in our library.

Methods

Noise in the library can be reduced through three complementary steps: (1) improving publicity, (2) shutting down and modifying our facilities, and (3) enforcing the quiet rules.

Improving Publicity

First, the library must publicize the noise problem. This assertive move will demonstrate the staff's concern. Publicity could include articles by staff members in the campus newspaper, leaflets distributed on campus, and a freshman library orientation that acknowledges the noise problem and asks the cooperation of new students. All forms of publicity should detail the steps the library is taking to solve the problem.

Shutting Down and Modifying Facilities

After notifying campus and local newspapers, you should close the library for one week. To minimize disruption, the shutdown should occur between the end of summer school and the beginning of the fall term.

During this period, you can convert the fixed tables on the fourth and fifth floors to cubicles by the use of temporary partitions (6 cubicles per table). You could later convert the cubicles to shelves as the need for book space increases.

Then you can take all unfixed tables from the third, fourth, and fifth floors to the first floor, and set up a space for group study. Plans already are underway for removing the computer terminals from the fourth and fifth floors.

Enforcing the Quiet Rules

Enforcement is the essential, long-term element of this plan. No one of any age is likely to follow all the rules all the time — unless they are enforced.

First, you can make new "Quiet" posters to replace the present, innocuous notices. A visual design student can be hired to draw up large, colorful posters that attract attention. Either the design student or the university print shop can take charge of poster production.

Next, through publicity, library patrons can be encouraged to demand quiet from noisy people. To support such patron demands, the library staff can begin monitoring the fourth and fifth floors, asking study groups to move to the first floor, and revoking library privileges of those who refuse. Patrons on the second and third floors can be asked to speak in whispers. Staff members should set an example by regulating their own voices.

Costs and Materials

The major cost would be for salaries of new staff members who would help monitor. Next year's library budget, however, will include an allocation for four new staff members.

A design student has offered to make up four different posters for $200. The university printing office can reproduce as many posters as needed at no extra cost.

Prefabricated cubicles for 26 tables sell for $150 apiece, for a total cost of $3,900.

Rearrangement of various floors can be handled by the library's custodians.

The Student Fee Allocations Committee and the Student Senate routinely reserve funds for improving student facilities. A request to these organizations would yield at least partial funding for the plan.

Personnel

The success of this plan ultimately depends on the willingness of the library administration to implement it. You can run the program itself by committees made up of students, staff, and faculty. This is yet another area where publicity is essential to persuade people that the problem is severe and that you need their help. To recruit committee members, you can offer students an added incentive through the untapped resource of Contract Learning credits.

The proposed committees include an Antinoise Committee overseeing the program, a Public Relations Committee, a Poster Committee, and an Enforcement Committee.

Feasibility

On March 15, 1983, I mailed survey letters to 25 New England colleges, inquiring about their methods for coping with noise in the library. Among the respondents, 16 stated that publicity and the administration's attitude toward enforcement were key elements in their success.

Improved publicity and enforcement certainly could work for us as well. Moreover, slight modifications in our facilities to concentrate group study on the busiest floors would automatically lighten the burden of enforcement.

Expected Results

Publicity will improve communication between the library and the campus. An assertive stance will show that the library is aware of its patrons' needs and willing to meet those needs. Offering the program for public inspection will draw the entire community into improvement efforts. Publicity, begun now, will pave the way for the formation of committees.

The library shutdown will have a dual effect: it will dramatize the problem to the community, and also provide time for the physical changes. (An antinoise program begun with carpentry noise in the quiet areas would hardly be effective.) The shutdown will be both a symbolic and a concrete measure, leading to the reopening of the library with a new philosophy and a new face.

Continued, strict enforcement will be the backbone of the program. It will prove that staff members are concerned enough about the atmosphere to jeopardize their own friendly image in the eyes of certain users and that the library is not afraid to enforce its rules.

Conclusion

The conclusion restates the need for the project and persuades readers to act. It answers the questions readers will ask:

> – How badly do we need this change?
> – Why should we accept your proposal?
> – How do we know this is the best plan?

End on a strong note, with a conclusion that is assertive, confident, and encouraging — and keep it short.

CONCLUSION AND RECOMMENDATION

> The noise in Margate University library has become embarrassing and annoying to the whole campus. Forceful steps are needed to improve the academic atmosphere.
>
> Aside from the intangible question of image, close inspection of the proposed plan will show that it will work if you take the recommended steps and — most important — if the enforcement of quiet becomes a part of the library's services.

In certain long proposals, especially those that begin with a comprehensive abstract, the conclusion can be omitted.

When a few lines or short paragraphs can answer the reader questions in each section, a short proposal will suffice, but a complex plan calls for a long, formal proposal. The following section illustrates a formal proposal accompanied by all necessary supplements.

APPLYING THE STEPS

The formal planning proposal in Figure 18–1 typifies the kind of specialized proposal that justifies a funding request.

AUDIENCE-AND-USE ANALYSIS

A university newspaper is struggling to meet rising costs. The paper's yearly budget is funded by a college allocations committee that disburses money to all student organizations. Because of tight money, the newspaper has received no funding increase in three years.

The author of the proposal, the paper's business manager, must justify his request for a 17 percent budget increase for the coming year. This proposal, solicited by the allocations committee, is both external and internal. The primary audience will be the committee members, who will evaluate the soundness of the plan and decide whether to grant the additional funds. The secondary audience will be the newspaper staff, who will implement the plan — if it is approved by the allocations committee.

To encourage reader acceptance, the writer must propose a realistic plan showing that the newspaper staff is sincere in its intention to cut operating costs. At a time when everyone is expected to get by with less, the writer has to make an especially strong case for salary increases because a section of the plan calls for a slight increase in staff salaries, to attract talented personnel.

When considering budget requests, readers invariably want to know exactly *how* the money will be spent, so the writer has to spell it out by itemizing every expense in the proposed budget. Thus the "Costs" section is the longest of the proposal.

To further justify the requested budget, the writer decides to show just how well the newspaper staff manages its funds. Under "Feasibility" he gives a detailed comparison of funding, expenditures, and size of newspaper in relation to the other four colleges in his area. This section is the clincher because these hard facts are most likely to convince the committee that the proposed plan is cost-effective. So as not to clutter his discussion, the writer adds an appendix containing a table of figures for the above comparisons.

A BUDGET PROPOSAL
FOR
THE SMU TORCH
(1983-84)

Prepared for

The Student Fee Allocations Committee
Southeastern Massachusetts University
North Dartmouth, Massachusetts

by

William Trippe
Torch Business Manager

May 1, 1983

FIGURE 18-1   A Formal Proposal

ii

The SMU <u>Torch</u>
Old Westport Road
North Dartmouth, Massachusetts

May 1, 1983

Charles Marcus, Chairperson
Student Fee Allocations Committee
Southeastern Massachusetts University
North Dartmouth, Massachusetts 02747

Dear Dean Marcus:

No one needs to be reminded about the staggering effects of
inflation on our campus community. We are all forced to make
do with less.

Accordingly, we at the <u>Torch</u> have spent long hours devising
a plan to cope with increased production costs -- without
compromising the newspaper's tradition of quality service.
I think you and your colleagues will agree that our plan is
realistic and feasible. Even the "bare-bones" operation that
will result from our proposed spending cuts, however, will
call for a $4,432.14 increase in our 1983-84 budget.

We have received no funding increases in three years. Our
present need is absolute. Without additional funds, the <u>Torch</u>
simply can function no longer as a professional newspaper.
Therefore, I submit the following budget proposal for your
consideration.

Respectfully,

William Trippe

William Trippe
Business Manager, SMU <u>Torch</u>

FIGURE 18-1 (*Continued*)

iii

TABLE OF CONTENTS

FIGURE 18–1 (*Continued*)

INFORMATIVE ABSTRACT

The student newspaper at Southeastern Massachusetts University is crippled by inadequate funding, having received no budget increases in three years. Inflation is the major problem facing the Torch. Increases in costs of printing, layout, and photographic supplies have called for a decrease in production. Moreover, our low salaries are inadequate to attract and retain qualified personnel. A nominal increase would make salaries more competitive.

Our staff plans to cut costs by reducing page count, saving on photographic paper, reducing circulation, and hiring a new printer. The only proposed cost increase (for staff salaries) is essential.

A detailed breakdown of projected costs establishes the need for a $4,432.14 budget increase to keep the paper running weekly and with an adequate page count and distribution to serve our campus.

Compared with other college newspapers, the Torch makes much better use of its money. This comparison certifies the cost effectiveness of our proposal.

FIGURE 18-1 (*Continued*)

INTRODUCTION

Subject and Purpose

Our campus newspaper faces the contradictory challenge of surviving ever-increasing production costs while maintaining its quality. The following proposal offers a realistic plan for meeting this crisis. The plan's ultimate success, however, depends on the Student Fee Allocations Committee's willingness to approve a long-overdue increase in our 1983-84 budget.

Background

In its ten years, the Torch has grown in size, scope, and quality. It is now the central means of communication on campus. Roughly 6,000 copies (24 pages/issue) are printed weekly during each 14-week semester.

Each week, the Torch prints news, features, editorials, sports articles, announcements, notices, classified ads, a calendar column, and letters to the editor. A vital part of university life, the paper disseminates information, ideas, and opinions -- all with the highest professionalism.

FIGURE 18-1   (*Continued*)

2

Statement of Problem

   The <u>Torch</u>, with much of its staff about to graduate,

faces next year with little money, rising costs in every

phase of production, and the need to replace expensive

equipment that is outdated and in constant need of repair.

   Our newspaper also suffers from a serious lack of student

involvement. Few students can be expected to devote full

time and energy to the paper without some kind of salary.

Most staff members do receive minimal weekly salaries: from

$10.00 for the distributor to $45.00 for the Editor in Chief,

but salaries averaging less than $1.00 per hour cannot

possibly compete with a minimum wage of $3.10 per hour. As

inflation forces more students to work, the <u>Torch</u> will have

to do what it can to make its salaries more competitive.

   The newspaper's operating expenses can be divided into

four categories: composing costs, salaries, printing costs,

and miscellaneous (office supplies, mail, etc.). The first

three categories account for nearly 90 percent of the budget.

In the past year, costs in all categories have increased from

as little as 3 percent for darkroom chemicals to as much as

60 percent for photographic paper and film. Printing costs

FIGURE 18-1 (*Continued*)

3

(roughly one-third of our total budget) rose by 9 percent in
the past year, and another price hike of 10 percent has just
become effective.

Need

Despite the steady increase in production costs, the
Torch has received no increase in its yearly allocation
($21,500) in three years. Given these inescapable cost
increases, inadequate funding virtually is crippling our
newspaper.

Scope

The plan offered below includes:

1. methods for reducing production costs while main-
    taining the quality of our staff

2. an itemized list of projected costs for equipment,
    material, salaries, and services in 1983-84

3. a demonstration of feasibility by showing our cost
    effectiveness

4. a summary of the attitudes shared by our staff

FIGURE 18-1   (*Continued*)

4

PROPOSED PLAN

The following plan is designed to trim operating costs -- without compromising the quality of our newspaper.

Methods

We can respond successfully to our budget and staffing crisis by taking the following steps:

Reducing Page Count

By condensing free notices for campus organizations from two pages to one page, abolishing "personals," and limiting press releases to one page, we can reduce our average page count per issue from 24 to 20. This reduction will mean nearly a 17 percent saving in production costs.

Saving on Photographic Paper

A newspaper with fewer pages will call for fewer photographs. Also, we can further save money by purchasing a year's supply of photographic paper within the next month, before the upcoming 15 percent increase becomes effective.

Reducing Circulation

By reducing circulation from 6,000 to 5,000 copies weekly, we barely will cover the number of full-time day students, but we will save nearly 17 percent in printing costs.

FIGURE 18-1 (*Continued*)

5

Hiring a New Press

We can save money by hiring Sadus Press of Worcester to do our printing. Bids from several other presses (including our present printer) were at least 25 percent higher than Sadus's price. Moreover, no other company offers the delivery service we will get from Sadus.

Increasing Staff Salaries

Although our staff seeks talented students who expect little money and much good experience, our salaries for all positions must increase by an average of $5.00 weekly. With the minimum wage at $3.10 per hour, any of our staff could make as much money elsewhere by working only one-third of the time. In fact, many students could make more than the minimum wage by working for local newspapers. To illustrate: The Standard Times pays $20.00 to $30.00 for a news article and $10.00 for a photo, while the Torch pays nothing for articles and $2.00 for a photo.

A most striking example of our low salaries is the $3.10 hourly we pay our typesetters. On the outside, trained typesetters like ours make from $8.00 to $12.00 per hour. Thus, our present typesetting cost of $3,038 easily could be

FIGURE 18–1 (*Continued*)

6

as much as $7,000 -- or even higher if we had the typesetting
done by an outside firm, as many colleges do.

In short, without this nominal salary increase, we cannot possibly hope to attract and retain qualified personnel.

Budget Request for 1983-84

Our proposed budget is itemized below, but the main
point is clear: if the Torch is to remain a viable newspaper, increased funding is essential.

Projected Costs   Yearly Total

Equipment Leasing (We must continue to meet the obligations
of a leasing arrangement with Chase Manhattan Leasing
Corporation for Compugraphic equipment.)

1.	Execuwriter	$106.25/month
2.	7200 L and hardware	$127.38/month
3.	Execuwriter II	$138.63/month

$372.26/month = $3,350.34/year

Composing Chemicals

1.	Activator	$105.00
2.	Stabilizer	180.00
3.	Processor Cleaner	55.00

$340.00   = $340.00/year

FIGURE 18-1 (*Continued*)

7

Photo Paper for Execuwriters

1. 3" x 150'	$711.00	
2. 7" x 150'	556.37	
3. Headline paper	732.63	
4. 8" x 150'	227.00	
	$2,227.00	= $2,227.00/year

Layout Supplies

1. Font strips	$160.00	
2. Layout boards	160.00	
3. Exacto knives and blades	80.00	
4. No-repro pens	30.00	
5. Rulers	12.00	
6. Scotch tape	50.00	
7. Folders	30.00	
8. Reduction wheels	20.00	
9. Wax	45.00	
10. Construction paper	10.00	
11. Border tape	50.00	
	$647.00	= $647.00/year

Photography

1. 10 $1.00 photos/issue	$280.00	
2. 10 $2.00 photos/issue	560.00	
	$840.00	= $840.00/year

Photography Supplies

1. Paper	$730.00	
2. Film	250.00	
3. Darkroom Supplies	100.00	
	$1,080.00	= $1,080.00/year

Typing Staff

1. 35 hrs. at $3.10/hr.	$108.50/week	= $3,038.00/year

FIGURE 18–1 (*Continued*)

8

Salaries

1.	Editor in Chief	$1,400.00/year	
2.	News Editor	840.00	
3.	Asst. News Editor	420.00	
4.	Features Editor	840.00	
5.	Head Writer (features)	420.00	
6.	Sports Editor	840.00	
7.	Head Writer (sports)	420.00	
8.	Advertising Manager	1,050.00	
9.	Advertising Designer	700.00	
10.	Free Ad Designer	280.00	
11.	Layout Editor	840.00	
12.	Art Director	560.00	
13.	Photo Editor	840.00	
14.	Business Manager	840.00	
		$10,290.00	= $10,290.00/year

Fixed Printing Costs

1.	5,000 copies/week x 28	$12,399.80	= $12,399.80/year

Miscellaneous Costs

1.	Graphics by SMU art students: 3/wk. at $5.00 each	$420.00	
2.	Mail costs	550.00	
3.	Telephone costs	500.00	
4.	Print shop costs	200.00	
5.	Copier fees	50.00	
		$1,720.00	= $1,720.00/year

TOTAL YEARLY COSTS $35,932.14

Expected Ad Revenue
    ($1,000.00/mo x 10) $10,000.00

Total Cost minus Ad Revenue: $25,932.14

TOTAL BUDGET REQUEST $25,932.14

FIGURE 18-1   (*Continued*)

9

Feasibility

Beyond exhibiting our need, we feel that the feasibility
of this proposal can be measured through an objective assessment of our cost effectiveness: In comparison to other school
newspapers, how well does the Torch use its funds?

In a survey of the other four colleges in our area, we
found that the Torch -- by a sometimes huge margin --makes
the best use of its money. Table 1 in Appendix A shows that,
of the five newspapers, the Torch costs the students least,
runs the most pages per week, and spends the least money per
page, despite having a circulation two to three times the
size of the other papers.

The most striking comparison is between the Torch and
the newspaper at Fallow State College (Appendix A). Each
student at FSC pays $12.33 yearly for a paper averaging 12
pages per issue. Here at SMU, each student pays $4.06 yearly
for a paper averaging 20 pages per issue. Thus, for 33 percent of FSC's cost, SMU students are getting 66 percent more
newspaper.

The Torch has the lowest yearly cost of all five papers,
despite the largest circulation. With the budget increase

FIGURE 18–1 (*Continued*)

10

requested above, the cost would rise only by $0.44, for a

yearly cost of $4.48 to each student. Although Alden Col-

lege's paper costs each student $4.29, it is published only

every third week, averages 12 pages per issue, and costs

nearly $50.00 yearly per page to print -- as opposed to our

yearly printing cost of $38.25 per page.

As the figures in Appendix A demonstrate, our cost

management is responsible and effective.

Personnel

Students on the Torch staff are unanimous in their de-

termination to maintain the highest professionalism. Many

are planning careers in journalism, writing, editing,

advertising, photography, or public relations. In any

issue, the balanced, enlightened coverage is evidence of

our judicious selection and treatment of articles and our

shared concern for quality.

CONCLUSION

As a forum for ideas and opinions, the Torch continues

to reflect a seriousness of purpose and a commitment to free

FIGURE 18-1 (*Continued*)

11

expression. Its place in the campus community is more vital
than ever in these troubled times.

There are increases and decreases in student fee allocations every year. Last year, for example, eight allocations
were increased by an average of $2,166. The Torch has re-
ceived no increase since 1979-80. Presumably, various increases materialize as priorities change and as special circumstances arise. The Torch staff urges the Allocation
Committee to respond to the paper's legitimate and proven
needs by increasing our 1983-84 allocation to $25,932.14.

FIGURE 18-1 (*Continued*)

12

APPENDIX A

Table 1.   A Comparison of Allocations and Performance of Five Massachusetts College Newspapers

	Stonehorse College	Alden College	Simms University	Fallow State	SMU
Enrollment	1,600	1,400	3,000	3,000	5,000
Fee paid (per year)	$65.00	$85.00	$35.00	$50.00	$65.00
Total fee budget	$88,000	$119,000	$105,000	$150,000	$334,429.28
Newspaper budget	$10,000	$6,000	$25,300	$37,000	$21,500
					$25,932.14[a]
Yearly cost per student	$6.25	$4.29	$8.43	$12.33	$4.06
					$5.20[a]
Format of paper	Weekly	Every third week	Weekly	Weekly	Weekly
Average no. of pages	8	12	18	12	20
Average total pages	224	120	504	336	560
					672[a]
Yearly cost per page	$44.50	$50.00	$50.50	$110.11	$38.25
					$38.69[a]

[a]These figures are next year's costs for the SMU Torch.

Source: Figures were quoted by newspaper business managers in April 1983.

FIGURE 18–1 (*Continued*)

CHAPTER SUMMARY

A proposal is an offer to do something or a suggestion for action. Three common types of proposals are the planning proposal, the research proposal, and the sales proposal. These can be internal or external, solicited or unsolicited. The planning proposal answers this central reader question:

- What are the benefits of following your suggestions for change?

The research proposal answers these questions:

- Why is this project worthwhile?
- What, exactly, qualifies you to undertake this project?
- What are its chances of success?

The sales proposal answers the question:

- How will you serve our needs better than your competitors?

Any successful proposal will answer applicable questions about *what, why, how, when,* and *how much.*

To ensure a good proposal, follow these guidelines:

1. Use the appropriate format and supplements.
2. Be sure your subject is focused and your purpose worthwhile.
3. Identify all related problems.
4. Offer realistic methods.
5. Provide concrete and specific information.
6. Use visuals whenever possible.
7. Maintain the appropriate level of technicality.
8. Create a tone that connects with your readers.

As you plan and write your proposal, work from a detailed outline that has a distinct introduction, body, and conclusion:

1. In the introduction, answer the *what* and *why,* and clarify the subject, background, and purpose of your proposal. Establish need and benefits, along with your qualifications. Identify data sources, any limitations of your plan, and its scope.

2. In the body, answer the *how, when,* and *how much.* Spell out your plan by enumerating methods, work schedules, materials and equipment, personnel, facilities, costs, expected results, and feasibility.

3. In the conclusion, summarize key points, and stimulate action.

REVISION CHECKLIST FOR PROPOSALS

Use this list to revise your proposal.

Format

1. Is the short, internal proposal a memo, and the short, external proposal a letter?
2. Is the long proposal a formal document with adequate supplements to serve the different needs of different readers?
3. Is the format professional in appearance?
4. Are headings logical and adequate?
5. Does the title forecast the proposal's subject and purpose?

Content

1. Is the subject original, concrete, and specific?
2. Is the purpose clear and worthwhile?
3. Does everything in the proposal support its stated purpose?
4. Does the proposal *show* as well as *tell* (opinions based on fact, assertions backed up by evidence, convincing reasons)?
5. Is the proposed plan, service, or product beneficial?
6. Are the proposed methods practical and realistic?
7. Are all related problems identified?
8. Is the proposal free from overstatement?
9. Are visuals used effectively and whenever possible?
10. Is the proposal's length appropriate to the complexity of the subject?

Arrangement

1. Are sections arranged in an introduction, body, and conclusion?
2. Does each section contain the *relevant* headings from the general outline on pages 435–436?
3. Does the introduction stimulate interest and answer the reader questions on page 436 that are applicable?
4. Does the body spell out and justify your plan by answering the applicable reader questions on page 438?
5. Does the conclusion encourage readers to accept the proposal?
6. Are there clear transitions between related ideas?

Style

1. Is the level of technicality appropriate for the intended primary audience?
2. Are the supplements written at a level of technicality that follows the guidelines on page 434?
3. Is the informative abstract sufficiently nontechnical to be understood by all intended readers?
4. Does the tone connect with readers (confident, encouraging, and diplomatic)?
5. Is each sentence clear, concise, and fluent (pages 43–57)?
6. Is the language convincing and precise (pages 57–72)?
7. Is the proposal in correct English (Appendix A)?

EXERCISES

1. After consultation with instructors or people on the job, write a memo to your instructor identifying the kinds of proposals most often written in your field. Are such proposals usually short or long, internal or external, solicited or unsolicited? Who are the decision makers in your field? Provide at least one detailed scenario of a work situation that calls for a proposal. Identify both the primary and secondary audience before you begin writing. Include your sources of information in your memo.

2. Assume the chairperson of your high school English Department has asked you, as a recent graduate, for suggestions about changing the English curriculum to better prepare students for writing. Write a proposal, based on your experience since high school, that will move readers to action. (Primary audience: the English Department chairperson and English faculty; secondary audience: school committee members.) In your external, solicited proposal be sure to identify problems, needs, and benefits and to spell out a realistic plan. Review the general outline on pages 435–436 before selecting specific headings for your proposal.

3. From a faculty member in your major, obtain a copy of a recent proposal for changes in the department's course offerings or staffing, or for solving a departmental problem (e.g., instituting a minor, adding new courses, dropping certain courses, hiring new faculty, or changing degree requirements for a major). Using the revision checklist, evaluate the proposal. Submit your findings in a memo to your instructor, or discuss your evaluation in class, using an opaque projector.

4. After identifying your primary and secondary audience, compose a

short planning proposal for improving an unsatisfactory situation in the classroom, on the job, or in your dorm or apartment (e.g., poor lighting, drab atmosphere, health hazards, poor seating arrangements). Choose a problem or situation whose resolution is more a matter of common sense and lucid observation than of intensive research. Be sure to (a) identify the problem clearly, give a brief background, and stimulate reader interest; (b) state clearly the methods proposed to solve the problem; and (c) conclude with a statement designed to gain reader support of your proposal.

5. Write a research proposal to your instructor (or an interested third party) requesting approval for the final term project (likely an analytical report or formal proposal). Identify the subject, background, purpose, and benefits of your planned inquiry, as well as the intended audience, scope of inquiry, data sources, methods of inquiry, and a task timetable. Be certain that adequate primary and secondary sources are available. Convince your reader of the soundness and usefulness of the project.

6. As an alternate term project to the formal analytical report (see Chapter 19), develop a long proposal for solving a major problem, improving a key situation, or satisfying an urgent need in your school, community, or job. Choose a subject of sufficient complexity to justify a formal proposal, a topic requiring a good deal of research (mostly primary). Identify an audience (other than your instructor) who will use your proposal for a specific purpose. Compose an audience-and-use analysis, using the sample on page 442 as a model.

Here are some possible subjects for your proposal:

– improving living conditions in your dorm or fraternity/sorority
– creating a student-operated advertising agency on campus
– creating a day care center on campus
– creating a new business or expanding an existing business
– saving labor, materials, or money on the job
– improving working conditions
– supplying a product or service to clients or customers
– improving campus facilities for the handicapped
– developing a community waste-recycling program
– developing a bicycle safety program in your community
– increasing tourist trade in your town
– eliminating traffic hazards in your neighborhood
– developing plans to handle parking problems (e.g., plans to handle campus parking and traffic patterns for sports events)
– reducing energy expenditures on the job
– improving security in dorms or in the college library
– improving in-house training or job-orientation programs
– deciding how a church or synagogue might best use a bequest of money
– creating a jogging path through the campus
– creating a bike path or special bike lane on main streets

- creating a new student organization within the government (say, one to handle Chicano affairs)
- finding ways for an organization to raise money
- improving faculty advisement of students
- purchasing new equipment
- improving the food service on campus
- funding the purchase of new uniforms for an athletic or musical organization
- easing freshmen students through the transition to college
- improving bomb-alert procedures in the student union building
- making word processors available to all advanced writing students
- establishing a college-wide computer fluency program
- persuading local manufacturers to disclose their toxic-waste dump sites
- changing the grading system at your school
- leasing microcomputers to students
- establishing more equitable computer terminal use
- offering a service to potential customers

19

Analyzing Data and Writing a Formal Report

ANALYSIS DEFINED

Analytical reports are question-answering or problem-solving reports. Analysis is basic to our thinking, and in this sense all the assignments you have written have involved analysis. A summary, for instance, requires an analysis of the original for essential points; expanded definition often includes an analysis of parts; description or process explanation requires analysis of the item or process. In each case, we divide the subject into its parts (partition) and group these parts within specific categories according to their similarities (classification).

The analysis of data is somewhat different from these earlier assignments: it is based not only on observation, but also on investigation and research. The subjects of earlier assignments had observable structures: for example, the stethoscope described in Chapter 10 has distinct parts and the tree-felling instructions in Chapter 11 have distinct steps. Your subject was wholly before you, intact, waiting only to be discussed.

In an analysis of data, however, your subject is not an item that needs describing or a process that needs explaining. Instead it is a question that needs answering, or a problem that needs solving. Your analysis might influence a major decision. Say, for example, you received the following assignment from your supervisor:

> Recommend the best method for removing the heavy-metal contamination from our company dump site.

Clearly, to get this job done, you will have to do more than observe your subject. Since high-level decisions will depend on your findings, you must *seek out* and *interpret* all data that will help you make the best recommendations. This is where you apply the research activity (where, say, you identified the *various* methods of removing heavy-metal contamination) discussed in Chapter 14, and it is where you face your greatest reporting challenge. Analysis almost always leads to recommendations.

PURPOSE OF ANALYSIS

On the job, you apply analytical skills to problems, proposals, and planning. A structural engineer, for example, analyzes material and design in evaluating plans for a suspension bridge or skyscraper. Before buying a stock, the wise investor analyzes economic trends, treasurers' reports, performance, dividend rates, and market conditions. A legal defense is built on the attorney's analysis and logical reconstruction of events in a case. Medical diagnosis relies on the precise identification and classification of symptoms, along with the communication of findings, interpretations, and recommendations for treatment.

As an employee you may be asked to evaluate a new assembly technique on the production line, or to locate and purchase the best equipment at the best price. You might have to identify the cause of a monthly drop in sales, the reasons for low employee morale, the causes of an accident, or the reasons for equipment failure. You might need to assess the feasibility of a proposal for company expansion or investment. The list is endless, but the process is basically the same: (1) making a plan, (2) finding the facts, (3) interpreting the findings, and (4) drawing conclusions and making recommendations.

TYPICAL ANALYTICAL PROBLEMS

The aim of an analytical report is to show how you arrived at your conclusions. Your approach will depend on your subject, purpose, and reader's needs. Here are some typical analytical problems:

"Will *X* Work for a Specific Purpose?"

Analysis can answer practical questions. Say your employer is concerned about the effects of stress on personnel. He or she may ask you to investigate the claim that transcendental meditation has therapeutic benefits — with an eye toward a TM program for employees. You would design your analysis to answer this question: "Does TM have therapeutic benefits?" The analysis would follow a *questions-answers-conclusions* structure. Because the report could lead to action, you probably would include recommendations based on your conclusions.

"Is *X* or *Y* Better for a Specific Purpose?"

Analysis is essential in comparisons of machines, processes, business locations, computer systems, or the like. Assume, for example, you manage a ski lodge and need to answer this question: Which of the two most popular types of ski binding is best for our rental skis? In a comparative analysis of the *Salamon 555* and the *Americana* bindings, you might assess the strengths and weaknesses of each in a point-by-point comparison: toe release, heel release, ease of adjustment, friction, weight, and cost. Or you might use an item-by-item comparison, discussing the first whole binding, then the next.

The comparative analysis follows a *questions-answers-conclusions* structure and is designed to help the reader make a choice. Examples are found in magazines such as *Consumer Report* and *Consumer's Digest*.

"Why Does *X* Happen?"

The problem-solving analysis is designed to answer questions like this: Why do independent television service businesses have a high failure rate? (See the sample report later in this chapter.) This kind of analysis follows a variation of the questions-answers-conclusions structure: namely, *problem-causes-solution*. Such an analysis has the following steps:

1. identifying the problem
2. examining possible and probable causes, and isolating definite ones
3. proposing solutions

An analysis of low employee morale would follow the same structure.

"What Are the Effects of X?"

An analysis of the consequences of an event or action would answer questions like this: How has air quality been affected by the local power plant's change from burning oil to coal? (See the sample report later in this chapter.)

Another kind of problem-solving analysis is done to predict an effect: "What are the consequences of my changing majors?" Here, the structure is *proposed action–probable effects–conclusions and recommendations.*

"Is X Practical in a Given Situation?"

The feasibility analysis assesses the practicality of an idea or plan: Will the consumer interests of Hicksville support a microcomputer store? In a variation of the question-answers-conclusions structure, a feasibility analysis uses *reasons for–reasons against,* with both sides supported by evidence. Business owners often use this type of analysis.

Combining Types of Analyses

These major types of analytical problems overlap considerably. Any one study may in fact require answers to two or more of these questions. For instance, the sample report on pages 508–516 is both a feasibility analysis and a comparative analysis. It is designed to answer these two questions: Is it wise to computerize our business? and If so, what computer system is best?

Although both might treat the same question or problem, the analytical report usually differs from the research report discussed in Chapter 15. The research report's primary purpose is simply to *inform* readers — not move them to action. Although the analytical report relies on the same research techniques, its purpose is both to *inform* and *recommend.* Analytical reports lead to action.

ELEMENTS OF A USEFUL ANALYSIS

The analytical report incorporates many writing strategies used in earlier assignments, along with the guidelines that follow.

Clearly Identified Problem or Question

Know what you're looking for. If your car's engine fails to turn over when you switch on your ignition, you would wisely check your battery and electrical connections before dismantling parts of the engine. Apply a similar focus to your report.

Earlier, a hypothetical employer posed this question: "Will transcendental meditation help reduce stress among my employees? The question obviously requires answers to three other questions: What are the therapeutic claims of TM? Are they valid? Will TM work in this situation? How transcendental meditation got established, how widespread it is, who practices it, and other such questions are not relevant to this problem (although some questions about background might be useful in the report's introduction). Always begin by defining clearly the central questions and thinking through any subordinate questions they may imply. Only then can you determine the data or evidence you need.

With the central questions identified, the writer of the TM report can formulate her statement of purpose:

> This report examines for the general reader some of the claims about therapeutic benefits made by practitioners of transcendental meditation.

The writer might have mistakenly begun instead with this statement:

> This report examines transcendental meditation and communicates its findings to the general reader.

Notice how the first version sharpens the focus by expressing the precise subject of the analysis: not transcendental meditation (a huge topic), but the alleged *therapeutic benefits* of TM.

Define your purpose by condensing your approach to a basic question: Does TM have therapeutic benefits? or Why have our sales dropped steadily for three months? Then restate the question as a declarative sentence in your purpose statement.

A Report with No Bias

Interpret evidence impartially. Stay on track by beginning with an unbiased title. Consider these two title versions:

> 1. The Accuracy of the Talmo Seismograph in Predicting Earth Tremors
> 2. An Analysis of the Accuracy of the Talmo Seismograph in Predicting Earth Tremors

If the Talmo Seismograph's accuracy were not yet confirmed, then the first title would be misleading. In that case, the second version would provide a more accurate forecast. Throughout your analysis, stick to your evidence. Do not force personal viewpoints on your material.

Accurate and Adequate Data

Do not distort the original data by excluding vital points. Say you are asked to recommend the best chainsaw for a logging company. Reviewing test reports, you come across this information:

> Of all six brands tested, the Bomarc chainsaw proved easiest to operate. However, it also had the fewest safety features.

If you cite these data, present both points, not simply the first — even though you may personally prefer the Bomarc brand. *Then* argue for the point you prefer.

Reserve personal comments or judgments for your conclusion. As space permits, include the full text of interviews or questionnaires in appendixes.

Fully Interpreted Data

Explain the significance of your data. Interpretation is the heart of the analytical report. You might, for example, interpret the chainsaw data this way:

> Our cutting crews often work suspended by harness, high above the ground. And much work is in remote areas. This means that safety features should be our first requirement in a chainsaw. Despite its ease of operation, the Bomarc saw does not meet our needs.

By saying "This means . . ." you engage in analysis — not mere information-sharing. *Merely listing your findings is not enough.* Spell out the meaning.

Valid Conclusions and Recommendations

A useful conclusion may appeal secondarily to *emotion* ("You will love this device") but it always appeals primarily to *reason* ("This device will best serve your needs"). When analyzing a controversial subject, try especially hard to remain impartial. Say you work in law enforcement administration and have been asked to study this question: Is the prisoner furlough system working in our state? Do justice to this topic by making sure your data gathering is complete, your interpretations are not shaded by prior opinion, and your conclusions and recommendations are based on the facts.

When you do reach definite conclusions, state them with assurance and authority. Avoid noncommittal statements ("It would seem that . . ." or "It looks as if . . ."). Be direct ("The earthquake danger at the proposed reactor site is critically high."). If, on the other hand, your analysis does not yield a definite conclusion, do not force a simplistic one on your material.

Clear and Careful Reasoning

The reporting process is not simply a mechanical process of collecting and recording information. If it were, machines could be programmed for the job. Each step of your analysis requires decisions about what to record, what to exclude, and where to go next. As you evaluate your data (Is this reliable and important?), interpret your evidence (What does it mean?), and make recommendations based on your conclusions (What action is needed?), you might have to adjust your original plan. You cannot know what you will find until you have searched. Remain flexible enough to revise in the light of new evidence.

Appropriate Length and Visuals

Depending on the problem or question, your analysis may range in length from a short memo to a long report with supplements. Make it just long enough to show readers how you arrived at your conclusions.

Use visuals generously (Chapter 13). Graphs are especially useful in an analysis of trends (rising or falling sales, radiation levels, etc.). Tables, charts, photographs, and diagrams work well in comparative analyses.

EVALUATING AND INTERPRETING DATA

Researching (Chapter 14) is only part of your task. As you sort material, you must evaluate the reliability of your evidence as well as interpret it. Facts or statistics out of context can be interpreted in many ways, but you are ethically bound to find answers that stand the greatest chance of being truthful. To ensure the validity of your report, choose reliable sources; distinguish hard from soft evidence; and avoid specious reasoning.

Choose Reliable Sources

Make sure each source is reputable, impartial, and authoritative. Say you are analyzing the alleged benefits of transcendental meditation. You could expect claims in a reputable professional journal, such as the *New England Journal of Medicine,* to be reliable. Also, a reputable magazine, such as *Scientific American,* would be a reliable source. Obviously, claims in glamour or movie magazines are highly suspect. Even claims in monthly "digests," which offer simplistic and largely undocumented "wisdom" to mass reading audiences, should be carefully verified.

Interview only those who have practiced TM extensively. Anyone who has practiced for only a few weeks could hardly assess bodily changes reliably. And to the extent you rely on personal experience reports, including those from long-time practitioners, you need a representative sample. Even reports from ten successful practitioners would be a small sample unless those reports were supported by laboratory data. On the other hand, with a hundred reports from people ranging from students to judges and doctors, you might not have "proved" anything, but your evidence would be somewhat persuasive.

Your own experience, also, is not a valid base for generalizing. We cannot tell whether our experience is representative, regardless of how long we might have practiced meditation. If you have practiced TM with success, interpret your experience only within the broader context of your collected sources.

Some issues (e.g., the nuclear-arms buildup or causes of inflation) are always controversial, and they will never be resolved. Although we can get verifiable data and can reason persuasively on some subjects, no amount of close reasoning by any expert and no supporting statistical analysis will "prove" anything about a controversial subject. Somebody's thesis that the balance of payments is at the root of the inflation problem cannot be proved. Likewise, one could only *argue* (more or less effectively) that federal funds would or would not alleviate poverty or unemployment. Some problems simply are more resistant to solution than others, no matter how reliable the sources.

Given these difficulties, resist the temptation to report hasty but unverified answers. Better to report no answers than misleading ones. Take no claims for granted, and cross-check all data.

Distinguish Hard from Soft Evidence

Hard evidence consists of observable facts. It can stand up under testing because it is verifiable. Soft evidence consists of uninformed opinion. It may collapse under testing unless the opinion is expert, authoritative, and unbiased.

Base your conclusions on hard evidence. Early in your analysis, you might read an article that makes positive claims about TM, but the article provides no data on measurements of pulse, blood pressure, or metabolic rates. Although your own experience and opinion might agree with the author's, do not hastily conclude that TM is beneficial to everyone. So far, you have only two opinions — yours and the author's — without any scientific support (e.g., tests of a cross-section under controlled conditions). Conclusions at this point would rest on soft evidence. Only after a full survey of reliable sources can you decide which conclusions are supported by the bulk of your evidence.

Until evidence proves itself hard, consider it soft. Say a local merchant claims that a nearby vacant shop is the best business location in town, but does not support that assertion with facts. Before you rent the shop, cross-check other sources, and observe customer traffic in adjoining businesses. On the other hand, if the previous owner gave you the same judgment, even though he did not support it with fact, you might take it seriously — particularly if you know he is honest and has vacated the shop because he has just purchased a $325,000 retirement home at age fifty-five.

Be careful not to confuse probable, possible, and definite causes in a problem-solving analysis. Sometimes a definite cause can be identified easily (e.g., "The engine's overheating is caused by a faulty radiator cap"), but usually a good deal of searching and thought are needed to isolate a specific cause. Say you tackle this question: Why are there no children's day care facilities on our state college campus? A brainstorming session yields this list of possible causes:

- lack of need among students
- lack of interest among students, faculty, and staff
- high cost of liability insurance
- lack of space and facilities on campus
- lack of trained personnel
- prohibition by state law
- lack of legislative funding for such a project

Assume you proceed with interviews, questionnaires, and research into state laws, insurance rates, and local availability of qualified personnel. You gradually rule out some items, but others emerge as probable causes. Specifically, you find a need among students, high campus interest, an abundance of qualified people for staffing, and no state laws prohibiting such a project. Three probable causes remain: lack of funding, high insurance rates, and lack of space. Further inquiry shows that lack of funding and high insurance rates *are* issues. These causes, however, can be eliminated by the creation of other sources of revenue: charging a fee for each child, holding fund-raising drives, or diverting funds from other campus organizations. Finally, after examining available campus space and speaking with school officials, you arrive at one definite cause: lack of space and facilities.[1]

Early in your analysis you might have based your conclusions hastily on soft evidence (say, an opinion — buttressed by a newspaper editorial — that the campus was apathetic). Now you can reason from a basis of solid, factual evidence. You have moved from a wide range of possible causes to a few probable causes, to the most likely cause. Because you have covered your ground well, your recommendations will have credibility and authority.

Sometimes, finding a single cause is impossible, but this reasoning process can be tailored to most problem-solving analyses. Although few, except the simplest, effects have one cause, usually one or more principal causes emerge. By narrowing the field, you can focus on the real issues.

Avoid Specious Reasoning

Specious reasoning is deceptive because it seems correct at first glance, but is faulty when scrutinized. Reasoning based on soft evidence is often specious. Conclusions you derive speciously fail under testing. Assume, for example, you are an education consultant. Your community has asked you to analyze the accuracy of IQ testing as a measure of intelligence and as a predictor of students' performance. Reviewing your collected evidence, you find a positive correlation between low IQ scores and low achievers, and vice versa. You then verify your own statistics by examining a solid cross-section of reliable sources. At this point you might feel justified in concluding that IQ tests do measure intelligence and predict performance. This conclusion, however, would be specious unless you could show that

1. Neither parents, teachers, nor the children tested had seen individual test scores and had thus been able to develop biased attitudes.
2. Children testing in all IQ categories had later been exposed to an identical curriculum at an identical pace. In other words, they were not channeled into programs on the basis of their scores.

Your total data could be interpreted only within the context of these two variables. Even hard evidence can be used to support specious reasoning, unless it is interpreted precisely, objectively, and within a context that accounts for all variables.

---

[1] Of course, one could argue that lack of space and facilities is somehow related to funding. And the fact that the college is unable to find funds or space may be related to the fact that student need is not sufficiently acute or interest sufficiently high to exert real pressure. Lack of space and facilities, however, emerges as the *immediate* cause.

PLANNING AND WRITING THE FORMAL REPORT

Remain Flexible

Your analysis will develop its own direction as it proceeds, depending on what you find at each point in your search. Because you will be writing and revising while searching, you will need clear points of reference to remain on track. At various points, pose and answer the following questions:

1. What am I looking for?
2. How should I structure my inquiry to obtain this information?
3. How will I best communicate my process of inquiry and my findings?

These are ongoing questions whose answers can change as you work. Initially, question 1 will be answered in your statement of purpose; question 2, in your tentative or working outline; and question 3, in your topic or sentence outline — the blueprint for your actual report. Be flexible enough to modify your approach in case of the unexpected. Here are a few possibilities:

1. Near the end of your research you uncover issues not summarized in your statement of purpose (e.g., that TM has no effect on certain people). So you modify your statement of purpose (to include counterclaims).

2. You think of new topics that are not in your working outline or find that information on one of your original topics is unavailable. Thus, your working outline needs revision.

3. As you write the first draft, your report seems disorganized. So you rearrange the outline before writing the next draft.

Your finished report will be the product of countless decisions and revisions. Revise and reshuffle as often as necessary.

Work from a Detailed Outline

You might outline *before* or *after* writing a first draft. In any case, the finished report depends on a good outline. The following model outline can be adapted to most analytical reports:

    I. INTRODUCTION
       A. Definition, Description, and Background of the Question, Issue, Problem, or Item
       B. Purpose of the Report, and Intended Audience
       C. Sources of Information
       D. Working Definitions (here or in a glossary)
       E. Limitations of the Study

      F. Scope of the Inquiry (with topics listed in ascending or descending order of importance)

II. COLLECTED DATA
    A. First Topic for Investigation
      1. Definition
      2. Findings
      3. Evaluation of findings
      4. Interpretation of findings
    B. Second Topic for Investigation
      1. First subtopic
        a. Definition
        b. Findings
        c. Evaluation of findings
        d. Interpretation of findings
      2. Second subtopic
        etc.

III. CONCLUSION
    A. Summary of Findings
    B. Comprehensive Interpretation of Findings
    C. Recommendations and Proposals (as needed)

This outline is only tentative. Modify it if necessary.

The three sample reports that follow are built on this model outline. The first report, "Survival Problems of Television Service Businesses," is not reproduced entirely; however, parts of each major section (introduction, body, and conclusion) are illustrated and explained. The second report (on air quality) is entirely reproduced. The third paper (on personal computers) is included for further study and discussion.

Each report responds to a slightly different question or problem. The first tackles the question, Why does *X* happen? The second, What are the effects of *X*? The third tackles two questions: Is *X* feasible? and Which version of *X* is better for our purposes? At least one of these reports should serve as a model for your own analysis.

Introduction

The introduction describes and defines the question or problem, and provides background. Identify your intended audience and discuss briefly your sources of data, along with reasons for omitting certain data (e.g., key person not available for interview). List working definitions, unless you have so many you need a glossary. If you do use a glossary and appendixes, refer to them at this point. Finally, define the scope of your report by listing all major topics discussed in the body.

SURVIVAL PROBLEMS OF TELEVISION
SERVICE BUSINESSES

INTRODUCTION

Definition, Description, and Background

A television service business specializes in repairing selected home entertainment products. The modern household contains TV sets, master antenna systems, and related electronic equipment, which require periodic service.

The "TV repairman" (or electronics technician) has become a household necessity. Like the family plumber, electrician, physician, and lawyer, the electronics technician is considered one of the professionals who keep the American home functioning.

Because of their recent arrival (TV has been in general use for just over 35 years), the professional personality of technicians escapes public analysis. Service costs regarded as excessive cause public distrust. Many feel the TV service field is lucrative and that the technicians are sometimes dishonest. Technicians are aware of this public image.

Coupled with the public relations problem is the history of a technology that constantly changes. Service technicians perennially struggle against obsolescence. "The state of the art" is the phrase constantly ringing in their ears, diverting attention from the need for sound business management. Poor public relations, efforts to keep up with everchanging technology, and poor financial management all contribute to the high rate of TV service business failure.

Purpose of Report, and Intended Audience

This analysis was prompted by the fact that TV service businesses rank second only to auto service stations in bankruptcies. The purpose of the report is to increase public understanding of this vital element of today's service technology, and to direct the attention of technician–shop owners to potentially fatal problems that may plague their businesses.

Sources of Information

The main source of information for this report is a study of a small business I've been familiar with for several years. An interview with the present owner is attached in Appendix A. A second interview with another local shop owner is attached in Appendix B. The Department of Labor Statistics and several trade magazines and journals round out the sources. Published material on this subject is understandably scarce.

Limitations of Study

This study is limited by the narrow sampling possibilities in our geographic area. Moreover, it is limited by the reluctance of many small business owners to "tell it like it is." Owners of TV service shops, specifically, are most hesitant to disclose financial records. The owner of a very large TV and appliance center (that does its own servicing) refused an interview. Unfortunately, it was the only large facility in our area.

Working Definitions

Television service shop: The conventional TV service shop is essentially a service center for home entertainment products. Items routinely serviced include TV sets, radios, stereo equipment, and antenna systems. Service businesses exist with or without sales. Some businesses refuse to sell equipment because of the added bookkeeping and inventory problems.

  Service technician: An electronics service technician is a person trained to service home entertainment products. Outside technicians make service calls, and inside technicians do the bench work. According to the Department of Labor, in 1965 there were 115,000 people engaged in radio and television service. One-third were self-employed. Of the remainder, two-thirds worked in TV shops, and the others were employed by manufacturers (1:321). In 1972, the number of technicians was reported at 140,000 with roughly the same breakdown as in 1965 (2:361). In 1974, technicians numbered 135,000 with the same breakdown (3:307). This last figure is striking because it shows a marked decrease in a field considered by the Department of Labor to have a scarcity of personnel. The rapid drop also occurred in a decade when the population and the number of items to be serviced increased greatly.

Scope

The topics of investigation in this report are, in ascending order of importance, the history of television service businesses, the problem of public relations, the demand for growing technical competence, and the problems of financial management.

Body

The body divides a complex subject into related topics and subtopics, ranked in order of their importance (ascending or descending). Partition the subject into its main parts and those, in turn, into subparts. Carry your division as far as you can, to make sense of the topic. In the sample that follows, the major topic, "The Problems of Financial Management," is divided into two subtopics: "Technician versus Manager" and "Cash Flow." This last subtopic, in turn, is divided into three sub-subtopics: "Sales," "Collection for Service," and "Credit Arrangements." These divisions and subdivisions keep the writer on track and help readers follow the inquiry. Be sure readers can draw conclusions identical to your own on the basis of your evidence. A problem-solving analysis requires that you discuss all possible causes, narrowing your focus to probable and then definite causes. Sift, evaluate, and interpret clues to reach a valid conclusion. The process might be diagrammed like this:

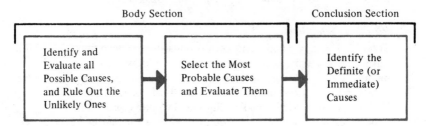

One major topic of the television service report follows. To save space, the other major topics are omitted.

COLLECTED DATA

The Problems of Financial Management

Technician versus Manager

Technician-owners cannot function productively as both full-time technicians and full-time business managers. Without improving their technical proficiency, they cannot keep the quality of their service up to professional standards. On the other hand, they must maintain the detailed financial records that are the backbone of any successful business. Shop owners usually attend evening courses and leave their workbenches for a part of the day to attend to the management of their business. A shop with several technicians can easily permit this arrangement, but, in a one- or two-man shop these multiple commitments can spell disaster.

Some more prosperous small shops have turned to outside bookkeeping and accounting agencies. Computerized methods have reduced the cost of these services, to the relief of the harried shop owner. Many shops showing only a marginal profit, however, cannot afford outside help. Their only recourse is to request assistance from the Small Business Administration, which provides advice, financial analysis, and small loans, but these services are offered only on a limited, first-come, first-served basis.

Cash Flow

The cash flow of a small service business is affected by sales, collection for service, and credit arrangements.

Sales. Some service shops also sell TV sets, stereos, and other home-entertainment products. In many cases, this policy has been their downfall because of the owner's lack of financial discipline. Most manufacturers permit "floor plans." TV sets, for example, are consigned to a seller for a small down payment (usually 10 percent of retail). The seller is required to pay for each item *as it is sold,* or within specified time limits, but many shop owners use this money to finance their operation. When payment is due, the money has been spent. This too-familiar situation has led to countless bankruptcies.

Collection for Service. Most shops generate their own capital. Therefore, strict collection for services rendered is the common practice. Signs on the walls of most shops ask for cash only — no charges and no checks. When the customer is unable or unwilling to pay, however, the technician must hold the set for collateral. This results in a shop cluttered with sets that are generating no capital.

Credit Arrangements. Most service work is done on a cash-on-delivery basis. Only warranty work (paid for by manufacturers) is done on credit. Manufacturers are dependable and usually pay for warranty work monthly. Checks totaling between $500 and $1,000 per month are quite welcome and generally cover most of the fixed monthly costs (rent, utilities, etc.) of a small service business.

Evaluation and Interpretation

Some owners use cash that they owe to manufacturers. Others have their shops cluttered with items held as collateral. The overriding problem in financial management, however, stems from inadequate record-keeping. This is where technician-owners were found to be weakest. They simply do not have time to leave the bench work to do "paperwork." And without good records they are unable to keep track of and manage their cash flow.

Conclusion

The conclusion culminates your report, and it is likely to be of most interest to readers because it answers the questions that sparked the analysis in the first place. (Many reports, therefore, are submitted with the conclusion *preceding* the introduction and body sections.)

Here you summarize, interpret, and recommend. Although you have interpreted evidence at each step of your analysis, your conclusion pulls the strands together in a broader interpretation and suggests action, where appropriate. Because of its importance, this final section must be consistent in three ways:

1. Your summary must reflect accurately the body of your report.
2. Your overall interpretation must be consistent with the findings in your summary.
3. Your recommendations must be consistent with the purpose of the report, the evidence presented, and the interpretations given.

Your summary and interpretations should lead logically to your recommendations. Here is the conclusion section of the television service report:

CONCLUSION

Summary of Findings

The History of Television Service Businesses

Many survival problems in today's service business derive from early habits and traditions. Virtually all early television technicians were military veterans trained as "radio mechanics." They moved into shops and went right to work with little business training. Their management expertise was confined to a working knowledge of a very active cash register.

Public Relations

Price gouging, which occasionally occurred in the early years, has been eliminated by the profession's development of ethical standards. Suspicions linger, however, largely because customers have little understanding of the electronic repairs needed for their sets. As a result, many customers turn to service departments in larger stores. Among the businesses studied, there seemed to be an embarrassing lack of public relations effort. The reasons given for this deficiency were lack of time and lack of advertising funds.

The Demand for Technical Competence

The shop owners studied were technicians of the highest caliber, who continually upgrade their skills. Realistically, a shop cannot endure without strong professional standards. Customers simply would not return. In shops with several technicians, individuals have begun to specialize. One shop has a TV department, an auto-radio and stereo department, and a commercial electronic department that services such items as microwave ovens and motel master systems for signal distribution. All technicians surveyed feel that specialization will improve individual competence.

Financial Management

Shop owners who insist on doing the bench work themselves and who neglect the daily management of the business are unable to maintain proper financial records or to stabilize their cash flow. With few exceptions, unsound financial management leads to business failure.

Comprehensive Interpretation of Findings

The areas of public relations and financial management pose threats to the welfare of independent TV service businesses.

Public Relations

Official and consumer attitude toward this profession still reflects the "radio mechanic" days of the postwar years. Technicians are still viewed as "TV repairmen" who have easily learned a simple trade and who keep customers uninformed so they can overcharge them. Consumers' lack of familiarity with the product and servicing techniques is compounded by technicians' unwillingness or inability to better inform their customers.

Financial Management

Unsound financial management is the largest cause of TV service business failures. When the books are not balanced, no business can survive for long.

Recommendations

Service technicians show a puzzling lack of initiative to erase the "radio mechanic" and "TV repairman" image. This attitude and the owner-technician's hesitancy to become a real "businessman" are problems requiring action.

Collective action by shop owners is the best way to strengthen public relations. Local technicians' guilds should design and sponsor advertising to inform consumers and to improve the technician's image. Even though work may be plentiful, technicians owe it to themselves to demonstrate their competence, reliability, and honesty.

Service shop owners must adopt more efficient business practices. They should enroll at local colleges in courses such as primary management, accounting, and business methods. Properly trained, they will understand such important items as balance sheets and income statements, as well as their responsibilities in the tax accounting of their business. Professionalism in management will show immediate results. Any problems with financing or cash flow will become obvious, enabling owners to take action.

Support the Text with Supplements

Submit your completed report with these supporting documents, in order:

- title page
- letter of transmittal
- table of contents
- table of figures
- abstract
- **report text (introduction-body-conclusion)**
- glossary (as needed)
- appendixes (as needed)
- works-cited page (or alphabetical or numbered list of references)

Each of these supplements is discussed in Chapter 12. Forms of documentation are discussed in Chapter 15.

APPLYING THE STEPS

The report in Figure 19–1, written for a general audience, is patterned after our model outline. The report answers this analytical question: "What are the effects of X"? For documentation, this writer uses a numbered list of references.

AUDIENCE-AND-USE ANALYSIS

One purpose of this report is to satisfy a course requirement; so one copy is accompanied by a letter of transmittal (not shown) to the instructor (the secondary audience). But the report has a primary audience as well: the writer's colleagues, all volunteers in an environmental group.

Ever since a local power plant changed from oil burning to coal, area residents have expressed concern about air quality and acid levels in rain and other precipitation. The area already has been damaged by acid rain, presumably caused by industrial emissions from several states away. Many people now fear that massive coal burning in the local plant will increase the pollution. Bill Kelly, member of Greenleaf Alliance, decides to research the effects of this change to coal.

Both Bill's primary and secondary audience can be considered laypersons. They want to know if the change to coal has further damaged the local environment. But before they can appreciate the author's conclusions, they will need answers to these questions:

– Exactly how does the burning of fossil fuel cause pollution?
– What specific pollutants are produced?
– What are the effects of these pollutants on human health and the environment?

Bill designs his introduction to answer these questions.

The report body and conclusion then go on to answer other questions that Bill anticipated from his audience:

– Why did the plant have to convert to coal, specifically, instead of, say, natural gas?
– What specific physical changes were involved?
– What, if any, pollution-control measures were taken?
– Have recent tests been done on plant emissions and local air quality? If so, how do the results compare with those of tests done *before* the conversion?

Bill answers these questions in the least complex ways. Because he is not writing for technical professionals, he leaves out specialized details (say, the differences in fixed-carbon, volatile, and moisture content among anthracite, bituminous, and lignitic coals). But he does, in his introduction, define certain specialized terms (*ash, BTU*) essential for nontechnical readers to understand his report.

To increase readability and interest, Bill uses visuals and headings generously.

AN ANALYSIS OF THE IMPACT ON AIR QUALITY
FROM
THE CHANGEOVER TO COAL BURNING
AT
NEW ENGLAND POWER'S BRAYTON POINT GENERATING STATION

Prepared for

Dr. John M. Lannon
Technical Writing Instructor
Southeastern Massachusetts University

by

William J. Kelly

March 31, 1984

FIGURE 19–1   An Analytical Report

16 Jefferson Court
Fall River, MA 02720
March 31, 1984

The Greenleaf Alliance
The Olympia Building, Rm. 318
Fall River, MA 02720

ATTENTION: Acid Rain Committee

As part of our country's drive for energy independence, many
power-generating plants have been converted from oil burning
to coal. Coal is a more available and stable source of
energy because it is the largest part of our domestic energy
reserves. The Brayton Point Generating Station in Somerset,
MA, has changed from oil to coal, and my course project has
been to see if the change has significantly affected air and
precipitation quality in the region.

As the enclosed paper shows, particulates, sulfur oxides, and
nitrogen oxides, all produced and released during the combustion of fossil fuels, are major environmental concerns.
Nitrogen oxides and sulfur oxides are particularly dangerous
because they go through a chemical change in the atmosphere
and return to earth as acid rain, a major threat to waterways and the life they support.

The evidence indicates that New England Power's change to
coal burning has had no significant effect. In fact, some
areas may have had a slight improvement. New England Power
has maintained air quality by using only low-sulfur coal and
installing in-stack anti-pollution devices.

My study suggests that conversion to a "dirtier" fuel can be
accomplished without significantly increasing pollution. In
a world already polluted, this finding is obviously good news.
Please call me should you have any questions.

Sincerely,

William J. Kelly

William J. Kelly

FIGURE 19–1   (*Continued*)

iii

TABLE OF CONTENTS

FIGURE 19–1 (*Continued*)

iv

TABLE OF FIGURES AND TABLES

FIGURE 19–1 (*Continued*)

INFORMATIVE ABSTRACT

New England Power has converted three power-generating units
at its Brayton Point Generating Station in Somerset, MA,
from oil burning to coal. Because coal is generally recognized as a "dirtier" fuel, there was concern that the change
would increase air pollution from particulate, nitrogen oxide,
and sulfur oxide emissions. Nitrogen oxides and sulfur
oxides are major threats to the environment because they combine with moisture in the air to form acid rain. The available information indicates, however, that the change has not
significantly affected our region's air quality. New England
Power's use of in-stack anti-pollution devices and their
selection of only low-sulfur coal apparently have offset the
increase normally expected in a switch from oil to coal.

FIGURE 19–1   (*Continued*)

INTRODUCTION

Background

The Brayton Point Generating Station, New England Power Company's Somerset, Massachusetts, power plant, currently burns coal in its three generators. These three units had burned oil before the conversion, which began in 1978 and was completed late in 1979. The conversion to coal was necessitated by the national movement from imported fuels toward more available domestic fuel.

Purpose and Audience

This report examines the question of whether coal burning at Brayton Point has increased acid rain locally. It is written for interested local citizens.

Description of Air Pollution from Fossil Fuels

The three major pollutants associated with the burning of fossil fuels are particulates, nitrogen oxides, and sulfur oxides.

Particulates, sometimes called fly ash, are the unburned remnants of fuel. The amount of particulates (also measured under the title of Total Suspended Particulates, or TSP)

FIGURE 19-1 (*Continued*)

2

produced by coal depends upon the ash, sulfur, and carbon

contents of coal. For instance, coal with a high ash content

can produce significant amounts of particulates, compounds

linked to ailments including asthma, emphysema, bronchitis,

and lung cancer (1:49).

Nitrogen oxide compounds also are released during the

burning of coal. They include nitric oxide (NO) and the more

toxic nitrogen dioxide (NO_2). Nitrogen oxides are a sig-

nificant problem because these compounds appear in greater

amounts with coal burning than with oil burning. In fact,

the original estimate for Brayton Point indicated an

expected increase of up to 73 percent (1:55).

Other pollutants produced during coal combustion are

sulfur oxides, including various sulfates and sulfur dioxide

(SO_2). Sulfur oxides cause eye and throat irritation, in-

crease respiration rates, and aggravate chronic respiratory

conditions. In addition, they damage vegetation and property.

They corrode paint and metal, as well as eroding various

types of stone (1:54).

When sulfur oxides combine with nitrogen oxides and the

water vapor in the atmosphere, the result is highly acid

FIGURE 19–1   (*Continued*)

3

precipitation. This phenomenon (acid rain) is one of the

biggest threats to regional waterways and the various forms

of life they support.

Data Sources

I compiled information for this report from government

studies, selected publications on energy, and an interview

with an environmental engineer for New England Power Company.

Working Definitions

Ash: the part of coal that remains in a solid form

after combustion and that does not contribute to

the heat value of the coal.

BTU: British Thermal Unit, the amount of heat

necessary to raise the temperature of one pound

of water one degree Fahrenheit; equal to 252

calories, 778 foot-pounds, 1055 joules, or 0.293

watt-hours.

Exceedance: violation of an air quality standard.

National Ambient Air Quality Standards: guidelines for

the levels of various pollutants in the surrounding

atmosphere.

FIGURE 19-1 (*Continued*)

4

OPEC: Organization of Petroleum Exporting Nations. A
        cartel of major oil producers that controls the
        amount of oil released to the international
        market.

Scope

Although New England Power has other pollution problems
from the changeover (run-off from the coal pile, increased
noise pollution, discharge of heated water into the surrounding rivers, etc.), this study focuses on how the change
from oil to coal burning has affected air and precipitation
quality in terms of particulates, nitrogen oxides, and
sulfur oxides.

COLLECTED DATA

The Background of the Changeover

The Need for Alternative Energy

In the three years following the 1973 oil embargo, OPEC
raised the price of oil some 366 percent (2:164). These increases, coupled with earlier hardships from oil shortages,
spurred a movement in the U.S. towards conservation and
energy independence. This meant looking for an alternative
source of energy.

FIGURE 19–1 (*Continued*)

5

Coal as an Energy Alternative

The logical answer to the energy question for the
United States is coal; as Figure 1 shows, 81 percent of the
U.S. energy reserves are coal. Further, as Figure 2 shows,
these reserves make up 31 percent of the total world reserves
of coal, indicating that imports would be unnecessary. Although many experts consider oil a superior fuel, the problem
the U.S. faces with oil is availability:

> Oil and natural gas currently provide about three-
> quarters of the total energy produced in the
> United States each year, but account for just
> 7 percent of the nation's energy reserves...
> [resulting] in the United States being dependent
> on imported crude and refined petroleum for nearly
> half of its total oil supply (2:176).

As Figure 2 shows, the vast reserves of coal represent one
way to decrease imported fuel.

Federal Restrictions on Oil Burning

Within a few years of the 1973 oil embargo, a national
movement to convert power plants from oil to coal burning
began. During the embargo, the Environmental Protection
Agency (EPA) had granted variances allowing all three units
at Brayton Point to burn coal in non-compliance with air-
quality standards from 1974-1975. Then, in 1977, the Federal

FIGURE 19-1 (*Continued*)

6

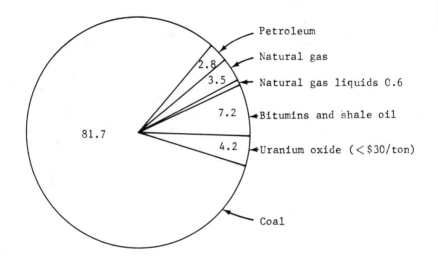

KNOWN RECOVERABLE U.S. ENERGY RESERVES
(1974 estimate)

Petroleum

Natural gas

2.8

3.5   Natural gas liquids 0.6

7.2   Bitumins and shale oil

81.7

4.2   Uranium oxide (<$30/ton)

Coal

Figure 1.   Percentages of U.S. Energy Reserves

Source: Figure appears in Coal Data Book, President's
          Commission on Coal, 1980: 65.

FIGURE 19–1   (*Continued*)

7

RECOVERABLE COAL RESERVES--WORLD
(1974 estimate)

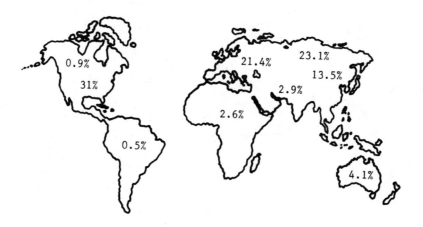

Figure 2. Percentages of World Coal Reserves

Source: Figure appears in Coal Data Book, President's
Commission on Coal, 1980: 63.

FIGURE 19-1 (*Continued*)

8

Energy Administration (FEA) issued a Prohibition Order,

blocking the burning of either oil or natural gas at Brayton

Point. At the same time, the Power Plant and Industrial Fuel

Act of 1978, Public Law 95-620, was instituted by the Depart-

ment of Energy. This law greatly limited the use of oil as a

primary fuel in power plants throughout the nation (3:211).

These actions effectively forced the conversion of many

generating stations, including Brayton Point, to coal burning.

The Changes Necessary for the Conversion

    Primary Changes

    The conversion to coal at Brayton Point required several

steps. Among them was the Draft Environmental Impact State-

ment (cited in this report) required by the agencies that

monitor air and environmental quality. This study outlined

the changes and problems a changeover creates. The document

was completed in 1978 and conversion begun shortly after.

    Some of the actual conversion was simple. For instance,

generating units one and two had originally been designed to

burn pulverized coal, so the return conversion was not dif-

ficult. Also, because unit three had been designed to burn

either oil or coal, no elaborate conversion was required.

FIGURE 19-1 (*Continued*)

9

Additional Pollution Controls

Added to the coal burning units were electrostatic precipitators, in-stack devices that trap particulates by
charging, or ionizing, the particles as the gas passes through
the stack. Although they do not completely eliminate dangerous particulates, they reduce them between 80 and 99 percent (4:80).

Other Changes

On-site changes were also required. For instance, the
harbor leading to the dock of the plant required dredging to
accommodate the coal ships. The coal storage area was expanded and adapted to minimize seepage and run-off from the
coal pile. Changes were made in the fuel-conveying system.
Altogether, the changes took about 18 months, and cost
roughly $180 million (5).

The Effects of the Changeover

Emissions Analysis

The changeover to coal was completed in 1979, and the
New England Power Company has continued to monitor the
effects of the switch. Principally their testing has measured
particulate, nitrogen oxide, and sulfur oxide emissions.

FIGURE 19-1 (*Continued*)

10

Emission Control

New England Power's pollution control centers on coal
selection. According to Nicole L. Letendre, Environmental
Engineer for New England Power, the utility exercises "front-
end control": they select only low-sulfur coal. Although
coal with 1.5 pounds of sulfur per million BTU's of fuel is
acceptable, the West Virginia coal used at Brayton Point
actually contains .8 pounds of sulfur per million BTU's.
Obtaining such low-sulfur coal is difficult for New England
Power, since the typical coal mined today contains about 2.5
pounds of sulfur per million BTU's (6:133).

Test Results

Testing Background

In 1982, Brayton Point underwent an EPA Compliance Stack
Test for their three power-generating units. Examined in the
study were the amounts of particulates, nitrogen oxides, and
sulfur oxides. This test differed from earlier studies in
that neither nitrogen oxides nor sulfur oxides had previously
been measured--only particulates.

Before the conversion, of course, New England Power had
burned oil at Brayton Point. Therefore, Table 1 shows amounts
of particulate emissions for oil burning.

FIGURE 19-1 (*Continued*)

11

Table 1. Results of 1978 Testing for Particulate Emissions
          from High-Sulfur Oil

GENERATING UNIT	PARTICULATES[a]
One	.05
Two	.083
Three	.025

[a]pounds per million BTU of fuel

Source: Nicole L. Letendre, Environmental Engineer, New
         England Power

Between 1973 and late 1979, Brayton Point burned coal
for about six months. However, the coal was a higher ash,
higher sulfur Polish coal. As Table 2 shows, there was a
significant increase in particulate emissions.

In 1980, New England Power began using a lower ash,
lower sulfur domestic coal. This switch meant smaller amounts
of particulate emissions, as Table 3 shows.

FIGURE 19-1 (*Continued*)

12

Table 2.   Results of Earlier 1980 Testing for Particulate
           Emissions from Higher-Sulfur, Higher-Ash Polish
           Coal

GENERATING UNIT	PARTICULATES[a]
One	.471
Two	.334
Three	(Still Burning Oil)

[a]pounds per million BTU of fuel

Source: Nicole L. Letendre

Table 3.   Results of Later 1980 Testing for Particulate
           Emissions from West Virginia Coal

GENERATING UNIT	PARTICULATES[a]
One	.228
Two	.183
Three	(Still Burning Oil)

[a]pounds per million BTU of fuel

FIGURE 19-1   (*Continued*)

13

Broader Testing

Tables 1, 2, and 3 show no figures for the nitrogen and sulfur oxides produced during the earlier oil-burning years. By 1982, however, when all three units were burning coal, the EPA required testing for these compounds. Table 4 shows the results.

Table 4. Results of 1982 EPA Compliance Stack Tests for Particulate, Nitrogen Oxide, and Sulfur Oxide Emissions

GENERATING UNIT	PARTICULATES[a]	NITROGEN OXIDES[a]	SULFUR OXIDES[a]
One	.088	.665	1.93
Two	.0157	.627	1.71
Three	.0104	1.49	1.89

[a]pounds per million BTU of fuel

Source: Nicole L. Letendre

These 1982 tests provide some insight into New England Power's attempts to control particulate, nitrogen oxide, and sulfur oxide emissions.

FIGURE 19-1 (*Continued*)

14

Particulates

New England Power has been largely successful in controlling particulate emissions. Although the 1982 and 1978
figures are similar, in 1978 Brayton Point was burning oil.
So, the conversion has not increased particulate pollution.

Nitrogen Oxides

Although no figures exist for nitrogen oxide emitted
during the earlier oil-burning years, the amounts are
generally higher for coal than for oil, according to Nicole
Letendre. In fact, the Draft Environmental Impact Statement
for Brayton Point predicted that nitrogen oxide emissions
could increase as much as 73 percent.

The EPA suggests a limit of .3 pounds per million BTU
burned for nitrogen oxides. The 1982 tests show that Brayton
Point exceeds the limit in all three units. However, this
exceedance is minimal. Further, because the largest producer
of nitrogen oxides is the automobile, the Brayton Point increase does not significantly affect local air quality. To
ensure that emissions remain at this level, New England Power
maintains monitoring stations in surrounding communities.
Current data show that the Southeastern Massachusetts and

FIGURE 19-1 (*Continued*)

15

Rhode Island areas at least meet national air standards for
nitrogen oxides.

Sulfur Oxides

As with nitrogen oxides, no figures exist for sulfur
oxides during the earlier oil-burning years. However, because of New England Power's front-end control (using only
low-sulfur coal), sulfur oxide emissions are probably less
than they were when New England Power burned high-sulfur oil
before the conversion.

Presently there is no established limit for sulfur
emissions. If Brayton Point, therefore, causes no exceedances
of the National Ambient Air Quality Standards for sulfur
oxides, the plant is considered in compliance. As with
nitrogen oxides, New England Power's monitoring stations
measure the air quality. Data show that Southeastern Massachusetts and Rhode Island have a "Better than National
Standards" rating for sulfur dioxide (1:35).

FIGURE 19-1 (*Continued*)

16

CONCLUSION

Summary of Findings

    Because coal is considered a "dirtier" fuel than oil,

Brayton Point's conversion from oil to coal burning raised

concerns about increased pollution from particulates, nitrogen

oxides, and sulfur oxides. The nitrogen and sulfur oxides

are a particular concern because, once airborne, they are

chemically transformed and combine with atmospheric moisture

to form acid rain.

    To help control particulate pollution, New England

Power uses electrostatic precipitators to trap particles

before they leave the stack. New England Power's efforts at

particulate control have been largely successful; today's

particulate emissions generally match 1978 oil-burning levels.

    Although figures are not available for nitrogen oxides

during the earlier oil-burning years, levels are considered

higher for coal than oil. All three units currently exceed

EPA limits; however, because automobiles produce most nitrogen

oxides, the exceedances are less significant--as long as

nitrogen oxide levels in the region at least meet national

standards. At the present time, Southeastern Massachusetts

FIGURE 19-1 (*Continued*)

and Rhode Island meet this minimum requirement.

No figures are available for sulfur oxides emissions
during the oil-burning years, but levels are considered lower
because the coal now used at Brayton Point has a lower percentage of sulfur than the oil used before the conversion.
Therefore, although levels of sulfur oxides are higher than
either particulates or nitrogen oxides, they are still seen
as an improvement over 1978 levels. Because no set limits
exist for sulfur emissions, Brayton Point is considered in
compliance as long as regional air quality meets national
standards. Southeastern Massachusetts and Rhode Island
currently have a "Better than National Standards" rating for
sulfur dioxide.

Comprehensive Interpretation of Findings

Levels of nitrogen oxides and sulfur oxides are particularly significant because these substances convert to
acid rain. However, since nitrogen oxides are near EPA
limits, and since the sulfur oxides level is presumed lower
with the low-sulfur coal than it was with the oil used before
the conversion, Brayton Point's conversion from oil burning
to coal does not appear to have increased the acid rain
problem.

FIGURE 19-1 (*Continued*)

18

Recommendations

　　1.　Local environmental groups should periodically review
the results of air quality monitoring and emissions testing
for all three generating units.

　　2.　The Ph (acid/base concentration) of regional precipitation should be routinely measured to ensure that the
acid rain problem becomes no more severe than it was before
the conversion.

FIGURE 19-1　(*Continued*)

19

REFERENCES

1. <u>Draft Environmental Impact Statement, Coal Conversion
   Program, New England Power Company, Brayton Point
   Generating Station, Plants 1, 2, and 3, Somerset,
   Bristol County, Massachusetts.</u> Washington, D.C.:
   U.S. Department of Energy, 1978: 35-54.

2. Ruedisili, Lon C., and Morris W. FireBaugh (eds.).
   <u>Perspectives on Energy</u>, second edition. New York:
   Oxford University Press, 1978: 164-176.

3. President's Commission on Coal. <u>Coal Data Book</u>. Washington, D.C.: U.S. Government Printing Office, 1980.

4. <u>McGraw-Hill Encyclopedia of Energy</u>. New York: McGraw-
   Hill, Inc., 1976.

5. Letendre, Nicole L., Environmental Engineer for New
   England Power Company. Interview about the cost of the
   conversion from oil burning to coal burning, August 1,
   1983.

6. Walker, Harry O. <u>Energy: Options and Issues</u>. University
   of California: Reproduced and Distributed Courtesy of
   Professor Harry O. Walker, University of California,
   Davis, Pacific Gas & Electric, San Francisco, California,
   Westinghouse Electric Corporation, 1977.

FIGURE 19-1 (*Continued*)

The next report combines a feasibility analysis with a comparative analysis. To illustrate yet another form of documentation, this writer uses the author-year system.

To save space, no front matter (title page, etc.) is reproduced here. The letter of transmittal for this report, however, is shown on page 244.

AUDIENCE-AND-USE ANALYSIS

One purpose of this report is to satisfy a requirement in a Technical Communication course, but the instructor is the secondary audience. Alan Greene, the author, addresses his remarks to his primary audience, the owners of ComTech, Beth Hall and Jeanne Smith. They asked their friend, Alan, who plans a career as a computer consultant, to analyze computer systems to see if one could be found that meet their needs and budget. Greene chose the topic, knowing he could help his friends while also establishing a worthwhile credential.

Smith and Hall run a part-time business (ComTech) writing résumés and cover letters for college students, and various business reports and sales brochures for small area businesses. They want to begin keeping books for ten small businesses they service. They wonder about the feasibility of computerizing these operations. Would it be a practical investment?

While talking with Hall and Smith to determine the content of his report, Alan got them to explain clearly the specific computer applications they desired. Because most of their work involves typing, Alan quickly determined that ComTech needs a computer with word processing capabilities. The business also needs a letter-quality printer that is compatible with their computer. Since they plan to expand the business to include bookkeeping, the system also must have software for accounts payable and receivable, general ledger, and, if possible, inventory control.

From discussions about his clients' computer needs, Alan chose the following eight features for analysis: (1) screen format, (2) numeric keypad, (3) memory capacity, (4) documentation, (5) monitor quality, (6) microprocessor capacity, (7) software availability, and (8) service and dealer support. So by learning how his readers planned to use the computer, he was able to focus his report and tailor it to their needs by comparing only those features relevant to them. And through his comparison, he was able to establish the feasibility of their investment.

While planning his report, Alan also considered its arrangement and style. Since the eight features were equally important to ComTech, he gave them equal ranking. And if the report were going to be useful to Hall and Smith, they would have to understand the concepts and terminology. Therefore whenever he used an unfamiliar term, he defined it parenthetically or in a sentence definition. He also used an informal tone and a "you" perspective throughout.

Like any worthwhile analysis, this report enabled Hall and Smith to make an *informed* decision, based on facts.

Because microcomputer technology advances so rapidly, some of Alan's conclusions will be outdated.

A COMPARATIVE ANALYSIS OF SIX PERSONAL COMPUTERS TO DETERMINE WHICH IS PREFERABLE FOR ComTech BUSINESS AND COMMUNICATION SERVICES

INTRODUCTION

Description of Problem

In 1977, Apple Computer began marketing its personal computer. By 1981, one hundred companies — with over 160 models — were competing for the personal computer market (*Fortune*, Oct. 1982). This phenomenal growth and competition have led to a 40 percent yearly increase in sales (projected through 1986) (*Fortune*, May 1982). For the consumer, this growth has meant better products, service, and prices. But choosing a personal computer is frustrating for people who must learn "computerese" before finding the system that best suits their needs.

Audience and Background

This analysis has been written for Beth Hall and Jeanne Smith who have a part-time business (ComTech) writing résumés and letters for students, and reports and sales brochures for small area businesses. They now want to an *informed* decision, based on facts.

Purpose

Hall and Smith need a versatile computer and letter-quality printer for business applications. This report intends to provide them with answers as to which, if any, computer system would best meet their needs and budget.

Sources of Information

Primary data come from interviews with local computer dealers. Secondary sources include business and computer magazines.

Working Definitions

Although computer jargon will be kept to a minimum, the following definitions are necessary for understanding this report:

1. daisy wheel printer: a printer with a flat circular typing element having the characters attached around the element on the ends of stalks. Such printers are commonly part of word processing systems since they produce solid, typewriter-like characters, in contrast to dot matrix printers, which produce characters made of a series of dots.

2. data: a general term meaning any information (facts, numbers, letters, symbols, etc.) that can be manipulated or produced by the computer.

3. disk drive: a device that reads from, or writes to, magnetic discs.

4. diskette: a 5¼" floppy disk that stores data and programs. The diskette is inserted in the disk drive.

5. memory: the part of the computer that stores data and programs.

6. program: an ordered list of instructions directing a computer to carry out a desired sequence of operations.

Scope

This report will compare six personal computer systems on the basis of the following eight features: (1) screen format, (2) numeric keypad, (3) memory capacity, (4) documentation, (5) monitor quality, (6) microprocessor capacity, (7) service and support, and (8) available software.

Limitations of Report

This report limits itself to systems in the price range cited: $3,000 to $5,000. Since you also want a system that can be bought and serviced locally, only the following computers will be considered: TRS-80 Color Computer, Commodore 64, TRS-80 Model III, Apple IIe, IBM Personal Computer (PC), and the Victor 9000.

Because you will be using software applications (e.g., word processing, accounting), you have no need for programming languages such as COBOL, or BASIC; so these languages will not be considered in my discussion of system capabilities.

System prices change often. So prices should be read as approximate.

COLLECTED DATA

Screen Format

Definition

Screen format represents the number of characters (numbers, letters, etc.) that can appear on the monitor's screen. An 80 x 25 format, for instance, means that 80 characters appear on one line, and 25 lines are displayed on the screen. Without 80-character format, word processing is difficult: the operator cannot see how text on the screen will look as hard copy (the printed document). With the 80-character format, what you see on the screen is what you get as hard copy— a necessity for effective document design.

Findings

The TRS-80 Color Computer has only 32 characters per line, while the Commodore and TRS-80 Model III have 40 and 64 characters respectively. Although the standard Apple IIe has only a 40-character format, it can be upgraded to an 80 x 24 format. Both the IBM PC and the Victor 9000 have an 80 x 25 format as standard equipment. (The 9000 also allows a 132-character format, a plus for accounting work requiring many number columns.)

Evaluation and Interpretation of Findings

For word processing, the TRS-80 and the Commodore 64 are inadequate. With practice, you could adapt to the Model III's 64-character format. The IIe, PC and 9000 have the desired format.

The accounting work you plan would also be tedious with the smaller formats, since you could not use many of the better accounting or spreadsheet programs. (An electronic spreadsheet facilitates many business functions such as cash-flow projections, sales projections, expense accounts, tax calculations, or project forecasting.) Some spreadsheets do work with screen formats of 40 columns or less, but an 80-column format lets you work with more information at once.

A Numeric Keypad

Definition

A numeric keypad has numbers arranged, as on an adding machine, to the right of the regular keyboard. Numbers can be entered much faster with the pad than by the number keys on the top row of the keyboard.

Findings

The Color Computer, the Commodore, and the IIe lack numeric keypads. A pad can be added to the IIe as a peripheral. The Model III, PC, and 9000 have attached numeric keypads.

Evaluation and Interpretation of Findings

Since both of you have experience as data entry clerks, you would have trouble adjusting to the number row on a regular keyboard for your accounting work. And, of course, you would lose data entry speed. Although you can add a pad to the IIe, it adds to workstation clutter and moves around the desk as you enter data. So for accounting and spreadsheets, the Model III, PC, and 9000 are best.

Monitor Quality

Definition

Monitor quality depends on the screen's resolution. Resolution refers to the number of dots (the matrix) used to form each image on the screen. A matrix of 5 x 7 dots (usually the minimum to form a single character) is minimally acceptable. Televisions used as display screens have a 5 x 7 format. Monitors with an 80-character format use a 7 x 9 matrix for better resolution.

Findings

The Color Computer uses a color TV as its monitor, with a 5 x 7 matrix. The Model III has an attached monitor (keyboard, computer, and monitor are one unit) with a 5 x 7 matrix. The IIe and Commodore use a standard 5 x 7 matrix, but these monitors are not attached; they can be replaced by higher-resolution monitors. The 7 x 9 matrix is standard for the PC and 9000, but the 9000's antiglare shield makes its monitor superior.

Evaluation and Interpretation of Findings

Higher-resolution monitors are expensive (the PC's $350 monitor, for example, equals the cost of the *entire* Color Computer system), but worth the price. In fact, in a survey conducted by Verbatim, the largest manufacturer of diskettes, 63.4 percent of respondents cited eyestrain as the worse computer-related health problem (Group Attitudes Corp., 1982).

For your word processing, spreadsheets, or accounting, you will need a good monitor. The PC or 9000 monitors are among the best, but you could upgrade the IIe or the Commodore.

Memory Capacity

Definition

Memory is measured in Kilobytes, abbreviated K. A kilobyte is 1024 characters of computer memory. Since data are stored in the computer's memory during processing, more K's means more data can be manipulated simultaneously.

Random access memory (RAM) is the principal memory *area* of a computer used to store programs and data during processing. The size of RAM determines how much information your computer can work with at once (*MicroSoftware Solutions*, 1983).

Findings

The Color Computer comes with 16K of RAM. The Color Computer can be expanded to 64K, but access to 32K of that total requires expensive add-on disk drives. The Model III, with 48K, is not expandable to include more memory. The Commodore, IIe, and PC come with 64K, but only the PC is expandable (to 512K). Standard for the 9000 is 128K, expandable to 896K.

Evaluation and Interpretation of Findings

Since most of the powerful software programs (e.g., WordStar, VisiCalc) require a minimum of 64K, neither Radio Shack computer (the Color Computer and the Model III) is adequate for your purposes. You should consider no less than 64K (expandable to 128K). Otherwise, your computer will be unable to grow with your business. Although the 9000 is expandable to 896K, this capacity is much more than you will need. The PC is your best bet since it is expandable to 512K: sufficient to meet your needs for many years, while still within your budget.

Microprocessor Capacity

Definition

A *microprocessor* is the controlling unit of a microcomputer (a term now synonymous with personal or small business computers like those being compared here). The microprocessor is a tiny silicon chip containing the micro-switches (or logical elements) for all data processing and software instructions. The 16-bit processor is second-generation personal computer technology; the 8-bit was the first.

Bit is an acronym for binary digit. Basically, computers work on a series of miniscule (micro) "on"/"off" switches, usually represented by a "0" or "1." For example, when you depress "a" on the keyboard, a series of 6 to 8 micro switches are activated to form the "a" on the monitor and in the computer's memory. The combined bits needed to form the character "a" are called a "byte."

Findings

Of the six systems compared here, only the PC and 9000 have 16-bit processors.

Evaluation and Interpretation of Findings

Because most popular small computers (e.g., IIe and Commodore) have 8-bit processors, they are limited. For instance, an 8-bit computer only works with 64K of RAM (without expensive design changes). In contrast, 16-bit computers like the PC or 9000 can work with 16 to 64 times more RAM (Hewin, 1982). More RAM leads to broader applications. With a spreadsheet program like VisiCalc, for example, more RAM means more rows and columns to work with. Or, for sorting lists, a 16-bit computer increases sorting speed by allowing you to manipulate all data within RAM rather than continually calling for blocks of data from disk storage (Hewin, 1982).

Other limitations of the 8-bit microprocessor include slower speeds for all applications, inability to do simultaneous tasks (e.g., word processing while printing another document), and the inability to link the computer to other terminals (Hewin, 1982). Although this last application may not concern you now, it could if your business grows. For example, you might want to add a terminal so both of you can access your main computer simultaneously. The PC and 9000 offer that capacity.

Documentation

Definition

Documentation includes all operating instructions for the hardware and software. Systems with clear documentation are "user friendly."

Findings

Of the six systems, the IIe and the Commodore have the clearest documentation.

Evaluation and Interpretation of Findings

Although documentation is included with all hardware and software, manufacturers too often give technical explanations instead of simple instructions. Some of the PC's documentation has this problem. The IIe, on the other hand, follows Apple's tradition for excellent documentation. As one expert notes, "the IIe's documentation package is . . . outstanding and certainly worthy of imitation . . ." (Edwards, 1983).

Radio Shack's documentation for the Color Computer and Model III is good. Owners I spoke with said the documentation is user friendly. *Popular Computing* calls the Commodore "The People's Computer" partly because of its user-friendly documentation (Roberts, 1983).

The 9000, the most powerful system in this comparison, is a business computer rather than a personal computer. As such, its documentation is written for a more technical audience. Of the six systems, the IIe and Commodore have the best documentation. There is hope for the PC, though, since most of its software is being written by third-party vendors.

Available Software

Definition

Documentation explains and instructs users on using hardware and software (programmed instructions that tell the computer what to do). Computers will not function without software.

All software for the computers compared here comes on 5¼" diskettes. Prices depend on the functions the software can perform. Sophisticated programs like Wordstar or VisiCalc cost $495. The accounting programs you need cost together roughly $1,500.

Findings

The IIe and Radio Shack models have the most software available, but the PC will soon lead.

Evaluation and Interpretation of Findings

Many experts tell their clients to find the software they need, then to find a system that will run it. Although this was sound advice a year ago, it no longer applies for the systems compared in this analysis. For instance, more software is available for Radio Shack's computers (Color Computer and Model III) than for any other systems. And since software for the IIe is compatible with earlier software for the Apple II+ and the Apple II, the IIe is second only to the Color Computer and Model III for available applications.

If the trend continues, the PC will have more available software than any other personal computer, since two hundred programs *a week* are being sent to IBM for review. Moreover, Quadram, an independent manufacturer, has developed a $600 plug-in module that allows PC owners to run all Apple software. (Peripherals [add-ons like telephone hookups or special printers] for the PC are reaching the market at an even faster rate.)

The Commodore has fewer software applications than the preceding four systems, but the company has announced plans for many new programs, all costing less than $100 (Roberts, 1983). Since the 9000 is the newest system, it has the fewest applications. But Victor is marketing its own spreadsheet (VictorCalc) and a word-processing program especially suited to its 16-bit processor (Vose, 1982). Software for the 9000 will be limited, however, since third-party vendors are not writing programs for it as they are for the more popular (and less costly) Radio Shack systems, the IIe, and the PC.

Service and Dealer Support

Definition

Service includes inspecting, adjusting, upgrading and repairing hardware. Support consists of helping buyers get their systems running, answering questions, explaining how specific software interacts with their systems, stocking supplies and peripherals, providing backup systems during servicing, and helping resolve minor crises that typically arise with new systems. For instance, if you cannot make a disk copy or if your printer begins chewing ribbons, a good dealer should be able to solve these problems quickly and, ideally, over the phone.

Findings

No system has a definite edge in dealer service and support.

Evaluation and Interpretation of Findings

While gathering data for this report, I went to the closest Commodore outlet (in a general department store). When I asked for a demonstration, neither clerk even knew how to turn the computer on.

PC dealers, on the other hand, must meet IBM's rigorous standards for technical competence, financial standing, and support. Because IBM is choosy about its dealers, support usually is good (Harbatkin, 1983).

The extensive Radio Shack network does offer buyers support and service for the Color Computer and Model III. But since Radio Shack has such an extensive network of stores, dealer support varies from store to store. For instance, the staffs of the three Radio Shacks I surveyed in this area seem to range from excellent to competent.

Victor and Apple have good dealer support, but expertise among sales people varied greatly within the same store. Fortunately, each store does have competent staff, so you usually can get the needed answers.

Dealer service and support vary greatly. Every computer owner I consulted had at least one horror story about poor service and support, regardless of the system. Their misfortunes, though, are minor compared with those of owners who bought by mail. Be sure to buy from a major area dealer, preferably part of a chain or franchise. That way, if your dealer cannot help you, he will call someone in the chain who can.

CONCLUSION

Summary of Findings

1. Only the 9000 and PC come with an 80 x 25 screen format, but the IIe can easily be upgraded.

2. A numeric keypad can be added to the IIe (as a peripheral), while an attached pad is standard for the Model III, 9000, and PC.

3. The 9000 and PC have the best monitors, but the IIe and Commodore can be upgraded.

4. The Color Computer has only 16K of RAM, the Model III, 48K. The Commodore, IIe, and PC come with 64K, but, of the three, only the PC is expandable (to 512K). Standard for the 9000 is 128K, expandable to 896K.

5. Of the six systems, only the PC and 9000 have 16-bit microprocessors.

6. User-friendly documentation is among the best features of the IIe and Commodore. Radio Shack's systems have good documentation, while the PC's is uneven, ranging from obscure to good. The 9000's manuals are for technical audiences.

7. The IIe and Radio Shack models have the most software, but the PC will soon lead.

8. Dealer support varies greatly, with no system having a definite edge.

Interpretation of Findings

Although well within your budget, the Color Computer, Commodore and Model III are inadequate for your needs. They lack expandable memory, 80 x 25 screen formats, and high-resolution monitors.

The Victor 9000 would be an excellent choice since it meets most of your criteria. But its $4,995 price is beyond your budget, since word processing software and a printer would add at least $2,500 before you bought accounting software.

The PC meets more of your needs than the IIe, but the IIe system would cost about $1,000 less — unless you decide to upgrade it with a numeric keypad, a better monitor, an 80 x 25 format, and more RAM.

TABLE 1   A Summary Comparison of the Apple IIe and IBM PC

Features	*Apple IIe*	*IBM PC*
Screen Format	80 x 25 [a]	80 x 25
Numeric Keypad	no [b]	yes
K's of RAM	64	64 [c]
Documentation	excellent	fair
Quality of monitor	good	excellent
Service and support	good	good
Availability of software	excellent	good [d]
Microprocessor	8-bit	16-bit

[a] The IIe must be upgraded to this format.

[b] A numeric keypad can be added as a peripheral.

[c] The PC is expandable to 512K RAM.

[d] The PC software library is growing rapidly. And Quadram offers an enhancement board that allows PC owners to run all Apple software.

Recommendations

Buy the IBM PC for your business. It has become the industry standard. As a result, hardware and software manufacturers are marketing PC-compatible products. This competition will not only lower prices, but also provide enormous choice.

In addition, the PC can grow with your business. The Apple cannot. Although 8-bit software will not become immediately obsolete, more and more programmers are writing 16-bit software because it is much faster, more powerful, and more adaptable to the applications you need.

The basic PC system (with one 320K disk drive) would cost around $3,000. A mid-range, letter-quality printer like the DaisyWriter would cost an additional $1,500, leaving $500 for software. This system will fill all your present needs, and in a few months you could add a second disk drive and accounting software.

REFERENCES

Edwards, John. 1983. "Apple IIe." *Popular Computing* (March): 108–110.

"Glossary." 1983. *MicroSoftware Solutions.* Catalogue: 31.

Harbatkin, Lisa. 1983. "A Dealer in Every Port Whether You Travel or Not." *PC Magazine* (July): 103–106.

Hewin, Larry and Duane Saylor. 1982. "The 8- versus 16-bit Debate." *Desktop Computing* (December): 68–70.

"Information Systems for Tomorrow's Office." 1982. *Fortune* (18 October): 18–82.

"Office Worker Views and Perceptions of New Technology in the Workplace: A Summary of Findings." 1982. Verbatim Survey, by Group Attitudes Corporation, New York.

Roberts, Bruce. 1983. "The Commodore 64: The People's Computer." *Popular Computing* (May): 92–98.

"Trends in Computing: Applications for the 80s." 1982. *Fortune* (31 May): 20–68.

Vose, Michael. 1982. "Victor Unveils Business System." *Desktop Computing* (September): 78–79.

CHAPTER SUMMARY

Analytical reports are question-answering or problem-solving reports. An analysis of data requires that you collect evidence from various sources and use this evidence to draw definite conclusions and make specific recommendations. As you plan your report consider which of these questions or combination of questions your analysis is intended to answer:

1. Will X work for a specific purpose?
2. Is X or Y better for a specific purpose?

3. Why does X happen?
4. What are the effects of X?
5. Is X practical in a given situation?

Condense your approach to a basic question, and restate it as a sentence in your statement of purpose. Sometimes you will combine these approaches.

After identifying the problem or question, interpret all data impartially and fully. Make the report detailed enough to show how you arrived at your conclusions. Use visuals and, except for a memo report, supplements.

As you sift data and write, choose the most reliable sources, distinguish hard from soft evidence, and avoid specious reasoning.

Follow these steps in planning and writing your analysis:

1. Make a detailed outline and develop your report from it:
   a. In your introduction, identify the subject; describe the problem or question; and give background. Identify your reader, sources of data and limitations. List working definitions. If you use a glossary or appendixes, mention them. Finally, list all major topics covered in the body.
   b. In your body, divide your subject into major topics and subtopics. At each level of division, define the topic; discuss your findings; and evaluate and interpret the findings.
   c. In your conclusion, summarize major findings; explain their overall meaning; and make recommendations based on your interpretation.

Revise the report, and submit the final draft with all needed supplements.

REVISION CHECKLIST FOR ANALYTICAL REPORTS

Use this list to refine the content, arrangement, and style of your report.

Content

1. Does the report grow from a clear statement of purpose (answering a practical question, examining quality, solving a problem, measuring feasibility?
2. Is the title unbiased and accurate?
3. Is the report's length adequate and appropriate?
4. Are all limitations of the report spelled out?
5. Is each topic defined before it is discussed?
6. Are visuals used effectively and whenever possible?

7. Is the analysis based on hard evidence (no unsupported opinions)?
8. Is it free from specious reasoning?
9. Are all data sources credible?
10. Are all data accurate?
11. Are all data unbiased?
12. Are all data complete?
13. Are all data fully interpreted?
14. Is the documentation adequate and correct?
15. Are recommendations based on impartial interpretations of data?

Arrangement

1. Does the report have a fully developed introduction, body, and conclusion?
2. Are headings appropriate and adequate?
3. Are there enough transitions between related ideas?
4. Does the report have all needed front matter (title page, letter of transmittal, table of contents, table of illustrations, informative abstract)?
5. Does the report have all needed end matter (glossary, list of references, appendix)?

Style

1. Is the level of technicality appropriate for the stated audience?
2. Are front-matter pages numbered in small roman numerals?
3. Are all sentences clear, concise, and fluent?
4. Is the language convincing and precise?
5. Is the report written in correct English (Appendix A)?

Now list those elements of the report that need improvement.

EXERCISES

1. *Self-help analysis:* Choose a problem in your life (low grades, poor love life, insufficient time for relaxation, depression, anxiety, etc.). Define the problem clearly in several sentences. Brainstorm for possible causes. After carefully evaluating each possible cause, compile a list of probable causes, thereby narrowing your original list. Now, evaluate each probable cause, to emerge with one or more definite causes. Summarize your findings;

draw conclusions from your evidence; and make recommendations for solving the problem.

2. *In class:* Divide into groups of about eight. Choose a subject for group analysis — preferably, a campus issue — and partition the topic through group brainstorming. Next, select major topics from your list and classify as many items as possible under each major topic. Finally, draw up a working outline that could be used for an analytical report on this subject.

3. Prepare a questionnaire based on your work in exercise 2 and administer it to members of your campus community. List the findings of your questionnaire and your conclusions in clear and logical form.

4. In the periodical section of your library, find examples of reports or articles that use each of the following types of analysis:

  a. answering a practical question (Will X work for a specific purpose?)
  b. comparing two or more items (Is X or Y better for a specific purpose?)
  c. solving a problem (Why does X happen?)
  d. determining consequences (What are the effects of X?)
  e. assessing feasibility (Is X practical in a given situation?)

Provide full bibliographical information, along with a *descriptive* abstract (see page 105) of each article.

5. In the periodical and newspaper section of your library, compile a list of sources, by title, ranked in general order of reliability: (a) List five highly reliable sources. (b) List five sources that are less reliable. Briefly explain the reason for each choice by discussing your criteria for judgment.

6. The statements below are followed by false or improbable conclusions. In order to prevent specious generalizations, what specific supporting data or evidence would be needed to justify each conclusion?

  a. Eighty percent of black voters in Mississippi voted for John Jones as governor. Therefore, he is not a racist.
  b. Fifty percent of last year's college graduates did not find desirable jobs. Therefore, college is a waste of time and money.
  c. Only 60 percent of incoming freshmen eventually graduate from this college. Therefore, the college is not doing its job.
  d. He never sees a doctor. Therefore, he is healthy.
  e. This house is expensive. Therefore, it must be well built.

7. Write an informative abstract of the report on pages 508–516. (You may wish to review Chapter 6.)

8. Write an analytical report, using the following guidelines (not necessarily in sequence):

  a. Choose a subject for analysis from the list at the end of this exercise, from your major, or from an area of interest.
  b. Identify the problem or question so you will know exactly what you are looking for.

   c. Restate the main question as a declarative sentence in your statement of purpose.

   d. Identify an audience — other than your instructor — who will use your information for a specific purpose.

   e. Hold a private brainstorming session to generate major topics and subtopics.

   f. Use the topics to make an outline based on the model outline in this chapter. Divide as far as necessary to identify all points of discussion.

   g. Make a list of all sources (primary and secondary) that you will investigate in your analysis.

   h. Write a proposal memo to your instructor (pages 437–438), describing the problem or question and your plan for analysis. Attach a working bibliography to your memo.

   i. Use your working outline as a guide to research and observation. Evaluate sources and evidence, and interpret all evidence fully. Modify your outline as needed.

   j. Submit a progress report to your instructor (pages 404–408), describing work completed, problems encountered, and work remaining.

   k. Write an analysis of your audience's needs and level of technical knowledge. (Use the samples on pages 482 and 507 as models.)

   l. Write the report for your stated audience, and include topic headings and transitions between topics. Work from a clear statement of purpose and be sure that your reasoning process is shown clearly. Verify that your evidence, conclusions, and recommendations are consistent.

  m. After writing your first draft, make any needed changes in the outline and revise your report according to the revision checklist. Include all necessary supplements.

   n. Exchange reports with another class member for further suggestions for revision.

Many of the research projects on pages 332–334 can be adapted to recommendation-type reports. Here are some other subjects for analysis:

- the effects of a vegetarian diet on physiological energy
- the causes of student disinterest in campus activities
- the student transportation problem to and from your college
- two or more brands of tools or equipment from your field
- noise pollution from nearby airport traffic
- the effects of toxic chemical dumping in your community
- the advisability of earning a graduate degree in your field
- the adequacy of veterans' benefits
- the effects of sun on skin
- the (causes, effects of) acid rain in your area
- the impact of a civic center or stadium on your community

- the adequacy of security on your campus or in your dorm
- the adequacy of your college's remediation program
- the feasibility of opening a particular business
- the best location for a new business
- causes of the high drop-out rate of students in your college
- the pros and cons of condominium ownership
- the feasibility of moving to a certain area of the country
- job opportunities in your field
- effects of the 200-mile limit on the fishing industry in your coastal area
- the effects of budget cuts on public higher education in your state
- the best nonprescription cold remedy
- the adequacy of zoning laws in your town
- effect of population increase on your local water supply
- adequacy of your student group-health-insurance policy
- the feasibility of using biological pest control as an alternative to pesticides
- the feasibility of large-scale desalination of sea water as a fresh water source
- effective water conservation measures that can be used in your area
- the effects of legalizing gambling in your state
- effective measures for relieving the property-tax burden in your town
- the causes of low morale in the company where you work part time
- causes of poor television reception in your area
- the effects of thermal pollution from a local power plant on marine life
- measures for the reclamation of strip-mined land
- the feasibility of using wood as an energy source
- two or more brands of a wood-burning stove
- the feasibility of converting your home to solar heating
- the adequacy of police protection in your town
- effective measures for improving the fire safety of your home
- the best energy-efficient, low-cost housing design for your area
- effective measures for improving productivity in your place of employment
- reasons for the success of a specific restaurant (or other business) in your area
- the feasibility of operating a campus food co-op
- the best investment (real estate, savings certificates, stocks and bonds, precious metals and stones, etc.) over the past ten years
- the problem of water supply and waste disposal for new subdivisions in your town
- the advisability of home birth (as opposed to hospital delivery)
- the adequacy of clean-up measures for toxic waste in your area
- the adequacy of the evacuation plan for your area in the event of a nuclear emergency or other disaster

20

Oral Reporting

ORAL REPORTS DEFINED

An oral report is any spoken statement requiring preparation and forethought — anything from a brief discussion to formal speeches or lectures. This chapter will discuss the more formal reporting situations, which require planning and preparation.

Like the written report, the oral report must be clear, informative, and technically appropriate for its audience, but the oral report has advantages as well as disadvantages. An oral report is more personal, may be more memorable, saves time, and elicits immediate response; in contrast, a written report is easier to refine and organize, can be more complex, can be studied at the reader's own pace, and is easily reviewed.

PURPOSE OF ORAL REPORTS

Your oral reports will vary in style, range, complexity, and formality in various situations. Your most formal oral reporting may include convention speeches, reports at national meetings, reports to officers and company personnel about a new project, speeches to community groups, and the like.

These formal talks may be designed to *inform* (e.g., to describe a new procedure for handling customer complaints), to *persuade* (e.g., to induce company officers to vote a pay raise), or to do both. The higher your status, on the job or in the community, the more you will give formal talks.

CHOOSING THE BEST TYPE OF REPORT

An oral report's effectiveness depends largely on *how* it is delivered. Here are the possibilities:

The Impromptu Delivery

The impromptu ("off-the-cuff") delivery generally is ineffective for formal reports. Being unified, coherent, fluent, and informative without preparation is nearly impossible. Don't delude yourself by thinking "It's all in my head." Get your plan down on paper.

The Memorized Delivery

In the memorized delivery, you write your report and memorize it. You might sound like a parrot, and forgetting a line can be disastrous. Because your personality and body language influence audience interest, you can hardly expect to seem natural while mechanically reciting your lines.

The Reading Delivery

Reading your report aloud is boring. Unless you maintain eye contact with your audience, and vary your gestures and tone, you might look like a robot. If you *do* plan to read the report, study it, and practice aloud to decrease reliance on the text. Otherwise, you might slur or mispronounce words, or keep your nose stuck to the text in fear of losing your place.

The Extemporaneous Delivery

An extemporaneous delivery is carefully planned, practiced, and based on notes that keep you on track. This is the most popular speaking technique. By following notes in sentence-outline form, you maintain control of your material. Also, you speak in a natural, conversational style, with only brief glances at your notes. Because an extemporaneous delivery is based on key ideas (represented by topic sentences to jog your memory) rather than fully developed paragraphs to be read or memorized, you can adjust your pace and diction as needed. Instead of coming across as a droning, mechanical voice, you are seen as a distinct personality. The one danger of an extemporaneous delivery is that you might get off track, unless you are thoroughly prepared.

PREPARING EXTEMPORANEOUS REPORTS

Plan your report step by step, to stay in control and build confidence. Some suggestions follow.

Know Your Subject

A sure way to appear foolish is to speak on something you don't fully understand or know too little about. Do your homework, *exhaustively*. Be prepared to explain and defend each assertion and statement of opinion with fact. When you know your subject, you can avoid tentative and equivocating statements that begin with "I feel," "I guess," "I suppose," or "I imagine." Your audience expects to hear a knowledgeable speaker. Don't disappoint them.

Identify Your Audience

Everyone has sat through boring or confusing lectures. Surely you don't want to impose this kind of agony on your audience. You expect any speaker to be interesting, informative, and clear; live up to these same expectations for your audience. At one extreme, speakers who oversimplify and belabor obvious or trivial points are boring; at the other extreme, those who speak in generalities or fail to explain complex information are confusing. Adjust the amount of detail and the level of technicality to your audience.

Many audiences contain people with varied technical backgrounds. So unless you know the background of each person, speak to a general audience (Chapter 2), as in a classroom of mixed majors.

As you plan the report, answer these questions about your audience:

1. What points do I want to make?
2. How can I make each point interesting and understandable?

If your subject is controversial, consider the average age, political views, educational level, and socioeconomic status of most audience members. Then decide how to speak candidly and persuasively without offending anyone.

Plan Your Report

Statement of Purpose

Formulate, on paper, a statement of purpose, no more than two or three sentences long. Why are you speaking on this subject? Who is your audience? What effect do you wish to have?

Homework

Begin gathering data well ahead of time. Use the summarizing techniques in Chapter 6 to identify and organize major points.

If the oral report is simply a spoken version of your written report, you'll need substantially less preparation. Simply expand your outline for the written report into a sentence outline. For our limited purposes, we will assume that your oral report is based on a written report.

Sentence Outline

Below is a typical sentence outline for a 15-minute oral report derived from a written report twenty pages long.

ORAL REPORT OUTLINE

Arnold Borthwick

Purpose: By informing Cape Cod residents about the dangers to the Cape's fresh water supply posed by rapid population growth, this report is intended to increase local interest in the problem.

I. INTRODUCTION

    A. Do you know what you are drinking when you turn on the tap and fill a glass?

    B. The quality of our water is high, but not guaranteed to last forever.

    C. The somewhat unique natural storage facility for our water supply creates a dangerous situation.

    D. Cape Cod's rapidly increasing population easily could pollute our water.

    E. In fact, pollution in some towns already has begun.

II. BODY

A. The groundwater is collected and held in an aquifer.
   1. This water-bearing rock formation forms a broad, continuous arch under the entire Cape (Visual #1).
   2. The lighter fresh water flows on top of the heavier salt water.
B. With increasing population, sewage and solid waste from landfill dumps invade the aquifer.
   1. As wastes flow naturally toward the sea, they can invade the drawing radii of town wells (Visual #2).
   2. The Cape's sandy soil causes rapid seepage of wastes into the groundwater in the aquifer.
C. Increased population also causes overdraw on some town wells, resulting in salt water intrusion.
D. Salt and calcium used in snow removal add to the problem by entering the aquifer from surface runoff.
E. The effects of continuing pollution of the Cape's water table will be far-reaching.
   1. Drinking water will have to be piped in over 100 miles from Quabbin Reservoir.
   2. The Cape's beautiful fresh water ponds will be unfit for swimming.
   3. Aquatic and aviary marsh life will be threatened.
   4. The sensitive ecological balance of Cape Cod's environment will be destroyed.
F. Such damage would, in turn, lead to economic disaster for Cape Cod's major industry — tourism.

III. CONCLUSION

A. In summary, from year to year, this problem becomes more real than theoretical.
B. The conclusion is obvious: If the Cape is to survive ecologically and financially, immediate steps must be taken to preserve our *only* water supply.
C. The following recommendations offer a starting point for effective action:
   1. Restrict population density in all Cape towns by creating larger building-lot requirements.
   2. Keep strict watch on proposed high-density apartment and condominium projects.
   3. Create a committee in each town to educate residents about the importance of conserving water, thereby reducing the draw on town wells.
   4. Prohibit salt, calcium, and other additives in sand spread on snow-covered roads.
   5. Identify alternatives to landfill dumps for solid waste disposal.

D. This crucial issue deserves the immediate attention of every Cape Cod resident.

E. Question-and-answer session.

Each sentence in this outline is a topic sentence for a paragraph the speaker will develop in detail. (Review outlining techniques in Chapter 9.)

Identify areas in your outline where visuals might help (Chapter 13). Flip charts and transparencies are especially effective for small (classroom-size) audiences.

Before practicing your delivery, transfer your outline to 3 x 5 notecards, which you can hold in one hand and shuffle as needed. Or insert the outline pages in a looseleaf binder for easy flipping. In either case, type or print clearly, leaving enough white space so you can locate material at a glance.

Visual Aids

Well-chosen visuals increase audience interest. But visuals are not a substitute for your report. Select visuals that will clarify and enhance your talk — without making you fade into the background.

Use visuals whenever possible to emphasize a point, when *showing* will be more effective than merely *telling* (Chapter 13). The map shown in Figure 20–1 (3' x 5' as used in the actual talk) is designed to reinforce Arnold Borthwick's explanation of how solid wastes contaminate wells. In this case, the *telling* is clarified greatly by the *showing*.

As you plan visuals for your report, ask yourself these questions:

1. What points do I want to highlight? What visual will provide the best emphasis in each case?

2. How large should the visual(s) be, to be seen clearly by the entire audience?

3. What hardware is available (slide projector, opaque projector, overhead projector, film projector, videotape, blackboard, computer terminal)? Which is best for my purpose and audience? How far in advance do I have to request this equipment?

4. Can I make (or have made) drawings, sketches, charts, graphs, or maps as needed? Can transparencies (for an overhead projector) be made up or slides collected? Do handouts have to be typed and reproduced?

Follow the suggestions below for using visuals in your oral report:

1. Keep visuals simple. The audience will have no time to study each one.

2. Limit their number (no more than four or five in a 20-minute talk). Otherwise, your presentation may seem like a media event. Be selective in choosing what to highlight with visuals.

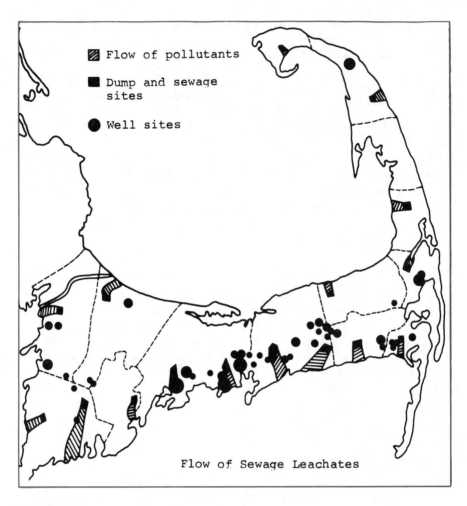

Flow of Sewage Leachates

FIGURE 20-1

3. Interpret each visual as you present it.

4. Stand aside when discussing a visual, so everyone can see it.

5. After discussing the visual, remove it, to refocus attention on you.

6. If you plan to do drawings on a blackboard, do them beforehand (in multicolored chalk). Otherwise, your audience will be twiddling their thumbs while you draw away.

7. If the audience has to remember certain material, prepare handouts for distribution *after* your talk.

8. Check the room beforehand to make sure you have the necessary space, electrical outlets, furniture, etc., for your equipment. If you will be addressing

a large audience by microphone, and are planning to point to certain features on your visuals, be sure the microphone is movable. Don't forget a pointer if you need one.

Delivery Time

Unless you have good reason to do otherwise, aim for a maximum delivery time of 20 minutes (the earlier outline is so designed). Longer talks may cause your audience to lose attention. Time yourself in practice sessions, and trim as needed.

Practice Your Delivery

Hold several practice sessions to learn the geography of your report. Then you won't fumble your actual delivery.

Feedback

Try to practice at least once before friends. Otherwise, use a full-length mirror and a tape recorder. Assess your rate of speaking from your friends' comments or from your taped voice (which, by the way, will sound high to you) and adjust your pace if necessary. Revise any parts that your friends find unclear. Have them evaluate your organization and tone.

Organization

Is your delivery unified, coherent, and logical? How well does it hang to-gether? Will the audience be able to follow your reasoning? Do you use enough transitional statements to reinforce the connections between ideas?

Tone

Maintain a conversational tone. Speaking as you do in the classroom should be appropriate.

Anticipate Audience Questions

Consider the parts of your report that might elicit questions and challenges from your audience. You might need to clarify or justify information that is new, controversial, disappointing, or surprising. Be prepared to support all conclusions, recommendations, assertions, and statements of opinion. If you doubt the validity of any statement, leave it unsaid. Try to predict your audience's responses so you'll be ready to field questions.

DELIVERING EXTEMPORANEOUS REPORTS

Capitalize on your good preparation by using the following guidelines for your delivery.

Use Natural Body Movements and Posture

If you move and gesture as you normally would in a conversation, your audience will be more relaxed. Nothing seems more pretentious than a speaker who works through a series of rehearsed moves and artificial gestures. Also, maintain a reasonable posture.

Speak with Confidence, Conviction, and Authority

Show your audience you believe in what you say. Avoid qualifiers ("I suppose," "I'm not sure," "but . . . ," "maybe," etc.). Also, clean up verbal tics ("er," "ah," "uuh," "mmm"). If you seem to be apologizing for your existence, you won't be impressive. Speaking with authority, however, is not the same as speaking like an authoritarian; no sermons, please.

Moderate Voice Volume, Tone, Pronunciation, and Speed

When using a microphone, people often speak too loudly. Without a microphone, they may speak too softly. Be sure you can be heard clearly, without shattering eardrums. Ask your audience about the sound and speed of your delivery after a few sentences. Your tone should be confident, sincere, friendly, and conversational.

Because nervousness can cause too-rapid speech and unclear or slurred pronunciation, pay close attention to your pace and pronunciation. Usually, the rate you feel is a bit slow will be just about right for your audience.

Maintain Eye Contact

Eye contact is vital in relating to your audience. Look directly into your listeners' eyes to hold their interest. With a small audience, eye contact is one of your best connectors. As you speak, establish eye contact with as many members of your audience as possible. With a large group, maintain eye contact with those in the first rows.

Read Audience Feedback

Addressing a live audience gives you the advantage of receiving immediate feedback on your delivery. Assess your audience's responses continually and make adjustments as needed. If, for example, you are laboring through a long list of facts, figures, examples, or statistical data, and you notice that people are dozing or moving restlessly, you might summarize the point you are making. Likewise, if frowns, raised eyebrows, or questioning looks indicate confusion, skepticism, or indignation, you can backtrack with a specific example or explanation. By tuning in to your audience's reactions, you can avoid leaving them confused, hostile, or simply bored.

Be Concise

Say what you came to say, then summarize and close — politely and on time. Don't punctuate your speech with "clever" digressions that pop into your head. Unless a specific anecdote was part of your original plan to clarify a point or increase interest, avoid excursions. Remember that each of us often finds what we have to say more interesting than our listeners do!

Summarize

Before ending, take a moment to summarize the major points and to reemphasize anything of special importance. As you conclude, thank your listeners.

Leave Time for Questions and Answers

As you begin, inform your audience that a question-and-answer period will follow. Announce a specific time limit (such as ten minutes), to avoid public debate. Then you can end the session gracefully, without making anyone feel cut off or excluded. Don't be afraid to admit ignorance. If you can't answer a question, say so, and move to the next question. End the session by saying "We have time for one more question," or the like.

CHAPTER SUMMARY

Compared to its written equivalent, an oral report elicits direct response, is more personal, may have a greater effect, and saves time. A written report is easier to refine and organize, can be complex, can be studied and reviewed.

When preparing a formal report, don't try to memorize the report, don't expect to read it to your audience, and don't try to cook it up on stage because you think "it's all in your head." Your typical situation will be extemporaneous, in which the report is by no means written out, but for which you have ample time for preparation and notes to rely on.

Follow these steps in preparing your delivery:

1. Know your subject matter so you can speak with authority.
2. Identify the technical level of your audience, and plan accordingly.
3. Plan your report by writing a statement of purpose, doing homework, making a sentence outline, and working out a delivery time.
4. Include well-chosen visuals.
5. Practice your delivery before friends or using a mirror and tape recorder.
6. Try to anticipate the kinds of questions your audience will ask.

Follow these guidelines in delivering your report:

1. Move and gesture as in a conversation; keep a natural posture.
2. Speak as though you know what you are talking about.
3. Keep tabs on your voice volume, tone, pronunciation, and speed.
4. Maintain eye contact, talking to your listeners, not at them.
5. Read audience feedback for any adjustments you might need to make.
6. Be concise; say what you came to say, and close.
7. Near closing, summarize your main points.
8. Conclude gracefully, leaving time for questions and answers.

REVISION CHECKLIST FOR ORAL REPORTS

Use this list to refine the content, arrangement, and style of your delivery.

Content

1. Is the report content suited to the makeup and needs of my audience?
2. Do I begin with a clear statement of purpose?
3. Does the report itself achieve my stated purpose (deliver what title and purpose statement promise)?
4. Do I know what kinds of questions to expect from my audience?

5. Do I have adequate and appropriate visuals?
6. Do I have enough facts to support my assertions?

Arrangement

1. Are the introduction-body-conclusion sections of my report clearly differentiated and fully developed?
2. Can I follow my outline with only brief glances?
3. Is my report coherent?
4. Do I stick to my purpose without digressing?
5. Do I summarize effectively, before concluding?

Style

1. Is my delivery relaxed and personable?
2. Am I comfortable with body movements and posture?
3. Do I speak with confidence and authority, but without pretention?
4. Do I pronounce all words distinctly?
5. Are the volume, tone, and speed of my delivery effective?
6. Do I maintain good eye contact with my audience/mirror reflection?
7. Is my delivery clear?

Now list those elements of your delivery that need improvement.

EXERCISES

1. In a memo to your instructor, identify and discuss the kinds of oral reporting duties you expect to encounter in your career.
2. In a memo to your instructor, identify your biggest fear about oral reporting. Discuss the reasons for this anxiety, describing it in detail. Finally, propose solutions to your problem.
3. Design an oral report for your technical writing class. (Base it on a written report.) Make a sentence outline, and include at least two visuals. Practice with a tape recorder and mirror or a friend. Use the revision checklist to evaluate your delivery.
4. Observe a lecture or speech, and evaluate it according to the revision checklist. Write a memo to your instructor (without naming the speaker), identifying weak areas and suggesting improvements. Identify strong areas as well.

Appendix A

Review of Grammar, Usage, and Mechanics

No matter how vital and informative a message may be, its credibility is damaged by basic errors. Any of these errors — an illogical, fragmented, or run-on sentence; faulty punctuation; or a poorly chosen word — stands out and mars otherwise good writing. Not only do such errors confuse and annoy the reader, but they speak badly for the writer's attention to detail and precision. Your career will make the same demands for good writing that your English classes do. The difference is that evaluation (grades) in professional situations usually shows in promotions, reputation, and salary.

None of the material here should be new to you. Everyone studies the ground rules of our language from the earliest grades onward. For reasons not yet understood, however, many writers continue to have trouble with "the basics." Although this appendix offers no overnight solutions to long-standing problems, it does provide a simple guide for basic repairs. Spend an occasional few minutes reviewing sections that fit your needs, and you may discover that some writing problems are easier to solve than you had realized.

Table A-1 contains the standard correction symbols along with their interpretation and page references. When your instructor or proofreader marks a symbol on your paper, turn to the appropriate section (here or in Chapters 3 and 4) for explanations and examples that will help you make corrections quickly and easily. This appendix is for your reference; use it when you need it, as you would a dictionary or thesaurus.

TABLE A-1 Correction Symbols

Symbol	Meaning	Page	Symbol	Meaning	Page
ab	abbreviation	573–575	, /	comma	557–563
agr p	pronoun/referent agreement	546–547	– – /	dashes	567–568
agr sv	subject/verb agreement	544–546	··· /	ellipses	566
appr	inappropriate diction	58–59	! /	exclamation point	555–556
bias	biased tone	70–72	– /	hyphen	568
ca	pronoun case	548–549	*ital*	italics	566
cap	capitalization	575–576	( ) /	parentheses	567
chop	choppy sentences	55–56, 542	. /	period	555
cl	clutter word	54	? /	question mark	555
coh	paragraph coherence	33–34	" / "	quotation	565–566
cont	contraction	564–565	; /	semicolon	556
coord	coordination	542–543	*qual*	needless qualifier	55
cs	comma splice	539–541	*red*	redundancy	51–52
dgl	dangling modifier	549–550	*rep*	needless repetition	52
euph	euphemism	62	*ref*	faulty reference	546–547
exact	inexact word	62–63	*ro*	run-on sentence	541
frag	sentence fragment	537–539	*seq*	sequence of development in a paragraph	34–39
gen	generalization	62	*shift*	sentence shift	552–553
jarg	needless jargon	59–60	*sp*	spelling	577
len	paragraph length	39	*st mod*	stacked modifiers	47–48
lev	level of technicality	11–16	*str*	paragraph structure	28
mng	meaning ambiguous	44–47	*sub*	subordination	543–544
mod	misplaced modifier	550–551	*th op*	"th" sentence openers	53
noun ad	noun addiction	54	*trans*	transition	569–572
om	omitted word	44	*trite*	triteness	61
over	overstatement	61–62	*un*	paragraph unity	33
par	parallelism	551–552	*v*	voice	49–51
pot	punctuation	553–569	*var*	sentence variety	56–57
ap /	apostrophe	563–565	*w*	wordiness	50–55
[ ] /	brackets	567	*wo*	word order	48
: /	colon	555	*ww*	wrong word	57–67
			#	numbers	576–577
			¶	begin new paragraph	27–28
			ts	topic sentence	29–30

COMMON SENTENCE ERRORS

Any piece of writing is only as good as each of its sentences. Here are the most common sentence errors, along with suggestions for easy repairs.

Sentence Fragment

As we said earlier, a sentence can be defined as the expression of a logically complete idea. Any complete idea must contain a subject and a verb and must not depend on another complete idea in order to make sense. Your sentence might contain several complete ideas, but it must contain at least one!

Although he was nervous, he grabbed the line, and he saved the sailboat.
(incomplete idea)   *(complete idea)*   *(complete idea)*

However long or short your sentence is, it should make sense to your reader. If the idea is not complete — if your reader is left wondering what you mean — you probably have left out an essential element (the subject, the verb, or another complete idea). Such a piece of a sentence is called a *fragment.*

Grabbed the line. (*a fragment because it lacks a subject*)

Although he was nervous. (*a fragment because — although it has a subject and a verb — it needs to be joined with a complete idea to make sense*)

The only exception to the rule for sentences is when we give a command (Run!) in which the subject (you) is understood. Because this is a logically complete statement, it can properly be called a sentence. So can this one:

Sam is an electronics technician.

Again the idea is logically complete. Although your readers may have some questions about Sam, they cannot fail to understand your meaning: somewhere there is a person; the person's name is Sam; the person is an electronics technician.

Suppose instead we write:

Sam an electronics technician.

This statement is not logically complete, therefore not a sentence. The reader is left asking, "What about Sam the electronics technician?" The verb — the word that makes things happen — is missing. By adding a verb we can easily change this fragment to a complete sentence.

Simple Verb   Sam **is** an electronics technician.

Verb plus Adverb   Sam, an electronics technician, **works hard.**

Dependent Clause,   **Although he is well paid,** Sam, an electronics tech-
Verb, and Subjective   nician, **is not happy.**
Complement

Do not, however, mistake the following statement — which seems to contain
a verb — for a complete sentence:

Sam being an electronics technician.

Such "ing" forms do not function as verbs unless accompanied by such other
verbs as *is, was,* and *will be.* Again, readers are left in a fog unless you complete your idea with an independent clause.

Sam, being an electronics technician, **was responsible for checking the
circuitry.**

Likewise, remember that the "to + verb" form does not function as a verb.

To become an electronics technician.

The meaning is unclear unless you complete the thought.

To become an electronics technician, **Sam had to complete a two-year
apprenticeship.**

Sometimes we can inadvertently create fragments by adding certain words
— *because, since, if, although, while, unless, until, when, where,* and others —
to an already complete sentence. We then change our independent clause
(complete sentence) to a dependent clause.

Although Sam is an electronics technician.

Such words subordinate the words that follow them so that an additional idea
becomes necessary to make the first statement complete. That is, they make
the statement dependent on an additional idea, which must itself contain a
subject and a verb and be a complete sentence. (See "Complex Sentences" and
"Faulty Subordination.") Now we have to round off the statement with a complete idea (an independent clause).

Although Sam is an electronics technician, **he hopes to become an electrical engineer.**

Note: Be careful not to use a semicolon or a period, instead of a comma, to
separate elements in the previous sentence. Because the incomplete idea (dependent clause) depends on the complete idea (independent clause) for its
meaning, you need only a *pause* (symbolized by a comma), not a *break* (symbolized by a semicolon), between these ideas. In fact, many fragments are
created when the writer uses too strong a mark of punctuation (period or

semicolon) between a dependent and an independent clause, thereby severing the needed connection. (See our later discussion of punctuation.)

Here are some fragments from student reports. Each is repaired in several ways. Can you think of any other ways of making these statements complete?

Fragment   She spent her first week on the job as a researcher. **Selecting and compiling technical information from digests and journals.**

Correct   She spent her first week on the job as a researcher, selecting and compiling technical information from digests and journals.

She spent her first week on the job as a researcher. She selected and compiled technical information from digests and journals.

She spent her first week on the job as a researcher by selecting and compiling technical information from digests and encyclopedias.

Fragment   **Because the operator was careless.** The new computer was damaged.

Correct   Because the operator was careless, the new computer was damaged.

The operator's carelessness resulted in damage to the new computer.

The operator was careless; as a result, the new computer was damaged.

Fragment   **When each spool is in place.** Advance your film.

Correct   When each spool is in place, advance your film.

Be sure that each spool is in place before advancing your film.

Comma Splice

In a sentence fragment, an incomplete statement is isolated from items on which it depends for its completion by too strong a punctuation mark: a period or semicolon is mistakenly used in place of a comma. In a comma splice, on the other hand, two complete ideas (independent clauses), which should be separated by a period or a semicolon, are incorrectly joined by a comma, as follows:

Sarah did a great job, she was promoted.

There are several possibilities for correcting this error:

  1. Substitute a period followed by a capital letter:

> Sarah did a great job. She was promoted.

  2. Substitute a semicolon to signal a relationship between the two items:

> Sarah did a great job; she was promoted.

  3. Use a semicolon with a connecting adverb (a transitional word):

> Sarah did a great job; **consequently** she was promoted.

  4. Use a subordinating word to make the less important sentence incomplete, thereby dependent on the other:

> **Because** Sarah did a great job, she was promoted.

  5. Add a connecting word after the comma:

> Sarah did a great job, **and** she was promoted.

Your choice of construction will depend, of course, on the exact meaning or tone you wish to convey. Each of the following comma splices can be repaired in the ways mentioned above.

Comma Splice   This is a fairly new technique, therefore, some people don't trust it.

Correct   This is a fairly new technique. Some people don't trust it.

This is a fairly new technique; therefore, some people don't trust it.

Because this is a fairly new technique, some people don't trust it.

This is a fairly new technique, **so** some people don't trust it.

Comma Splice   Ms. Jones was a strict supervisor, she was well liked by her employees.

Correct   Ms. Jones was a strict supervisor. She was well liked by her employees.

Ms. Jones was a strict supervisor; **however,** she was well liked by her employees.

Although Ms. Jones was a strict supervisor, she was well liked by her employees.

Ms. Jones was a strict supervisor, **but** she was well liked by her employees.

Ms. Jones was a strict supervisor; she was well liked by her employees.

Comma Splice A current is placed on the wires entering and leaving the meter, the magnetic field generated by the current moves the coil.

Correct A current is placed on the wires entering and leaving the meter. The magnetic field generated by the current moves the coil.

A current is placed on the wires entering and leaving the meter; the magnetic field generated by the current moves the coil.

A current is placed on the wires entering and leaving the meter; **consequently** the magnetic field generated by the current moves the coil.

When a current is placed on the wires entering and leaving the meter, the magnetic field generated by the current moves the coil.

A current is placed on the wires entering and leaving the meter, **and** the magnetic field generated by the current moves the coil.

Run-on Sentence

The run-on sentence, a cousin to the comma splice, crams too many ideas together without providing needed breaks or pauses between thoughts.

Run-on The hourglass is more accurate than the waterclock for the water in a water clock must always be of the same temperature in order to flow with the same speed since water evaporates it must be replenished at regular intervals thus not being as effective in measuring time as the hourglass.

Like a runaway train, this statement is out of control. Here is a corrected version:

Revised The hourglass is more accurate than the waterclock because water in a water clock must always be of the same temperature to flow at the same speed. Also, water evaporates and must be replenished at regular intervals. These temperature and volume problems make the waterclock less effective than the hourglass in measuring time.

Choppy Sentences

It is possible to write grammatically correct sentences that, nonetheless, read like Dick-and-Jane sentences in a third-grade reader. Short, choppy sentences cause tedious reading for your audience and bad publicity for you.

> Choppy   Brass-plated prongs are not desirable. They do not always make a good contact in the outlet. They also rust or corrode. Some of the cheaper plugs also have no terminal screws. The conductors in the cord are soldered to the prongs. Sometimes they are just wrapped around them. These types often come as original equipment on small lamps and appliances. They are not worth repairing when a wire comes loose.

Correct this problem by combining related ideas within single sentences and by using transitions to increase coherence.

> Revised   Brass-plated prongs are not desirable because they do not always make a good contact in the outlet and they rust and corrode. Furthermore, some of the cheaper plugs, which often come as original equipment on small lamps and appliances, have no terminal screws: the conductors in the cord are either soldered to the prongs or just wrapped around them. Therefore, these plugs are not worth repairing when a wire comes loose.

Notice that the original eight sentences have been replaced by three. (See also pages 55–57.)

Faulty Coordination

Give ideas of equal importance equal emphasis by joining them with coordinating conjunctions: *and, but, or, nor, for, so,* and *yet.*

> This course is difficult **but** worthwhile.
> My horse is old **and** gray.
> We must decide to support **or** reject the dean's proposal.

Do not coordinate excessively.

> Excessive   The climax in jogging comes after a few miles **and** I
> Coordination   can no longer feel stride after stride **and** it seems as if I am floating **and** jogging becomes almost a reflex **and** my arms **and** legs continue to move **and** my mind no longer has to control their actions.

Revised   The climax in jogging comes after a few miles when I can no longer feel stride after stride. By then I am jogging almost by reflex, nearly floating, my arms and legs still moving, my mind no longer having to control their actions.

Avoid coordinating ideas that can't be sensibly connected:

Faulty   John had a weight problem and he dropped out of school.

Revised   John's weight problem made him so depressed that he couldn't study, so he quit school.

Faulty   I was late for work and wrecked my car.

Revised   Late for work, I backed out of the driveway too quickly, hit a truck, and wrecked my car.

Don't use "try and." Use "try to" instead.

Faulty   I will try and help you.

Revised   I will try to help you.

Faulty   Bill promised to try and be on time.

Revised   Bill promised to try to be on time.

Faulty Subordination

When you subordinate, you make the less important ideas dependent on the most important idea. Through subordination you can combine related short sentences and also emphasize the most important idea. Consider, for instance, these two ideas:

Joe studies hard. He has a learning disability.

Because these ideas are each expressed as a simple sentence, they appear to be coordinate (equal in importance). But if you wanted to express your opinion about Joe's chances of succeeding, you would need a third sentence: "His handicap probably will make success impossible"; or "His will power will help him succeed." An easier and more concise way to inject your opinion is by combining ideas and subordinating the one that deserves less emphasis:

Despite his learning disability (*subordinate idea*), Joe studies hard (*independent idea*).

This first version suggests that Joe will succeed. Below, subordination is used to suggest the opposite:

Despite his diligent studying (*subordinate idea*), Joe has a learning disability (*independent idea*).

Be sure to place the idea you want emphasized in the independent clause; don't write

> Although Alfred is receiving excellent medical treatment, he is seriously ill.

if you mean to suggest that Alfred has a good chance of recovering.

Don't coordinate when you should subordinate:

> Weak — Television viewers can relate to an athlete they idolize and they feel obligated to buy the product endorsed by their hero.

Of the two ideas in the above sentence, one is the cause, the other the effect. Emphasize this relationship through subordination:

> Revised — Because television viewers can relate to an athlete they idolize, they feel obligated to buy the product endorsed by their hero.

When combining several ideas within a sentence, decide which is most important, and make the other ideas subordinate to it — don't simply coordinate:

> Faulty — This employee is often late for work, and he writes illogical reports, and he is a poor manager, and he should be fired.

> Revised — Because this employee is often late for work, writes illogical reports, and has poor management skills, **he should be fired.** (*The last clause is independent.*)

Don't overstuff sentences by subordinating excessively:

> Overstuffed — This job, which I took when I graduated from college, while I waited for a better one to come along, which is boring, where I've gained no useful experience, makes me anxious to quit.

> Revised — Upon college graduation, I took this job while waiting for a better one to come along. Because I find it boring, and have gained no useful experience, I am anxious to quit.

Faulty Agreement — Subject and Verb

Failure to make the subject of a sentence agree in number with the verb is a common error. Happily, it's an error easily corrected and avoided. We are not likely to use faulty agreement in short sentences, where subject and verb are not far apart. Thus we are not likely to say "Jack eat too much" instead of "Jack eats too much," but in more complicated sentences — in which the subject is separated from its verb by other words — we sometimes lose track of the subject-verb relationship.

Faulty   The lion's **share** of diesels **are** sold in Europe.

Although "diesels" is closest to the verb, the subject is "share," a singular subject that must agree with a singular verb.

Correct   The lion's **share** of diesels **is** sold in Europe.

Agreement errors are easy to correct when the subject and verb are identified.

Faulty   There **is** an estimated 29,000 **women** living in our county.

Correct   There **are** an estimated 29,000 **women** living in our county.

Faulty   A **system** of lines **extend** horizontally to form a grid.

Correct   A **system** of lines **extends** horizontally to form a grid.

A second situation that causes trouble in subject-verb agreement occurs when we use indefinite pronouns such as *each, everyone, anybody,* and *somebody.* They function as subjects and usually take a singular verb.

Faulty   **Each** of the crew members **were** injured.

Correct   **Each** of the crew members **was** injured.

Faulty   **Everyone** in the group **have** practiced long hours.

Correct   **Everyone** in the group **has** practiced long hours.

Sometimes agreement problems can be caused by collective nouns such as *herd, family, union, group, army, team, committee,* and *board.* They can call for a singular or plural verb — depending on your intended meaning. When denoting the group as a whole, use a singular verb.

Correct   The **committee meets** weekly to discuss new business.

The editorial **board** of this magazine **has** high standards.

To denote individual members of the group, however, use a plural verb.

Correct   The **committee have** voted unanimously to hire Jim.

The editorial **board are** all published authors.

Yet another problem occurs when two subjects are joined by *either . . . or* or *neither . . . nor.* Here, the verb is singular if both subjects are singular, and plural if both subjects are plural. If one subject is plural and one is singular, the verb agrees with the one that is closer.

Correct   Neither **John** nor **Bill works** regularly.

Either **apples** or **oranges are** good vitamin sources.

Either Felix or his **friends are** crazy.

Neither the boys nor their **father likes** the home team.

If, on the other hand, two subjects (singular, plural, or mixed) are joined by *both . . . and,* the verb will be plural. Whereas *or* suggests "one or the other," *and* suggests both subjects, thereby requiring a plural verb.

Correct   **Both** Joe and Bill **are** resigning.
The **book and** the **briefcase appear** expensive.

Faulty Agreement — Pronoun and Referent

A pronoun can make sense only if it refers to a specific noun (its referent or antecedent), with which it must agree in gender and number. It is easy enough to make most pronouns agree with their respective referents.

Correct   **Joe** lost **his** blueprints.
The **workers** complained that **they** were treated unfairly.

Some cases, however, are not so obvious. When, for example, an indefinite pronoun like *each, everyone, anybody, someone,* and *none* serves as the pronoun referent, the pronoun itself is singular.

Correct   **Anyone** can get **his** degree from that college.
Anyone can get **his or her** degree from that college.
Each candidate described **her** plans in detail.

Faulty Pronoun Reference

Whenever a pronoun is used, it must refer to one clearly identified referent; otherwise, your message will be vague and confusing.

Ambiguous   **Sally** told **Sarah** that **she** was obsessed with her job.

Does "she" refer to Sally or Sarah? Several interpretations are possible:

1. Sally is obsessed with her job.
2. Sally thinks Sarah is obsessed with her (Sally's) job. (Sarah is envious.)
3. Sally thinks that Sarah is obsessed with her own job.
4. Sally is obsessed with Sarah's job.
5. Sally thinks that someone else is obsessed with her (Sally's) job.

6. Sally thinks someone else is obsessed with Sarah's job.
7. Sally thinks someone else is obsessed with some other person's job.
8. Sally thinks someone else is obsessed with her own job.

Correct   Sally told Sarah, "I'm obsessed with my job."
Sally told Sarah, "I'm obsessed with your job."
Sally told Sarah, "You're obsessed with [your, my] job."
Sally told Sarah, "She's obsessed with [her, my, your] job."

Avoid using *this, that,* or *it* — especially to begin a sentence — unless the pronoun refers to a specific antecedent (referent).

Vague   As he drove away from his menial **job,** boring **lifestyle,** and damp **apartment,** he was happy to be leaving **it** behind.

Correct   As he drove away, he was happy to be leaving his menial job, boring lifestyle, and damp apartment behind.

Vague   The problem with our **defective machinery** is only compounded by the new **operator's incompetence. That** makes me angry!

Correct   I am angered by the problem with our defective machinery as well as by the new operator's incompetence.

Vague   Water boils at 212°F and freezes at 32°F, which makes it usable as a coolant in most parts of the country. Antifreeze is required to prevent **this.** Some manufacturers recommend **this** on cars with air conditioning because of the possibility of the heater core freezing.

Notice that the first "this" has no specific referent, and the second seems to refer to "antifreeze" but is placed too far from its referent. The meaning of the message is obscured by such vague construction.

Inaccurate   Increased blood pressure is caused by the narrowing of the blood vessels, making the pressure higher as **it** attempts to flow through the blood vessels.

Here, "it" seems to refer to "pressure," which is absurd.

Correct   Increased blood pressure is caused by the narrowing of the blood vessels, making the pressure higher as the blood attempts to flow through the vessels.

Faulty Pronoun Case

The case of a pronoun — nominative, objective, or possessive — is determined by its role in the sentence: as subject, object, or indicator of possession.

If the pronoun serves as the subject of a sentence (*I, we, you, she, he, it, they, who*), its case is *nominative*.

> **She** completed her graduate program in record time.
> **Who** broke the chair?

When a pronoun follows a version of the verb *to be* (a linking verb), it further explains (complements) the subject, and thus its case is nominative.

> It was **she**.
> The chemist who perfected our new distillation process is **he**.

If the pronoun serves as the object of a verb or a preposition (*me, us, you, her, him, it, them, whom*), its case is *objective*.

Object of the Verb	The employees gave **her** a parting gift.
Object of the Preposition	Several colleagues left with **him**.
	To **whom** do you wish to complain?

If a pronoun indicates possession (*my, mine, our, ours, your, yours, his, her, hers, its, their, theirs, whose*), its case is *possessive*.

> The brown briefcase is **mine**.
> **Her** offer was accepted.
> **Whose** opinion do you value most?

Here are some of the most frequent errors made in pronoun case:

Faulty	**Whom** is responsible to **who**? (*The subject should be nominative and the object should be objective.*)
Correct	**Who** is responsible to **whom**?
Faulty	The debate was between Marsha and **I**. (*As object of the preposition, the pronoun should be objective.*)
Correct	The debate was between Marsha and **me**.
Faulty	**Us** board members are accountable for our decisions. (*The pronoun accompanies the subject, "board members," and thus should be nominative.*)
Correct	**We** board members are accountable for our decisions.

Faulty   A group of **we** managers will fly to the convention.
(The pronoun accompanies the object of the preposition, "managers," and thus should be objective.)

Correct   A group of **us** managers will fly to the convention.

Hint: By deleting the accompanying noun from the two latter examples, we can identify easily the correct pronoun case ("We . . . are accountable . . ."; "A group of us . . . will fly . . .").

Faulty Modification

The word order (syntax) of a sentence determines its effectiveness and meaning. Certain words or groups of words are modified (i.e., explained or defined) by other words or groups of words. Prepositional phrases, for example, usually define or limit adjacent words:

> the foundation **with the cracked wall**
> the repair job **on the old Ford**
> the journey **to the moon**
> the party **for our manager**

As do phrases with "ing" verb forms:

> the student **painting the portrait**
> **Opening the door,** we entered quietly.

Or phrases with "to + verb" form:

> **To succeed,** one must work hard.

Or certain clauses:

> the man **who came to dinner**
> the job **that I recently accepted**

When using modifying phrases to begin sentences, we can get into trouble unless we read our sentences carefully.

Dangling   **Answering the telephone,** the cat ran out the open
Modifier   door.

Here, the introductory phrase signals the reader that the noun beginning the main clause (its subject) is what or who is answering the telephone. The absurd message occurs because the opening phrase has no proper subject to modify; it *dangles*.

Correct   As Mary answered the telephone, the cat ran out the open door.

A dangling modifier can also obscure the meaning of your message.

Dangling Modifier	**After completing the student financial aid application form,** the Financial Aid Office will forward it to the appropriate state agency.

Who completes the form — the student or the financial aid office? Here are some other dangling modifiers that make the message confusing, inaccurate, or downright absurd:

Dangling Modifier	**While walking,** a cold chill ran through my body.
Correct	While **I** walked, a cold chill ran through my body.
Dangling Modifier	**After a night of worry,** the lights came on.
Correct	**After we had worried all night,** the lights came on.
Dangling Modifier	Impurities have entered our bodies **by eating chemically processed foods.**
Correct	Impurities have entered our bodies by **our** eating chemically processed foods.
Dangling Modifier	**By planting different varieties of crops,** the pests were unable to adapt.
Correct	By planting different varieties of crops, **farmers** prevented the pests from adapting.

The word order of adjectives and adverbs in a sentence is as important as the order of modifying phrases and clauses. Notice how changing word order affects the meaning of these sentences:

I **often** remind myself of the need to balance my checkbook.
I remind myself of the need to balance my checkbook **often.**

Be sure that modifiers and the words they modify follow a word order that reflects your meaning.

Misplaced Modifier	Harry typed another memo on our new electric typewriter **that was useless.** (*Was the typewriter or the memo useless?*)
Correct	Harry typed another useless memo on our new electric typewriter.
	or
	Harry typed another memo on our new, useless electric typewriter.

Misplaced   He read a report on the use of nonchemical pesti-
Modifier   cides **in our conference room.** (*Are the pesticides to be used in the conference room?*)

Correct   In our conference room, he read a report on the use of nonchemical pesticides.

Misplaced   She volunteered **immediately** to deliver the radioac-
Modifier   tive shipment. (*Volunteering immediately, or delivering immediately?*)

Correct   She immediately volunteered to deliver. . . .
or
She volunteered to deliver immediately. . . .

Faulty Parallelism

Express items of the same importance in the same grammatical form:

Correct   We here highly resolve . . . that government **of the people, by the people, for the people** shall not perish from the earth.

The above statement describes the government by using three modifiers of equal importance. Because the first modifier is a prepositional phrase, the others must be also. Otherwise, the message would be garbled, like this:

Faulty   We here highly resolve . . . that government **of the people, which the people created and maintain, serving the people** shall not perish from the earth.

If you begin the series with a noun, use nouns throughout the series; likewise for adjectives, adverbs, and specific types of clauses and phrases.

Faulty   The new apprentice is **enthusiastic, skilled,** and **you can depend on her.**

Correct   The new apprentice is **enthusiastic, skilled,** and **dependable.** (*all subjective complements*)

Faulty   In his new job, he felt **lonely** and **without a friend.**

Correct   In his new job, he felt **lonely** and **friendless.** (*both adjectives*)

Faulty   She plans **to study** all this month and **on scoring well** in her licensing examination.

Correct   She plans **to study** all this month and **to score** well in her licensing examination. (*both infinitive phrases*)

Faulty   She **sleeps** well and **jogs** daily, **as well as eating** high-protein foods.

Correct   She **sleeps** well, **jogs** daily, and **eats** high-protein foods. (*all verbs*)

To improve coherence in long sentences, repeat words that introduce parallel expressions:

Faulty   Before buying this property, you should decide whether you plan to settle down and raise a family, travel for a few years, or pursue a graduate degree.

Correct   Before buying this property, you should decide whether you plan **to settle** down and raise a family, **to travel** for a few years, **or to pursue** a graduate degree.

Make all headings parallel in your table of contents, your formal outline, and your report.

Faulty   A. Picking the Fruit
              1. When to pick
              2. Packing
              3. Suitable temperature
              4. Transport with care

The logical connection between steps in that sequence is obscured because each heading is phrased in a different grammatical form.

Correct   A. Picking the Fruit
              1. Choose the best time
              2. Pack the fruit loosely
              3. Store at a suitable temperature
              4. Transport with care

Other forms of phrasing would also be correct here, as long as each item is expressed in a form parallel to all other items in the series.

Sentence Shifts

Shifts in point of view damage coherence. If you begin a sentence or paragraph with one subject or person, don't shift courses.

Shift in Person   When **you** finish such a great book, **one** will have a sense of achievement.

Correct   When **you** finish such a great book, **you** will have a sense of achievement.

Shift in Number	**One** should sift the flour before **they** make the pie.
Correct	**One** should sift the flour before **one** makes the pie. (Or better: Sift the flour before making the pie.)

Don't begin a sentence in the active voice and then shift to the passive voice.

Shift in Voice	**He delivered** the plans for the apartment complex, and the building site **was also inspected by him.**
Correct	**He delivered** the plans for the apartment complex and also **inspected** the building site.

Don't shift tenses without good reason.

Shift in Tense	She **delivered** the blueprints, **inspected** the foundation, **wrote** her report, and **takes** the afternoon off.
Correct	She **delivered** the blueprints, **inspected** the foundation, **wrote** her report, and **took** the afternoon off.

Don't shift from one mood to another (e.g., from imperative to indicative mood in a set of instructions).

Shift in Mood	**Unscrew** the valve and then steel wool **should be used** to clean the fitting.
Correct	**Unscrew** the valve and then **use** steel wool to clean the fitting.

Don't shift from indirect to direct discourse within the same sentence.

Shift in Discourse	Jim wonders **if he will get the job** and **will he like it?**
Correct	Jim wonders **if he will get the job** and **if he will like it.**
	Will Jim get the job, and will he like it?

EFFECTIVE PUNCTUATION

Punctuation marks are like road signs and traffic signals. They govern reading speed and provide clues for navigation through a network of ideas; they mark intersections, detours, and road repairs; they draw attention to points of interest along the route; and they mark geographic boundaries. In short, punctuation marks provide us with a practical and simple way of making ourselves understood. They take up the slack created when the spoken message (made clear by the speaker's tone, pitch, volume, speaking rate, pauses, body movements, and facial expressions) is transposed into a written message (silent, static words on a page). In fact, effective punctuation can often make the written message clearer than its spoken equivalent.

As an experiment, copy a paragraph — without the punctuation — from any book, and try to read it clearly.

Before we discuss individual punctuation marks in detail, a common sense review of the relationship among the four used most often (period, semicolon, colon, and comma) might help. These marks can be ranked in order of their relative strengths.

1. *Period.* The strongest mark. A period signals a complete stop at the end of an independent idea (independent clause). The first word in the idea following the period begins with a capital letter.

> Jack is a fat cat. His friends urge him to diet.

2. *Semicolon.* Weaker than a period but stronger than a comma. A semicolon signals a brief stop after an independent idea but does not end the sentence; instead, it provides advance notice that the independent idea that follows is *closely related* to the previous idea.

> Jack is a fat cat; he eats too much.

3. *Colon.* Weaker than a period but stronger than a comma. A colon usually follows an independent idea and, like the semicolon, signals a brief stop but does not end the sentence. The colon and semicolon, however, are never interchangeable. A colon provides an important cue: it symbolizes "explanation to follow." Information after the colon (which need not be an independent idea) explains or clarifies the idea expressed before the colon.

> Jack is a fat cat: he weighs forty pounds. (*The information after the colon answers "How fat?"*)

or

> Jack is a fat cat: forty pounds worth! (*The second clause is not independent.*)

Note: As long as any two adjacent ideas are independent they may correctly be separated by a period. Sometimes, a colon or a semicolon may be more appropriate for illustrating the logical relationship between two given ideas. When in doubt, however, use a period.

4. *Comma.* The weakest of these marks. A comma does not signal a stop at the end of an independent idea but only a pause within or between ideas in the sentence. A comma often indicates that the word, phrase, or clause set off from the independent idea cannot stand alone but must rely on the independent idea for its meaning.

> Jack, a fat cat, is jolly. (*In that sentence, the phrase within commas depends on the independent idea for its meaning.*)

> **Although he diets often,** Jack is a fat cat. (*Because the first clause depends on the second, any stronger mark would create a fragment.*)

A comma is rarely appropriate between two independent clauses unless there is a conjunction.

> Comma Splice   Jack is a fat cat, he eats too much.

So we see that punctuation marks, like words, convey specific meanings to the reader. These meanings are further discussed in the sections that follow.

End Punctuation

The three marks of end punctuation — period, question mark, and exclamation point — work like a red traffic light by signaling a complete stop.

Period

A period ends a sentence. Periods end some abbreviations.

> Ms.   Assn.   N.Y.
> M.D.   Inc.   B.A.

Periods serve as decimal points for figures.

> $15.95
> 21.4%

Question Mark

A question mark ends a sentence asking a direct question.

> Where is the balance sheet?

Do not use a question mark to end an indirect question.

> Faulty   He asked if all students had failed the test?
>
> Correct   He asked if all students had failed the test.
> *or*
> He asked, "Did all students fail the test?"

Exclamation Point

Because exclamation points mean that you are excited or adamant, don't overuse them. Otherwise you might seem hysterical or insincere.

> Correct   Oh, no!
> Pay up!
> My pants are missing!

Use an exclamation point only when the expression of strong feeling is appropriate.

Semicolon

A semicolon usually works like a blinking red traffic light at a deserted intersection by signaling a brief but definite stop.

Semicolon Separating Independent Clauses

Most commonly, semicolons separate independent clauses (logically complete ideas) whose contents are closely related.

> The project was finally completed; we had done a good week's work.

The semicolon can replace the conjunction-comma combination that joins two independent ideas.

> The project was finally completed, and we were elated.
> The project was finally completed; we were elated.

The second version emphasizes the sense of elation.

Semicolons Used with Adverbs as Conjunctions and Other Transitional Expressions

Semicolons must accompany adverbs and other expressions that connect related independent ideas (*besides, otherwise, still, however, furthermore, moreover, consequently, therefore, on the other hand, in contrast, in fact,* etc.).

> The job is filled; however, we will keep your résumé on file.

> Your background is impressive; in fact, it is the best among our applicants.

Semicolons Separating Items in a Series

When items in a series contain internal commas, semicolons provide clear separations between items.

> We are opening branch offices in the following cities: Santa Fe, New Mexico; Albany, New York; Montgomery, Alabama; and Moscow, Idaho.

> Members of the survey crew were John Jones, a geologist; Hector Lightweight, a draftsman; and Mary Shelley, a graduate student.

Colon

A colon works like a flare in the middle of the road. It signals you to stop and then proceed paying close attention to the situation ahead, the details of which will be revealed as you move ahead. Usually, a colon follows an introductory statement that requires a follow-up explanation.

> We need the following equipment immediately: a voltmeter, a portable generator, and three pairs of insulated gloves.

> She is an ideal colleague: honest, reliable, and competent.

> Two candidates are clearly superior: John and Marsha.

In most cases — with the exception of *Dear Sir:* and other salutations in formal correspondence — colons follow independent statements (logically and grammatically complete). Because colons, like end punctuation and semicolons, signal a full stop, they are never used to fragment a complete statement.

> Faulty   My plans include: finishing college, traveling for two years, and settling down in Boston.

No punctuation should follow "include."
Colons can introduce quotations.

> The supervisor's message was clear enough: "You're fired."

As shown on page 554, a colon normally replaces a semicolon in separating two related, complete statements, when the second statement directly explains or amplifies the first.

> His reason for accepting the lowest-paying job offer was simple: he had always wanted to live in the Northwest.

The statement following the colon explains the "reason" mentioned in the statement preceding the colon.

Comma

The comma is the most frequently used — and abused — punctuation mark. Unlike the period, semicolon, and colon, which signal a full stop, the comma signals a *brief pause*. Thus, the comma works like a blinking yellow traffic light for which you slow down without stopping. As we said earlier, a comma should never be used to signal a *break* between independent ideas; it is not strong enough.

Comma as a Pause Between Complete Ideas

In a compound sentence where a coordinating conjunction (*and, or, nor, for, but*) connects equal (independent) statements, a comma is usually placed immediately before the conjunction.

> This is a high-paying job, but the stress is high.

> This vacant shop is just large enough for our boutique, and the location is excellent for walk-in customer traffic.

Without the conjunction, each of the above statements would suffer from a comma splice, unless the comma were changed to a semicolon or a period.

Comma as a Pause Between an Incomplete and a Complete Idea

A comma is usually placed between a complete and an incomplete statement in a complex sentence to show that the incomplete statement depends for its meaning on the complete statement (i.e., the incomplete statement cannot stand alone, separated by a break symbol such as a semicolon, colon, or period).

> **Because he is a fat cat,** Jack diets often.
> **When he eats too much,** Jack gains weight.

Above, the first idea is made incomplete by a subordinating conjunction (*since, when, because, although, where, while, if, until,* etc.), which here connects a dependent with an independent statement. The first (incomplete) idea depends on the second (complete) for wholeness. When the order is reversed (complete idea followed by incomplete), the comma can usually be omitted.

> Jack diets often **because he is a fat cat.**
> Jack gains weight **when he eats too much.**

Because commas take the place of speech signals, reading a sentence aloud should tell you whether or not to pause (and use a comma).

Commas Separating Items (Words, Phrases, or Clauses) in a Series

Use a comma to separate items in a series.

> **Sam, Joe, Marsha,** and **John** are joining us on the hydroelectric project.

> The office was **yellow, orange,** and **red.**

> He works hard at **home, on the job,** and even **during his vacation.**

> The new employee complained **that the hours were long, that the pay was low,** that the work was boring, and **that the foreman was paranoid.**

> **She came, she saw,** and **she conquered.**

Do not use commas when *or* or *and* is used between all items in the series.

> She is willing to work in San Francisco or Seattle or even in Anchorage.

Add a comma when *or* or *and* is used only before the final item in the series.

> Our luncheon special for Thursday will be coffee, rolls, steak, beans, and ice cream.

Without the comma, that sentence might cause the reader to conclude that beans and ice cream are an exotic new dessert.

Comma Setting off Introductory Phrases

Infinitive, prepositional, or verbal phrases introducing a sentence usually are set off by commas.

Infinitive Phrase	**To be or not to be,** that is the question.
Prepositional Phrase	**In Rome,** do as the Romans do. **In other words,** you're fired. **In fact,** the finish work was superb.
Participial Phrase	**Being fat,** Jack was a slow runner. **Moving quickly,** the army surrounded the enemy.

When an interjection introduces a sentence, it is set off by a comma.

> **Oh,** is that the final verdict?

When a direct address introduces a sentence, it is set off by a comma.

> **Mary,** you've done a great job.

Comma Used to Avoid Ambiguity

By signaling a pause that you would make in speaking, a comma can increase the clarity of your statement.

Ambiguous	Outside the office building was colorful. (*Was the exterior of the building or the area surrounding the building colorful?*)
Clear	Outside, the office building was colorful.
Ambiguous	For Bill Smith's advice was a lifesaver. (*Was Bill Smith's advice a lifesaver, or was Smith's advice to Bill a lifesaver?*)
Clear	For Bill, Smith's advice was a lifesaver.

Read your sentences aloud to reveal any such ambiguities in your own writing.

Commas Setting off Nonrestrictive Elements

A restrictive phrase or clause describes, defines, or limits its subject in such a way that it could not be deleted without changing the meaning of the sentence.

All candidates **who have work experience** will receive preference.

The clause, "who have work experience," defines "candidates" and is essential to the meaning of the sentence. Without this clause, the meaning would be entirely different.

All candidates will receive preference.

The following sentence also contains a restriction.

All candidates **with work experience** will receive preference.

The phrase, "with work experience," defines "candidates" and thus specifies the meaning of the sentence. Because these elements *restrict* the subject by limiting the category, "candidates," each forms an integral part of the sentence and is thus not separated from the rest of the sentence by commas.

In contrast, a nonrestrictive phrase or clause does not limit or define the subject; such an element is optional because it could be deleted without changing the essential meaning of the sentence.

Our draftsperson, **who has only six weeks' experience,** is highly competent.

This house, **riddled with carpenter ants,** is falling apart.

In each of those sentences, the modifying phrase or clause does not restrict the subject; each could be deleted as follows:

Our new draftsperson is highly competent.
This house is falling apart.

Unlike a restrictive modifier, the nonrestrictive modifier does not supply the essential meaning to the sentence; therefore, it is set off by commas from the rest of the sentence.

To appreciate how the use of simple commas can affect meaning, consider the following statements:

Restrictive   Office workers **who drink martinis with lunch** have slow afternoons.

Because the restrictive clause limits the subject, "office workers," we interpret that statement as follows: some office workers drink martinis with lunch, and these have slow afternoons. In contrast, we could write

Nonrestrictive   Office workers, **who drink martinis with lunch,** have slow afternoons.

Here the subject, "office workers," is not limited or defined. Thus we interpret that all office workers drink martinis with lunch and have slow afternoons.

Commas Setting off Parenthetical Elements

Items that interrupt the flow of a sentence are called parenthetical and are enclosed by commas. Expressions such as *of course, as a result, as I recall,* and *however* are parenthetical and may denote emphasis, afterthought, clarification, or transition.

Emphasis	This deluxe model, **of course,** is more expensive.
Afterthought	Your report format, **by the way,** was impeccable.
Clarification	The loss of my job was, **in a way,** a blessing.
Transition	Our warranty, **however,** does not cover tire damage.

So is a direct address.

Listen, **my children,** and you shall hear . . .

A parenthetical expression at the beginning or the end of a sentence is set off with a comma.

Naturally, we will expect a full guarantee.
My friends, I think we have a problem.
You've done a good job, **Jim.**
Yes, you may use my name in your advertisement.

Commas Setting off Quoted Material

Quoted items included within a sentence are often set off by commas.

The customer said, **"I'll take it,"** as soon as he laid eyes on our new model.

Commas Setting off Appositives

An appositive, a word or words explaining a noun and placed immediately after it, is set off by commas.

Martha Jones, **our new president,** is overhauling all personnel policies.

The new Mercedes, **my dream car,** is priced far beyond my budget.

Alpha waves, **the most prominent of the brain waves,** are typically recorded in a waking subject whose eyes are closed.

Please make all checks payable to Sam Sawbuck, **company treasurer.**

Sarah, **my colleague,** has arrived.

Notice that the commas clarify the meaning of the last sentence. Without commas, the sentence would be ambiguous.

> Sarah my colleague has arrived.

Are you telling Sarah that your colleague has arrived?

> **Sarah,** my colleague has arrived.

Or are you saying that your colleague, Sarah, has arrived (as in the first version)?

Commas Used in Common Practice

Commas are used to set off the day of the month from the year, in a date.

> May 10, 1984

They are used to set off numbers in three-digit intervals.

> 11,215
> 6,463,657

They are also used to set off street, city, and state in an address.

> The bill was sent to John Smith, 184 Sea Street, Albany, New York 01642.

When the address is written vertically, however, the commas that are omitted are those that would otherwise occur at the end of each address line.

> John Smith
> 184 Sea Street
> Albany, New York 01642

Use commas to set off an address or date in a sentence.

> Room 3C, Margate Complex, is the site of our newest office.
> December 15, 1977, is my retirement date.

Use them to set off degrees and titles from proper names.

> Roger P. Cayer, M.D.
> Gordon Browne, Jr.
> Marsha Mello, Ph.D.

Commas Used Erroneously

Don't be a comma philanthropist. Avoid sprinkling commas where they are not needed or simply do not belong. In fact, you are probably safer using too few

commas than too many. Again, overuse of commas generally can be avoided
if you read your sentences aloud.

Faulty   As I opened the door, he told me, that I was late.
(separates the indirect from the direct object)

The most universal symptom of the suicide impulse,
is depression. *(separates the subject from its verb)*

This has been a long, difficult, project. *(separates the
final adjective from its noun)*

John, Bill, and Sally, are joining us on the design
phase of this project. *(separates the final subject
from its verb)*

An employee, who expects rapid promotion, must
quickly prove his worth. *(separates a modifier that
should be restrictive)*

I spoke in a conference call with John, and Marsha.
*(separates two words linked by a coordinating con-
junction)*

The room was, eighteen feet long. *(separates the
linking verb from the subjective complement)*

We painted the room, red. *(separates the object from
its complement)*

Apostrophe

Apostrophes are used for three purposes: to indicate the possessive, to indicate
a contraction, and to indicate the plural of numbers, letters, and figures.

Apostrophe Indicating the Possessive

At the end of a singular word, or of a plural word that does not end in *s*, add
an apostrophe plus an *s* to indicate the possessive.

The people's candidate won.
The chainsaw was Bill's.
The men's locker room burned.
The car's paint job was ruined by the hail storm.
I borrowed Doris's book.
Have you heard Ray Charles's song?

In some cases convention will require that you not add an *s*.

Correct   Moses' death
for conscience' sake

Do not use an apostrophe to indicate the possessive form of either singular or plural pronouns:

Correct   The book was hers.
           Ours is the best sales record.
           The fault was theirs.

At the end of a plural word that ends in *s*, add an apostrophe only.

Correct   the cows' water supply
           the Jacksons' wine cellar

At the end of a compound noun, add an apostrophe plus an *s*.

Correct   my father-in-law's false teeth

At the end of the last word in nouns of joint possession, add an apostrophe plus *s* if both own one item.

Correct   Joe and Sam's lakefront cottage

and an apostrophe plus *s* to both nouns if each owns specific items.

Correct   Joe's and Sam's passports

Apostrophe Indicating a Contraction

An apostrophe shows that you have omitted one or more letters in a phrase that is usually a combination of a pronoun and a verb.

Correct   I'm   they're
           he's   you'd
           we'll   who's
           you're   who'll

Don't confuse *they're* with *their* or *there*.

Faulty   there books
           their now leaving
           living their

Correct   their books
           they're now leaving
           living there

Remember the distinction this way:

Correct   Their boss knows they're there.

Don't confuse *it's* and *its*. *It's* means "it is." *Its* is the possessive.

Correct   It's watching its reflection in the pond.

Don't confuse *who's* and *whose*. *Who's* means "who is," whereas *whose* indicates the possessive.

Correct   Who's interrupting whose work?

Other contractions are formed from the verb and the negative.

Correct   isn't   can't
           don't   haven't
           won't   wasn't

Apostrophe Indicating the Plural of Numbers, Letters, and Figures

There are no buts about it: The 6's on this new typewriter look like smudged G's, 9's are illegible, and the %'s are unclear.

Quotation Marks

Quotation marks set off the exact words borrowed from another speaker or writer. At the end of a quotation the period or comma is placed within the quotation marks.

Correct   "Hurry up," he whispered.
           She told me, "I'm depressed."

The colon or semicolon is always placed outside the quotation marks:

Correct   Our contract clearly defines "middle-management personnel"; however, it does not clearly state the promotional procedures for this group.

           You know what to expect when Honest John offers you a "bargain": a piece of junk.

Sometimes a question mark is used within a quotation that is part of a larger sentence.

Correct   "Can we stop the flooding?" inquired the foreman.

When the question mark or exclamation point is part of the quote, it is placed within the quotation marks, thereby replacing the comma or period.

Correct   "Help!" he screamed.
           He asked John, "Can't we agree about anything?"

If, however, the question mark or exclamation point is meant to denote the attitude of the quoter instead of the quotee, it is placed outside.

Correct   Why did he wink and tell me, "It's a big secret"?
           He actually accused me of being an "elitist"!

When quoting a passage of fifty words or longer, indent the entire passage five spaces and single space between the lines of the passage to set it off from the text. Do not enclose the passage in quotation marks.

Use quotation marks around titles of articles, paintings, book chapters, and poems.

> Correct   The enclosed article, "The Job Market for College Graduates," should provide some helpful insights.

The title of a bound volume — book, journal, or newspaper — should be underlined to represent italics.

Finally, use quotation marks to indicate your ironic use of a word.

> Correct   He is some "friend"!

Ellipses

Use three dots in a row (...) to indicate that you have left some material out of a quotation. If the omitted words come at the end of the original sentence, a fourth dot indicates the period. Use several dots centered in a line to indicate that a paragraph or more has been left out. Ellipses help you save time and zero in on the important material within a quote.

> Correct   "Three dots ... indicate that you have left some material out. ... A fourth dot indicates the period. Several dots centered in a line ... indicate ... a paragraph or more. ... Ellipses help you ... zero in on the important material. ..."

Italics

In typing or longhand writing, indicate italics by <u>underlining</u>. Use italics for titles of books, periodicals, films, newspapers, and plays; for the names of ships; for foreign words or scientific names; for emphasizing a word (used sparingly); for indicating the special use of a word.

> *The Oxford English Dictionary* is a handy reference tool.
>
> The *Lusitania* sank rapidly.
>
> He reads *The Boston Globe* often.
>
> My only advice is *caveat emptor*.
>
> *Bacillus anthracis* is a highly virulent organism.
>
> *Do not* inhale these spores, under any circumstances!
>
> Our contract defines a *full-time employee* as one who works a minimum of thirty-five hours weekly.

Parentheses

Use commas normally to set off parenthetical elements, dashes to give some emphasis to the material that is set off, and parentheses to enclose material that defines or explains the preceding statements.

> An anaerobic (airless) environment must be maintained for the cultivation of this organism.

> The cost of manufacturing our Beta II transistors has increased by 10 percent in one year (see Appendix A for full cost breakdown).

> This new three-colored model (made by Ilco Corporation) is selling well.

Notice that material between parentheses, like all other parenthetical material discussed earlier, can be deleted without harming the logical and grammatical structure of the sentence.

Also, use parentheses to enclose numbers or letters that segment items of information in a series.

> There are three basic steps to this process: (1) . . . , (2) . . . , and (3). . . .

Brackets

Use brackets within a quotation to add material that was not in the original quote but that is needed for clarification. Sometimes a bracketed word will provide an antecedent (or referent) for a pronoun.

> "She [Jones] was the outstanding candidate for the job."

Brackets can enclose information taken from some other location within the context of the quotation.

> "It was in early spring [April 2, to be exact] that the tornado hit."

Use brackets to correct a quotation.

> "His report was [full] of mistakes."

Use *sic* ("thus so") when quoting a mistake in spelling, usage, or logic.

> His secretary's comment was clear: "He don't [sic] want any of these."

Dashes

Dashes are effective — as long as they are not overused. Make dashes on your typewriter by placing two hyphens side by side. While parentheses tend to deemphasize the enclosed material, dashes emphasize it.

Used selectively, dashes can provide dramatic emphasis for a statement, but they should not be used flagrantly as a substitute for all other forms of punctuation. In other words, when in doubt, do not use a dash!

Dashes can be used to denote an afterthought

> Have a good vacation — but don't get sunstroke.

or to enclose an interruption in the middle of a sentence.

> The designer of this building — I think it was Wright — was, above all, an artist.

> Our new team — Jones, Smith, and Brown — is already compiling outstanding statistics.

Although they can often be used interchangeably with commas, dashes dramatize a parenthetical statement more than commas do.

> Mary, a true friend, spent hours helping me rehearse for my interview.
> Mary — a true friend — spent hours helping me rehearse for my interview.

Notice the added emphasis in the second version.

Hyphen

Use a hyphen to divide a word at your right-hand margin. Consult your dictionary for the correct syllable breakdown:

> com-puter
> comput-er

Actually, it is best to avoid altogether this practice of word division at the ends of lines in a typewritten text.

Use a hyphen to join compound modifiers (two or more words preceding the noun as a single adjective).

> the rough-hewn wood
> the well-written report
> the all-too-human error

Do not hyphenate these same words if they *follow* the noun.

> The wood was rough hewn.
> The report is well written.
> The error was all too human.

Hyphenate an adverb-participle compound preceding a noun.

> the high-flying glider

Do not hyphenate compound modifiers if the adverb ends in *ly*.

> the finely tuned engine

Hyphenate all words that begin with the prefix *self*.

> self-reliance
> self-discipline
> self-actualizing

Hyphenate to avoid ambiguity.

> re-creation (*a new creation*)
> recreation (*leisure activity*)

Hyphenate words that begin with *ex* only if *ex* means "past."

> ex-foreman
> expectant

Hyphenate all fractions, along with ratios that are used as adjectives and precede the noun.

> a two-thirds majority
> In a four-to-one vote they defeated the proposal.

Do not hyphenate ratios if they do not immediately precede the noun.

> The proposal was voted down four to one.

Hyphenate compound numbers from twenty-one through ninety-nine.

> Thirty-eight windows were broken.

Hyphenate a series of compound adjectives preceding a noun.

> The subjects for the motivation experiment were fourteen-, fifteen-, and sixteen-year-old students.

TRANSITIONS AND OTHER CONNECTORS

Transitions help us make our meaning clear — to signal the reader that we are in a certain time frame or a certain place, that we are giving an example, showing a contrast, shifting gears, concluding our discussion, or the like. Here is a list of the most commonly used transitions and the relationships they signal:

Addition	I am majoring in naval architecture; **furthermore,** I spent three years crewing on a racing yawl.

> moreover   and
> in addition   again
> also   as well as

Place	Here is the switch that turns on the stage lights. **To the right** is the switch that dims them.

> beyond   to the left

over	nearby
under	adjacent to
opposite to	next to
beneath	where
inside	

Time The crew will mow the ball field this morning; **immediately afterward** we will clean the dugouts.

first	the next day
next	in the meantime
second	in turn
then	subsequently
meanwhile	while
at length	since
later	before
now	after

Comparison Our reservoir is drying up because of the drought; **similarly,** water supplies in neighboring towns are dangerously low.

likewise
in the same way
in comparison

Contrast or Alternative Felix worked hard; **however,** he received poor grades. **Although** Jack worked hard, he was never promoted.

however	but
nevertheless	on the other hand
yet	to the contrary
still	notwithstanding
in contrast	conversely
otherwise	

Results Jack fooled around; **consequently** he was shot by a jealous husband.
Because he was fat, he was jolly.

thus	thereupon
hence	as a result
therefore	so
accordingly	as a consequence

Example Competition for part-time jobs is fierce; **for example,** eighty students applied for the clerk's job at Sears.

for instance	namely
to illustrate	specifically

<table>
<tr><td>Explanation</td><td>She had a terrible semester; **that is,** she flunked four courses.</td></tr>
</table>

in other words	in fact
simply stated	put another way

<table>
<tr><td>Summary or Conclusion</td><td>Our credit is destroyed, our bank account is overdrawn, and our debts are piling up; **in short,** we are bankrupt.</td></tr>
</table>

in closing	to sum up
to conclude	all in all
to summarize	on the whole
in brief	in retrospect
in summary	in conclusion

Pronouns serve as connectors, because a pronoun refers back to a noun that you have used in a preceding clause or sentence.

As the **crew** neared the end of the project, **they** were all willing to work overtime to get the job done.

Low employee morale is damaging our productivity. **This** problem needs immediate attention.

Synonyms (words meaning the same as other words) can also be effective connectors. In the whaling intelligence paragraph on page 31, for example, "huge and impressive mammals" (a synonym for "whales") helps tie the ideas together.

Repetition of key words or phrases is another good connecting device — as long as it is not overdone. Here is another example of effective repetition:

Overuse and drought conditions have depleted our water supply to a critical level. Because of our **depleted water supply,** we will need to enforce strict **water-**conservation measures.

Here, the repetition also emphasizes a critical problem.

The following paragraph lacks adequate transitions, and the sentences seem choppy and awkward:

<table>
<tr><td>Choppy</td><td>Technical writing is a difficult but important skill to master. It requires long hours of work and concentration. This time and effort are well spent. Writing is indispensable for success. Good writers derive great pride and satisfaction from their effort. A highly disciplined writing course should be a part of every student's curriculum.</td></tr>
</table>

Here is the same paragraph rewritten to improve coherence:

Revised   Technical writing is a difficult but important skill to master. **Thus** it requires long hours of work and concentration. This time and effort, **however**, are well spent **because** writing is an indispensable tool for success. **Moreover,** good writers derive great pride and satisfaction from their effort. A highly disciplined writing course, **therefore**, should be a part of every student's curriculum.

Besides increasing coherence *within* a paragraph, transitions and other connectors can underscore relationships *between* paragraphs by linking related groups of ideas. Here are two transitional sentences that could serve as the concluding sentences for certain paragraphs or as topic sentences for paragraphs that would follow; or they could stand alone for emphasis as single-sentence paragraphs:

> Because the A-12 filter has decreased overall engine wear by 15 percent, it should be included as a standard item in all our new models.
>
> With the camera activated and the watertight cover sealed, the diving bell is ready to be submerged.

These sentences look both ahead and back, thereby providing a clear direction for continuing discussion.

Topic headings, like those used in this book, are another device for increasing coherence. A topic heading is both a link and a separation between related, yet distinct, groups of ideas.

Finally, a whole paragraph can serve as a connector between major sections of your report. Assume, for instance, that you have just completed a section of a report on the advantages of a new oil filter and are now moving to a section on selling the idea to the buying public. Here is a paragraph you might write to link the two sections:

> Because the A-12 filter has decreased overall engine wear by 15 percent, it should be included as a standard item in all our new models. Because tooling and installation adjustments will add roughly $100 to the list price of each model, however, we have to explain the filter's long-range advantages to the customers. Let's look at ways of explaining these advantages.

The effective use of transitional devices in your revisions is one way of transforming a piece of writing from adequate to excellent.

EFFECTIVE MECHANICS

Correctness in abbreviation, capitalization, use of numbers, and spelling is an important sign of your attention to detail. Don't ignore these conventions. (See Chapter 12 for format conventions.)

Abbreviations

(For correct abbreviations in documentation, see pages 317–322.)

In using abbreviations consider your audience; never use one that might confuse your reader. Often, abbreviations are not appropriate in formal writing. When in doubt, write the word out in full.

Abbreviate certain words and titles when they precede or immediately follow a proper name.

Correct	Mr. Jones	Raymond Dumont, Jr.
	Dr. Jekyll	Warren Weary, Ph.D.
	St. Simeon	

However, do not write abbreviations such as the following:

Faulty   He is a Dr.
           Pray, and you might become a St.

In general, do not abbreviate military, religious, and political titles.

Correct   Reverend Ormsby
           Captain Hook
           President Reagan

Abbreviate time designations only when they are used with actual times.

Correct   400 B.C.
           5:15 A.M.

Do not abbreviate these designations when they are used alone.

Faulty   Plato lived sometime in the B.C. period.
           She arrived in the A.M.

In formal writing, don't abbreviate days of the week, months, words such as *street* and *road* or names of disciplines such as *English*. Avoid abbreviating states, such as *Me.* for *Maine;* countries, such as *U.S.* for *United States;* and book parts such as *Chap.* for *Chapter, pg.* for *page,* and *fig.* for *figure.*

Use *no.* for *number* only when the actual number is given.

Correct   Check switch No. 3.

Abbreviate units of measurement only when they are used often in your report and are written out in full on first use. Use only those abbreviations that you are sure your reader will understand. Abbreviate items in a visual aid only if you need to save space.

Here is a list of common technical abbreviations for units of measurement:

ac	alternating current	kw	kilowatt
amp	ampere	kwh	kilowatt hour
A	angstrom	l	liter
az	azimuth	lat	latitude
bbl	barrel	lb	pound
BTU	British Thermal Unit	lin	linear
C	Centigrade	long	longitude
Cal	calorie	log	logarithm
cc	cubic centimeter	m	meter
circ	circumference	max	maximum
cm	centimeter	mg	milligram
cps	cycles per second	min	minute
cu ft	cubic foot	ml	milliliter
db	decibel	mm	millimeter
dc	direct current	mo	month
dm	decimeter	mph	miles per hour
doz	dozen	no	number
dp	dewpoint	oct	octane
F	Farenheit	oz	ounce
f	farad	psf	pounds per square foot
fbm	foot board measure	psi	pounds per square inch
fl oz	fluid ounce	qt	quart
FM	frequency modulation	r	roentgen
fp	foot pound	rpm	revolutions per minute
fpm	feet per minute	rps	revolutions per second
freq	frequency	sec	second
ft	foot	sp gr	specific gravity
g	gram	sq	square
gal	gallon	t	ton
gpm	gallons per minute	temp	temperature
gr	gram	tol	tolerance
hp	horsepower	ts	tensile strength
hr	hour	v	volt
in	inch	va	volt ampere
iu	international unit	w	watt
j	joule	wk	week
ke	kinetic energy	wl	wavelength
kg	kilogram	yd	yard
km	kilometer	yr	year

Here are some common abbreviations for reference in manuscripts:

anon.	anonymous	fig.	figure
app.	appendix	i.e.	that is
b.	born	illus.	illustrated
bull.	bulletin	jour.	journal
©	copyright	l., ll.	line(s)
c., ca.	about (c. 1963)	ms., mss.	manuscript(s)
cf.	compare	n.	note
ch.	chapter	no.	number
col.	column	p., pp.	page(s)
d.	died	pt., pts.	part(s)
diss.	dissertation	rev.	revised or review
ed.	editor	rep.	reprint
e.g.	for example	sec.	section
esp.	especially	sic	thus, so (to cite an error in the quote)
et al.	and others		
etc.	and so forth	trans.	translation
ex.	example	viz.	namely
f. or ff.	the following page or pages	vol.	volume

For abbreviations of other words, consult your dictionary. Most dictionaries have a list of abbreviations at the front or rear or include them alphabetically with the appropriate word entry.

Capitalization

Capitalize the following: proper names, titles of people, books and chapters, languages, days of the week, the months, holidays, names of organizations or groups, races and nationalities, historical events, important documents, and names of structures or vehicles. In titles of books, films, etc., capitalize the first word and all those following except articles or prepositions.

Joe Schmoe
A Tale of Two Cities
Protestant
Wednesday
the *Queen Mary*
the Statue of Liberty
April
Chicago
the Bill of Rights
the Chevrolet Vega
Russian
Labor Day
Dupont Chemical Company

Senator John Pasteur
France
The War of 1812
Daughters of the American Revolution
The Emancipation Proclamation
Jewish
Gone With the Wind

Don't capitalize the seasons, names of college classes (*freshman, junior*), or general groups (*the younger generation,* or *the leisure class*).

Capitalize adjectives that are derived from proper nouns.

Chaucerian English

Capitalize titles preceding a proper noun but not those following.

State Senator Marsha Smith
Marsha Smith, state senator

Capitalize words like *street, road, corporation, college* only when they accompany a proper name.

Bob Jones University
High Street
The Rand Corporation

Capitalize *north, south, east,* and *west* when they denote specific locations, not when they are simply directions.

the South
the Northwest
Turn east at the next set of lights.

Begin all sentences with capitals.

Use of Numbers

As a rule, write out numbers that can be expressed in two words or less.

fourteen	five
eighty-one	two million
ninety-nine	

Use numerals for all others.

4,364	2,800,357
543	3¼

Use numerals to express decimals, precise technical figures, or any other exact measurements.

50 kilowatts

14.3 milligrams

15 pounds of pressure

4,000 rpm

Express the following in numerals: dates, census figures, addresses, page numbers, exact units of measurement, percentages, ages, and times with A.M. or P.M. designations, monetary and mileage figures.

page 14

18.4 pounds

115 miles

the 9-year-old tractor

15.1 percent

1:15 P.M.

9 feet

12 gallons

$15

Do not begin a sentence with a numeral.

Six hundred students applied for the 102 available jobs.

If your figure consumes more than two words, revise your word order.

The 102 available jobs brought 780 applicants.

Do not use numerals to express approximate figures, time not stated with A.M. or P.M. designations, or streets named by numbers below 100.

about seven hundred fifty

four fifteen

108 East Forty-second Street

If one number immediately precedes another, spell out the first and use a numeral to express the second:

Please deliver twelve 18-foot rafters.

Only in contracts and other legal documents should a number be stated both in numerals and in words:

The tenant agrees to pay a rental fee of three hundred seventy-five dollars ($375.00) monthly.

Spelling

Spelling problems will not cure themselves. If you are bothered by certain spelling weaknesses, take the time to use your dictionary, *religiously*, for all writing assignments. As you read, note the spelling of the words that have given you trouble. You might even compile a list of troublesome words. Moreover, your college may have a learning laboratory or a skills resources center where you can get professional assistance with spelling problems. If not, your instructor might suggest several self-teaching books with instructions and exercises for spelling improvement.

Appendix B

Writers and Audiences on the Job

This appendix contains the full (and largely unedited) text of interviews with four working professionals who write daily. The first two can be classified as part-time writers; the latter two, full-time. The varied reporting tasks of these four respondents typify the range of primary audiences and information needs that writers face on the job:

- Blair Cordasco's audience needs instructions for improving their job performance.
- Glenn Tarullo's audience needs to make major investment and marketing decisions.
- Bill Trippe's audience needs to make major government policy decisions.
- Pam Herbert's audience needs clear documentation to understand and use her company's mainframe computer.

As the interviews show, each of these writers has secondary audiences as well.

Each respondent here was asked an identical set of eight questions about the amount and types of writing, audiences, writing challenges, deadlines, individual writing processes, and advice to students. Besides offering a vivid picture of real-world expectations, the collected responses contain much useful advice for anyone whose career will depend in some way on good writing.

FIRST RESPONDENT — WRITING FOR COLLEAGUES

Blair Cordasco is a training specialist for an international bank. Her main job is to develop instructional programs and procedures that help improve all levels of employee performance.

Question 1: What percentage of your job is spent writing, editing, and dealing with written communication?
Answer: Roughly 75 percent.

Question 2: What types of writing/editing do you do?
Answer: I do a few different types:
1. I write memos informing management and recommending types of training needed in various divisions. I also write status (periodic) reports on the types and evaluation of training that has been given. Basically, I keep management informed about the activities of the training department.
2. I write materials to be used for training (lectures, procedures, and manuals). Also, I rewrite technical manuals that accompany high-tech equipment so that nonspecialized managers can understand the operating principles of the equipment, to better supervise the staff using the equipment.
3. I edit my staff's material for clarity, conciseness, fluency, and basic correctness. Reading their material from the viewpoint of the person who has to learn from it, I adjust the level of technicality and detail for the intended audience. Then the edited material goes to the content experts who review the simplified versions for technical accuracy. Organization and format are *very* important. Word processors make the formatting much easier, and we use them for our training manuals.

Question 3: Who are the audiences for your writing, and what are their needs?
Answer: I write for two audiences:
1. Management staff (B.A.'s in finance, economics, etc. and M.B.A.'s), who range from college trainees all the way to upper management, with varying degrees of technical/business experience. Senior members of the audience need executive briefings (overviews, summaries), not a lot of detail, with technical material as appropriate. The junior members of the audience need training materials either for skills (interviewing, quality-control techniques, etc.) or for knowledge (overview of international banking policy, or fundamentals of data processing, etc.).
2. Clerical staff (high-school education), such as secretaries and terminal operators, who need to learn specific skills (word processing, key punching, etc.) they can apply directly to the job. The material has to be at their level of comprehension.

Question 4: What does your audience expect from your documents?
Answer: Upper management wants to know, "How does it affect the bottom line? How can I use this information to increase productivity and quality?" Lower management and trainees ask, "What's in it for me? Can it help me do *my* job better?"

I spend much of my time trying to convince people that the information we offer can help them. Managers have to be persuaded that your recommendations are worthwhile and the best way to go.

Clerical staff wants to know, "Will this help me keep my job?" Their lifeline to survival is the information and training we offer. An underlying assumption in training clericals is that the training will increase their self-esteem and sense of belonging: "Will this make me feel like a jerk or help me feel I can accomplish something?" Is the tone going to be condescending, or the material too technical? Or will the tone be friendly and engaging, and the material at the right level of technicality?

Question 5: What is your biggest writing challenge on the job?
Answer: Aside from the persuasive challenge I just described, the biggest challenge is taking a body of information and packaging it *accurately* in different ways for different audiences.

For instance, a new piece of equipment arrives, and the vendor documentation is too technical for operators (how to) and managers (overview). We have to take the vendor's information and rewrite (how to) for operators. Then, for managers, we have to explain the operating principles, point out potential problems and solutions, and identify major problem areas. So, from one technical manual, we have to extract instructions for the operators, and explanations and overviews for the managers.

A trainer really is dealing with all kinds of information and all kinds of vehicles, and trying to reduce them to the lowest common denominator — to make the meaning absolutely clear.

Question 6: What about deadlines?
Answer: Deadlines are a way of life. Nothing is more essential than getting information to the right people at the right time.

Question 7: Do you follow a standard and predictable process when writing?
Answer: Deadlines affect how we approach the writing process. With plenty of time, we can afford the luxury of the whole process: the careful decisions about our audience, purpose, content, organization, and style (as you would put it) — and plenty of revisions. At times, we have to take shortcuts.

Each project presents its own problem, so we have no pat way of writing training materials. With limited time, we have limited revisions — so you have to think on your feet.

Say, for instance, we receive word November 1 that a new check processing system will be installed and operating by December 1. Operators will begin training on this equipment one week before the system is implemented (last week of November). In this case, we might write backwards: the basic "How to" comes first, and the more elaborate documentation (for management) would come whenever we could complete it.

Question 8: What advice do you have for students?

Answer: Make sure that *whatever* you're writing is clear *to you* first, and that the way you've organized it makes sense *to you*. Then take that material and become more objective. Try to understand how your audience thinks: "How can I make this logical to my audience? Will they understand what I want them to understand?"

Organizing is the key. Develop whatever type of outlining or listing or brainstorming tool that works best for you, but *find* one that works, and use it consistently. Then you'll be comfortable with that general strategy whenever you sit down to write, especially under rigid deadline.

SECOND RESPONDENT — WRITING FOR CLIENTS

Glenn Tarullo is a senior project manager for a market research firm. Most of his writing is done for clients who will use it to make investment decisions based on the feasibility and strategy for marketing various new products.

Question 1: What percentage of your job is spent writing, editing, and dealing with written communication?
Answer: From 50 to 60 percent.

Question 2: What types of writing/editing do you do?
Answer: Aside from an occasional internal memo, my writing is designed for two purposes: (1) to gather, analyze, and interpret information from respondents, and (2) to report our findings to clients.

For respondents, I have to translate the market research (or information) needs of clients into precise questions that *cannot* be misinterpreted. The questions have to be so precise and unambiguous that the respondent knows *exactly* what we're asking.

Reports, on the other hand (of findings gathered from questionnaires), have to allow the clients to make marketing decisions based on their understanding of complex data analyses presented in a concise, *nontechnical* way. These people are not in any way experts. They want to know, "What do I do next?" Our reports interpret the findings ("What do they mean?"), and offer recommendations for marketing strategy ("What should I do?").

Question 3: Who are the audiences for your writing, and what are their needs?
Answer: Among questionnaire respondents, the range is vast: from scientists making a breakthrough in robotics, to homemakers choosing a particular brand of coffee off a supermarket shelf.

Clients range from an express-mail carrier who's spending $3 million to assess the feasibility of expanding into international express service, to an individual who wants to market a new brand of salad dressing or low-calorie chocolate.

Both respondents and clients need material that is concise, clear, and unambiguous.

Question 4: What does your audience expect from your documents?
Answer: Respondents expect questions they can interpret accurately — that is, in only *one* way.

Clients expect enough information to make basic investment and marketing decisions. Some clients need answers to very specific statistical questions: "How many overnight letters are sent daily from England to the U.S.? How many potentially could be sent daily — classified by type of industry (pharmaceutical, aerospace, etc.) and size of company?" Other clients need answers to more general questions:

- "How does one break into the salad dressing market?"
- "What are your recommendations for establishing a new brand of salad dressing?"
- "How can I make any money with my low-calorie chocolate?"
- "I have a million, and want to market this chocolate. What can you do for me?"

For this latter group of clients, we might research the taste, packaging, and price-point (say, $2/pound for chocolate) that is most popular for the particular product.

So, basically our audience's expectations range from those of real experts who know exactly what they want — down to the *n*th decimal point — to somebody who calls with a vague idea and asks, "What can you do for me?"

Question 5: What is your biggest writing challenge on the job?
Answer: My writing has to have *one* interpretation *only.* I have to be certain that respondents are answering *exactly* the question I had in mind, not inventing their own version of the question. Then I have to take this data and translate it into accurate interpretations and sensible recommendations for our clients.

Question 6: What about deadlines?
Answer: The priority is to *meet* the deadline, regardless of the quality of the report. You never have the time you need, so the people who survive are those who *are able to write it* — no matter what — with enough quality to satisfy client expectations.

The differences between real-world and college writing is that, in the real world, the deadlines are *never* extended — never any excuses for not getting the report done on time.

Whether you keep the client depends on how well you can write under impossible time pressures.

Question 7: Do you follow a standard and predictable process when writing?
Answer: I look at data and move to some kind of prose that gets revised as often as time allows. For long reports, I rely on outlines. The final product is a recommendation, but the process begins with my looking at a computer printout.

Even when we devise questions that we're sure are unambiguous to our intended respondents, we pretest them — often finding that the questions have been misinterpreted. So we write them and test them again, until we get it right.

The basic element in my writing process is checking and rechecking for ambiguous messages, and revising as often as time allows.

Question 8: What advice do you have for students?
Answer: Learn to write as quickly, accurately, and unambiguously as you can in a given amount of time. The issue is to have it there when it's expected — maybe not perfect, but to develop a style so that the document is out and it gets the job done.

Organize your time in such a way that you can always meet the deadline. Prepare to labor under Murphy's Law: there will always be some crisis. The survivors are those who anticipate the crises, and are able to deliver nonetheless.

THIRD RESPONDENT — WRITING FOR GOVERNMENT DECISION MAKERS

Bill Trippe is a communications specialist with a military contract company. His job is to provide all-around communications support for a large group of engineers who perform various technical analyses for the government.

Question 1: What percentage of your job is spent writing, editing, and dealing with written communication?
Answer: Writing is probably 30 to 40 percent of my job, with editing tasks taking up another 50 percent, and training and tutoring accounting for the remaining 10 to 20 percent.

Question 2: What types of writing/editing do you do?
Answer: The writing I do falls into four categories:
1. Semitechnical reports from my managers to upper managers and to our military sponsors. These are usually progress reports on our larger projects, and run from 1 to 20 pages in length.

2. Some sections of highly technical (engineer-to-engineer type) reports. The engineers and scientists will usually write the "body" of highly technical reports, and I will write the abstract, introduction, executive summary, acknowledgments, conclusions, lists of references and bibliography. I will also fill out certain sections that need more material — add figures and tables, expand on explanations for equations and other raw material.
3. Miscellaneous reports and articles for a general audience. These include articles and announcements for the company newspapers, project updates for the board of directors, and short papers about our work, to be used for recruiting new employees.
4. Training materials for engineers. I'm currently writing short essays on grammar, audience analysis, and techniques for oral presentations. I've also put together my own editorial style sheet that I constantly update and expand.

Question 3: Who are the audiences for your writing, and what are their needs?
Answer: The typical primary audience is either our military sponsors or our management.

Military

Specifically, the military sponsor means the Project Officer: a Colonel or Lt. Colonel who is the commanding officer in charge of procuring a military system. He usually has help from other commissioned officers and civilian engineers, who coordinate testing, management and technical reviews, quality assurance, and other specific areas of the program. The Project Officer is the decision maker — the primary reader — and his supporting staff are the secondary readers.

Their needs? Usually, they've asked our company to perform a specific technical analysis. Our formal products to them are a technical report, regular progress reports, and related briefings. They want thorough analysis that relates the state of the art and our company's expertise to their specific program. For example, if they want to build a radar to operate in extreme cold, we'll tell them if this can be done, how reliable the radar would be, and how much it would probably cost.

Their uses for the papers? As a basis for making decisions, or — once the decision has been made — as a means of persuading their superiors (right on up through the Pentagon and Congress) to go along with their recommendations.

Company Management

The progress reports that I ghost write for my bosses are usually written for what I would call "middle managers" — my bosses are the low-level managers and their superiors (the readers) are just above them and just below vice-presidents. They are all ex-engineers-turned-managers, so the technical content can be anywhere from general to complex. They are looking at how the project team is handling potential problem and risk areas — difficult technical tasks, areas where cost overruns are possible, and any problems with scheduling. I write weekly reports about every project, and other reports as necessary.

Their needs? They are a fairly easy audience to write for, mainly because they can read at almost any level. They are, however, a critical audience — almost to the point of hostility. They are tough on their underlings probably because their superiors are then tough on them, and the stakes get higher up there. But, as I said above, they are looking for highlights about management concerns: cost, schedule, and technical risk. They are also more concerned with the quality of the writing and presentation because they are conscious of company image and how the customer and the public perceive us.

Their use? They use all the information given them as a means for making sure the individual projects and tasks are running smoothly. They put together a picture of their area that they can then present to their superiors and to the customers.

For the other writing I do, the audience is everyone from the readers of the company newspaper to the board of directors: general readers, I guess. Everything in this category is really written for information purposes, for entertainment, or as a bit of fluff.

Question 4: What does your audience expect from your documents?
Answer: The military audience expects a technically complete report with a rigorous organization, a good balance of detail, strong graphics and tables, and a clear and readable writing style. Because not all the military people are engineers, certain sections of the report (the introduction, executive summary, and conclusions) have to be semitechnical. On the other hand, to satisfy the various interests of the specialists in the project office, many areas have to be covered in great detail, but without bogging down the basic paper. For this, we resort to appendices — lots of them. Occasionally we will publish completely separate reports: an executive summary, the analysis, and supporting appendices; this type of report usually goes over very well.

I've pretty well covered the expectations of our management audience in my response to question #3. But I should add that these readers really expect the writing to be tailored to their needs as an audience. When they ask for a one-page memo, they expect a one-page memo. The same is true for briefings; they ask for and get a three-minute presentation.

Question 5: What is your biggest writing challenge on the job?
Answer: All my writing tasks are a big challenge because they are highly visible and fall under the tightest deadlines.

The greatest pressure comes from the writing tasks because they are my product alone, whereas the editing I do is the author's product I've helped along. If the author's paper is unsatisfactory, I don't hear about it (unless there's an editorial problem, but that hasn't happened). However, if my reports are late or contain a mistake, then my bosses come directly to me. So I watch what I write very closely, and rewrite and proofread everything. I also work very closely with the typists and artists, to see that everything is exactly how I want it.

In a corporation, you are invisible unless you do something spectacular or make a mistake. Since the spectacular rarely happens, management tends not to notice you until you make a mistake.

Question 6:   What about deadlines?
Answer: I rarely miss a deadline.

I know this isn't my own rule, but I use it anyway: when I need to estimate how much time it will take me to do something, I make a reasonable estimate and then double it. Then when I know something is due, I use my doubled estimate and go to work on it, leaving everything else aside. This usually works, unless the computer goes down and I lose an afternoon's writing to that great electronic void, or a secretary is out sick, or everyone that I need to see is in a meeting.

The important thing is juggling jobs and deadlines. I work for six projects and really have eight bosses. If everyone is screaming for work, then I go to everyone, explain my schedule, and see if they can come up with a new deadline. If conflicts persist, then I explain the problem to the big boss, and he tells me what to work on first. My negotiations are usually enough; I've only had to involve the big boss once or twice.

I've brought work home, stayed late, gotten extra help from secretaries and co-op students, and done jobs faster than I've wanted to. But by reasonably estimating how long a job will take and by being assertive about how to handle conflicting deadlines, I haven't had to do any of these other things to a point of discomfort.

Question 7:   Do you follow a standard and predictable process when writing?
Answer: The process I follow depends on the length of the piece.

For short pieces (say, less than 3 pages) I outline in my head, draft, and revise. On my first draft of a short piece, I spend 40 to 50 percent of the time on the first one or two paragraphs, and crank out the rest quickly. Then I revise substantively two or three times, and tinker with the mechanics and format right up until printing. I use a word processor for everything I write and edit, so revising is easy: I'm never afraid to change something because of the extra work mechanical typing would create. I usually don't bother to make a printout until I'm pretty close to a final product, so it's very easy for me to revise something short several times in a day, tinker with it a half-dozen times, and then make a printout and tinker with it some more once I see what it looks like on paper.

For longer pieces, I try to follow a rigorous pattern:

a. Interview knowledgeable people first for ideas and inspiration
b. Write a paragraph-long statement of purpose and formal sentence outline
c. Review the outline and thesis with the boss of the project
d. Write a first draft, paying particular attention to the introduction
e. Have the draft reviewed for technical accuracy
f. Revise for accuracy and begin revising for editorial quality
g. Revise (from two to five substantive revisions, depending on deadline)
h. Correct the mechanics and format

Question 8: What advice do you have for students?
Answer: Whew! I could say a million things here.

The following paragraphs are for non-engineers interested in technical writing as a career.

First of all, technical writing is as difficult a writing job as is newspaper writing, advertising writing, creative writing, and any other job where your principal function is to work with words. So I think to go into this field you have to be fully competent as a writer: you need to have an inquisitive mind and critical abilities, have an eagerness to interview and do research, be a wide and thoughtful reader, have a college background that includes many writing courses beyond freshman english, have experience writing against deadlines, have at least a strong interest in and an ability to learn about technical areas (several courses in math, science, and computers would give you a great advantage), have the social ability to deal with many kinds of people of all ages and backgrounds, and have the instincts and composure to deal with corporate life.

Get practical experience before graduation. I combined my undergraduate writing program with an assortment of part-time writing and publication jobs. I was on the staffs of the college newspaper, literary magazine, yearbook, and radio station. I wrote for a local daily newspaper and did other freelance writing for newspapers and magazines. In short, I packed as much practical experience in as possible — and learned a great deal about writing, editing, production, graphic design, and photography. Despite the pressure you might feel to learn as much about the technical areas as possible, keep in mind that the jobs of technical writer and editor are to be a publications specialist and not an engineer.

I also had a strong academic program. I had an excellent writing program in college, minored in philosophy, and took an assortment of interesting and challenging courses. In this job, I have to think on my feet and learn complex things quickly. But I've never felt at a loss — and I've gained the respect of my peers and bosses — by being able to tackle a subject and get things down on paper. My liberal arts background — with its emphasis on reading and writing — is an asset in this field. On the other hand, you need some math and science knowledge and ability. My only college science courses were in biology and chemistry, but my high school program included physics and trigonometry. In interviews for technical writing jobs, I had to convince the people interviewing me that I had a strong aptitude for technical subjects. I would recommend getting a few technical courses under your belt, and you won't need anything else after that. Since graduating, I've taken many writing and editing courses (I'm in a Master's Degree program in writing), and I've taken graphic design and management courses, but I've only taken one technical course — an in-company course on using personal computers.

Another point: Learn what the established industry procedures are for writing, editing, production, art, and printing. Specifically:

— Learn how to use the editorial style books. *The U.S. Government Printing Office Style Manual* and *The University of Chicago Press Manual of Style* are the two most widely used.

— Learn everything you can about production; a little book published by the International Paper Company, *Pocket Pal*, is stuffed with information about production and printing. A part-time job on the school newspaper will get you experience in basic paste-up. Learn typing, word processing, and typesetting. This may be your biggest practical concern on the job. If you know what can be done and how long it takes to do it, then you will be able to reduce your production problems by at least half. Learn what you can about how artwork is prepared because illustrations are as much as 50 percent of a technical document; I work closely with two illustrators. Learn also what makes a good graphic and what doesn't. Learn something about how maps are made.

— Learn photography (including darkroom work — not just snapshots). I work with photographs all the time, and though I don't have to take pictures myself, I assign photographers to take pictures, work with them when they are taking pictures, choose pictures for papers and for audio-visual presentations, and work them into page layouts, presentations, and three-dimensional displays.

— Learn printing. Read what you can about it in *Pocket Pal*, and then look for a large printer in your area that offers tours and try to arrange one for a group from your school. If you know printing, then you will know a great deal about production — what looks good and what doesn't, what works and what doesn't. I deal with the print shop every day. I often have to rush work through there against the shop's normal schedule; if I weren't reasonable in my demands of these guys, they'd have my head.

And a word or two for engineers. According to most surveys I've read, you'll spend about 25 percent of your time on the job writing. And another 25 percent giving formal and informal oral presentations on your work. After your technical knowledge, your single most important skill will be your ability to communicate. Your career will soar or die based on your ability to communicate your technical knowledge to other people (especially your bosses). I see this happening all the time.

FOURTH RESPONDENT — WRITING FOR COMPUTER USERS

Pamela Herbert is a technical writer for a nationwide distributor. She writes user documentation (manuals for using the company's mainframe computer in various applications in various company locations), and upgrades existing manuals.

Question 1: What percentage of your time is spent writing, editing, and dealing with written communication?
Answer: The actual physical process of writing constitutes maybe 30 percent of my worktime. I spend more time coordinating research (I have to catch the field support people, who are my information sources, when they are in town), collecting information, organizing notes, planning a manual's general structure, preparing visuals to go with the text, and keeping track of which reviewer has which draft. Because I rely on others' feedback, I circulate materials often. So I write memos fairly often, about 3–5 times per week.

Question 2: What types of writing/editing do you do?
Answer: My writing consists of

— user documentation: procedure manuals written for the Hewlett-Packard 3000, a mainframe computer. (Our plants use computer applications such as Accounts Payable, Accounts Receivable, Inventory Control, Vending, etc. The manuals for these applications are called "control procedures." They are written according to a very structured format.)
— supplements or updates to existing documents. (I'm currently writing a user guide for a "report-writer" program to be used in our plants. The guide provides examples specific to our business, so that users can have models. Also, it defines terms and summarizes important information from the original manual for the software.)
— 3–5 memos per week, requesting information or reporting on my work.

My editing consists of work on

— revisions of articles written for our company newsletter.
— revisions of procedures written by programmers to be used by computer operators. These have less detail and less supplementary information than the

user documentation. They are "skeleton" procedures that simply need rewriting for clarity.

Question 3: Who are the audiences for your writing, and what are their needs?
Answer: My audience varies according to the type of writing involved. User documentation is called "control procedures"; the procedures are written both for upper management (so they can understand the system in general) and for data-entry clerks (so they can use the system). Therefore, control procedures are written as *tasks* required to perform a particular function with the computer.

System users in our branch offices are not, for the most part, data processing people. They are clerical workers using the computer to do the tasks they once did manually; they see the computer as a "black box," and so they need very specific, nontechnical instructions. Managers and other administrators simply can read the task overviews and outlines for a good picture of the process going on.

When researching a topic, I try to anticipate my audience's needs, and I ask the technical source person (usually a programmer or systems analyst) specific questions keyed to my audience's needs: Who performs the task? What materials are required? What does the task accomplish? What can go wrong? Otherwise, I would waste time soaking up like a sponge any and all information the source person feels like rattling off, whether it's important or not.

For memos, I try to keep the style consistent regardless of who will be reading it. Anything worth writing as a memo (instead of a scribbled note or phone call) is always in my plainest English — even though my boss initially balked at my writing "here it is. . . ." instead of "Enclosed please find. . . ." I have a hard time with the business-ese that so many people use. If they ask me to proofread memos or letters, I can't resist striking all the heretofore's and pleased-be-advised-that-pursuant-to-your-request's. People dislike surrendering those, though. So I'm rarely asked to proofread those things anymore!

Question 4: What does your audience expect from your documents?
Answer: As I mentioned earlier, my manuals have to be specific and straightforward.

Also, because I'm a member of the Quality Assurance group, one function of my job is to detect potential problems with the computer system: inconsistencies or "user-unfriendliness" that could result in a service call from a branch office to our systems support staff here. I am expected to document a system as it is; then, while reviewing my document, the department managers decide whether something in the system should be changed.

Also, I am expected to submit written reports of any Quality Assurance issues I encounter while researching the applications. Readers expect a crystal-clear description of the problem. Considered a "naive user," I am a fairly good test of how understandable a procedure will be to the average data-entry person. So I am expected to clarify any procedure that strikes me as unclear.

Question 5: What is your biggest writing challenge on the job?
Answer: The biggest challenge in my job is staying on top of a project. It is too easy to procrastinate, put off sifting through those notes, sitting down with that programmer, or making those phone calls to Jacksonville. My boss doesn't watch over me to make sure I'm working, but if a project is unusually delayed, he wants to know why.

The biggest *writing* challenge is getting my reviewers to *review!!* They hang on to a draft for weeks, and only after I hound them do they drag the thing out and review it. Then they hound *me* to finish the thing and get it into the field. I have to be diplomatic, of course, but it is frustrating to work hard to meet a deadline and then have the draft stagnating on someone's desk. All I can do is document the delays, so that I'm not held responsible for the lengthy production time on a document.

Question 6: What about deadlines?
Answer: Although I construct project timetables with deadlines, I always end up revising the charts and extending the deadlines. I know pretty much how long it will take to research, organize, and draft a manual. What I can't predict are reviewers who drop my work when they are suddenly called out of town, "emergency" projects that cut into my time, source people who are on the road for weeks, and downtime on our system here. Sometimes I spend a week at a branch office myself to do research, rather than wait for a technical person to become available at the home office. At any rate, I try to meet a project's deadline, but I do not lose sleep over it if I can't do it.

Question 7: Do you follow a standard and predictable process when writing?
Answer: The process involved in the production of a typical manual is

a. Once the project is assigned, I set up research time with the technical person.
b. I compile notes, collect report samples, and roughly sketch any data entry screens I'll have to document.
c. I analyze the notes, organizing information into tasks, procedure steps, knowledge topics, nice-to-knows, rules, cautions, etc.
d. I then write a draft (longhand) and from there enter it into the word processor.

    e. I assemble a complete draft, which is sent around on its in-house review. Major problems are ironed out, and I revise the first draft.

    f. The revised draft is mailed to a user who is familiar enough with the system to review the document competently.

    g. I usually assemble "dummy data" to run the procedure on the system myself, as a final test of the accuracy of what I've written.

    h. After "cosmetic" editing, the document is issued.

Deadlines do affect the process. If I'm really strapped for time, I'll scrap the field review (step f). I usually will make sure that the draft is as accurate as possible before it goes out on its first review in-house. If time gets really tight, though, I won't bother to double-check every issue with the first draft. I'll rap it out quickly and let any errors be caught in the review process. I try not to do this very often; shortcuts never save much time, and the glaring gaps in the text make me look like a sloppy writer.

Question 8:   What advice do you have for students?
Answer: DON'T let yourself slide into the jargon/business-ese trap. Hold fast to the way you learned to write — correctly.

    DO be flexible, though, when it comes to other people's writing. You can't preach English grammar to adults if they don't want to hear it.

    DO contribute your ideas. Many times I've found that the only reason something hadn't been done a better way was that no one had bothered to suggest it.

    DO keep lists. They help you to keep tabs on which document is at what stage, who is reviewing what, etc. It's much easier to concentrate on writing when you have everything written down neatly and don't have to commit it all to memory.

    DO try to get hands-on experience with what you're writing about, if possible. If you're writing a procedure, perform it yourself. If you're describing the parts of a device, get your hands on one. Knowing your subject thoroughly can make the writing 100 percent easier.

    DO keep up-to-date with the technical writing field. Subscribe to documentation newsletters (I get lots of info from the Society for Technical Communication, as well as a neat newsletter called *Simply Stated,* from the Document Design Center). Try to keep learning more about different documentation developments and techniques. You'll be more useful in your present job, and more marketable if you decide to look for a new one.

Index